Vibro-Acoustics
Fundamentals and Applications

Vibro-Acoustics
Fundamentals and Applications

Dhanesh N. Manik
Professor, Mechanical Engineering
IIT Bombay
Powai, Mumbai, India

CRC Press
Taylor & Francis Group
Boca Raton London New York

CRC Press is an imprint of the
Taylor & Francis Group, an **informa** business

CRC Press
Taylor & Francis Group
6000 Broken Sound Parkway NW, Suite 300
Boca Raton, FL 33487-2742

First issued in paperback 2020

ISBN-13: 978-1-4665-8093-0 (hbk)
ISBN-13: 978-0-367-73630-9 (pbk)

Library of Congress Cataloging-in-Publication Data

Names: Manik, Dhanesh N., author.
Title: Vibro-acoustics : fundamentals and applications / Dhanesh N. Manik.
Description: Boca Raton : CRC Press, 2017. | Includes bibliographical references and index.
Identifiers: LCCN 2016043113| ISBN 9781466580930 (hardback : alk. paper) |
ISBN 9781466580947 (ebook)
Subjects: LCSH: Sound-waves--Damping. | Damping (Mechanics) | Machine design. | Noise control.
Classification: LCC TA355 .M295 2017 | DDC 620.3/7--dc23
LC record available at https://lccn.loc.gov/2016043113

Visit the Taylor & Francis Web site at
http://www.taylorandfrancis.com

and the CRC Press Web site at
http://www.crcpress.com

Dedicated to my wife, Nisha

Contents

Preface

This book is mainly focused on the viewpoint of those interested in modeling vibro-acoustic systems for designing quiet machines at the drawing-board stage. It is well known that modeling is always cost-effective and is much superior to cut-and-try empirical methods that are not expected to give definite results. The subject matter of vibro-acoustics is very vast and is, therefore, difficult to compress it into about 500 pages of print. But the main objective is to provide all the necessary fundamentals for the reader who can very easily move to other reference books and then comprehend material that is available in the literature; in short, the main objective is to give an easier learning curve for those interested in vibro-acoustics, especially those students at the advanced undergraduate stage.

Background in the following areas is expected: mechanics, circuit theory, fluid mechanics, and mathematics. Therefore, those from any of the engineering disciplines should be able to follow this book. The style adapted in writing this book is pedagogical and hence very easy to follow without any secondary reading. The entire book can be taught as a semester course or some parts of this book can be used to teach a course on noise control.

Most parts of this book have evolved based on the postgraduate elective on vibro-acoustics I have taught at the Indian Institute of Technology (IIT) Bombay for the past 5 years. Therefore, it has undergone many useful changes over a period of time, based on the feedback from students. The course, taught over a period of 14 weeks with tutorials, quizzes, exams, and so on, generated content for this book. Hence, sufficient care has been taken to write this book so the student can easily comprehend the basics of vibro-acoustics through independent reading, supplemented by classroom teaching. I have also used most parts of this book to teach continuing education programs for practicing engineers. Therefore, practicing engineers can also benefit from using this as a reference book, as there are many illustrated practical applications.

The contents have been carefully selected based on my extensive experience of teaching industry professionals and students for the past 23 years. I would have rather preferred to directly use an existing textbook to teach vibro-acoustics. But unfortunately, it was hard to find a book that could give all the necessary information that is required for a beginner to master vibro-acoustics. In addition, many of the authors of reference material in vibro-acoustics are from a physics or electrical engineering background and, therefore, have a different perception of their audience. Due to these reasons, I decided to write this book so that someone interested in vibro-acoustics can get most of it from a cover to cover reading.

Another important reason for writing this book is the division among those working in various aspects of vibration and noise; some of them use only the modal approach and others the wave approach. There are a chosen few who use both of them, for example, Norton. Since each of these approaches has their own advantages, I have used both approaches extensively wherever relevant, so the reader feels much more comfortable to expand the horizons of learning without any difficulty. This approach, therefore, will help to spread the knowledge of vibro-acoustics to a wider audience, rather than being restricted to a chosen few.

The book is divided into 10 chapters. Beginning with a single-degree-of-freedom system, the vibratory behavior of multidegree-of-freedom systems and continuous systems are discussed using both the modal and wave approaches. The basic principles of airborne sound are then discussed along with their measurement techniques. Random vibration, which is commonly encountered in many vibro-acoustic systems, is discussed to emphasize the basic concept of broadband excitation. Acoustic sources like monopoles, dipoles, and baffled pistons are discussed, which will help in objective description of the source strength. This will also help relating how vibration can produce airborne sound and vice versa. The physics of sound–structure interaction with respect to flexible vibrating structure is clearly explained without using too much mathematics. The basic information for studying information on statistical energy analysis (SEA) is organized systematically in sequence in the first 9 chapters to prepare for the final chapter. In Chapter 10, the basics of SEA are presented along with many applications.

The book is organized in such a way that information related to vibration and noise is logically presented so that they can be eventually unified to determine the extent of noise radiation due to vibration. The sequence is such that all the chapters can be sequentially studied without having to go back and forth. The physics of dynamic systems is emphasized throughout so that the reader in not just lost in mathematics. Therefore, the information presented here will be very useful for modeling vibro-acoustic systems. After going through this book, many other reference books and journal papers in this area can be easily studied. The main objective of this book is to prepare a new graduate student in vibro-acoustics.

Dhanesh N. Manik
Mumbai, India

Acknowledgments

I thank my PhD supervisor at Auburn University, Dr. Malcolm J. Crocker, for introducing me to these wonderful subjects of vibration and acoustics. I thank my students who had taken my elective course on vibro-acoustics all these years for their valuable feedback. I thank IIT Bombay for providing support in preparing the manuscript.

Author

Dhanesh N. Manik was born in Mysore, India. He graduated with a bachelor's degree in mechanical engineering from Mysore University in 1982; a master's degree in mechanical engineering from Indian Institute of Science (IISc), Bangalore, in 1985; and a PhD in mechanical engineering from Auburn University in Alabama in 1991. He also briefly worked at Hindustan Aeronautics (Helicopter Division), Bangalore, from 1985 to 1986. He joined the faculty of Indian Institute of Technology Bombay in 1992 and is currently a professor in mechanical engineering. His main areas of research are statistical energy analysis (SEA) and machinery diagnostics. He is a member of the International Institute of Acoustics and Vibration; Acoustical Society of India; and National Committee on Noise Pollution Control, Central Pollution Control Board, India. He is a consultant to many projects from industry related to vibration and noise, and regularly conducts continuing education programs to industry professionals on vibration and noise. He has authored a book on control systems (Cengage India, 2012).

1

Single-Degree-of-Freedom (SDOF) System

A single-degree-of-freedom (SDOF) spring–mass–damper system is the basic building block of dynamic systems related to vibration and noise. Basic concepts of vibration and noise analysis can be very easily understood by studying such systems. In addition, the dynamic characteristics of transducers used in vibration and noise measurement are also very similar to those of an SDOF. Furthermore, most complex dynamical systems encountered in vibration and noise analysis can be modeled as a combination of such SDOF systems. Therefore, the study of an SDOF system is considered first in the study of vibration and noise. The main emphasis in this chapter is to study dynamic characteristics of SDOF systems in greater detail that are relevant and fundamental to the study of vibration and noise.

An SDOF system is shown in Figure 1.1. It consists of a spring of stiffness k (N/m), dashpot of viscous damping coefficient R_m (N-s/m), and a mass m (kg). Mass stores kinetic energy, spring stores potential energy, and the dashpot dissipates energy in the form of heat. Although the actual mechanism of energy dissipation could be much different in vibro-acoustic systems, a viscous damping model is assumed for mathematical convenience. The above elements are connected in an arrangement as shown and the system is assumed to have a single degree of freedom in the vertical direction and the displacement response is assumed to be measured from an equilibrium position of the mass. The movement of the mass can occur either due to initial conditions or due to an external excitation force that will be discussed later.

There are two important aspects to the study of a dynamic system. First is the response of the system due to the initial conditions imposed on it, known as the free vibration. Second is response of the system due to the force excitation applied on it, known as the forced vibration. Although free vibration brings out unique characteristic features of a dynamic system, the response due to force excitation depends on these characteristic features. Therefore, free vibration is studied first before studying forced vibration. The same sequence will be followed in this chapter and also for an SDOF.

The following sequence is followed in this chapter. Since the neglect of damping simplifies the discussion of dynamic systems, without losing much information on its characteristic features, the free vibration of an undamped system is studied first. The differential equation of an undamped SDOF is studied by using the method of differential operators as well as the complex exponential. It is then followed by the study of free vibration of a damped SDOF using a complex exponential. Then, important concepts of logarithmic decrement and impulse response function are discussed, followed by the convolution integral. Once the characteristic features of an SDOF, undamped natural frequency, and damping factor are known, the frequency response of the SDOF to harmonic excitation is discussed. Once free vibration of an SDOF is well understood, the forced response is discussed as follows.

The forced response is expressed in terms of impedance of an SDOF, and the frequency-dependent properties of magnitude and phase of the impedance are discussed in greater detail. This impedance is identified as mechanical impedance to differentiate it from electrical impedance. The electromechanical analogy is then discussed, which gives equivalent

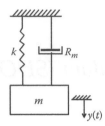

FIGURE 1.1
Single-degree-of-freedom spring–mass–damper system.

terms between electrical and mechanical systems. The power that is input to a mechanical system due to harmonic force excitation is discussed and is shown to be a function of frequency. From this, frequency regions are clearly demarcated that show the dominant contribution of stiffness, damping, and mass toward mechanical impedance in certain regions. The frequency regions are also presented on a Nyquist plot. The magnitude and phase response of displacement, velocity, and acceleration are discussed to understand the similarities with the frequency response based on impedance. The impedance concept is further extended to define admittance, which is a useful parameter in many applications. Finally, vibration transducers are discussed that will help measure vibration.

An SDOF will be discussed from different viewpoints that will help in understanding various concepts presented in later chapters.

1.1 Undamped Single-Degree-of-Freedom (SDOF)

To begin with, let us first consider an SDOF system without damping, as shown in Figure 1.2, which is the same as Figure 1.1 but without the damping element; the mass is assumed to move in the vertical direction. Although all practical systems have some amount of damping, an undamped model enables us to study the basic concept of exchange of kinetic and potential energy, which is mainly responsible for its dynamic motion. Damping is a mechanism that resists motion, and in many cases it may be very small. Therefore, the characteristic motion of many dynamic systems can be accurately studied by neglecting damping. Once the basic concepts are very clear, it would then be easier to study a realistic model by introducing damping later.

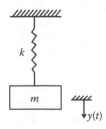

FIGURE 1.2
Undamped single-degree-of-freedom system.

1.1.1 Free Vibration

Consider a system as shown in Figure 1.2. When the mass is pulled down a certain distance (y_o) and let go, the entire system performs to and fro motion about the equilibrium position of the mass, defined by the initial deflection due to its weight. The physical mechanism of the to and fro motion, which can be termed as vibration, can be explained as follows. When the spring is elongated during the initial displacement of the mass, it stores a certain amount of potential energy. When the mass is let go, under the action of the restoring force of the spring, it attempts to move upward toward its equilibrium position. During this action, the mass is set into motion. It utilizes the potential energy that is stored in the spring, thus resulting in a decrease of the potential energy of the spring and an increase in the kinetic energy of the mass. This process of conversion of potential energy to kinetic energy continues until the mass reaches its static equilibrium position. At this position, the entire potential energy of the spring is exhausted in the process of its conversion to kinetic energy. The inertia of the mass, which now tries to move up, compresses the spring, thus enabling the spring to store potential energy. This results in the conversion of kinetic energy into potential energy. This process continues until all of the kinetic energy of the mass is exhausted. The mass now would be at a distance of y_o from its equilibrium position, which incidentally is equal to the initial displacement given to the mass. The spring that is fully compressed now tries to push the mass down, reversing its direction.

The cycle continues resulting in the to and fro motion of the mass. The vibratory motion continues forever, at least theoretically, in the absence of damping. In practice, however, due to the damping provided by air or due to the internal structural damping of the spring, the motion comes to a stop after some time. So far in our discussion, the mass of the spring was neglected. However, it can be accommodated, if it is of a significant value, by adding one-third of the spring mass to the main inertial element.

The motion of the aforementioned undamped system is known as free vibration. Free vibration is the to and fro motion of the body after it is given an initial displacement or velocity. The frequency of free vibration is known as the *undamped natural frequency* or *resonance frequency* of the system. It is dependent on the system parameters—mass and stiffness—which can respectively store kinetic and potential energy. The physical mechanism of free vibration when the spring is pulled a certain distance was described earlier. A similar motion takes place with a constant exchange of kinetic and potential energy if the mass were to be given an initial velocity, by striking it with an impulsive force.

Having understood the physical mechanism of free vibration, now the differential equation of motion for the undamped system can be formulated. The formulation of the differential equation will enable us to qualitatively and quantitatively assess the dynamic behavior of an SDOF.

1.1.2 Differential Equations of Undamped Free Vibration

The force acting on the mass at any instant of time, during free vibration, depends on the restoring force of the spring. This restoring force, f_r, is proportional to the displacement of the mass from its initial position and is given by

$$f_r = -ky \tag{1.1}$$

where k is the stiffness of the spring that is the proportionality constant between force and displacement, and y is the displacement of the mass from its equilibrium position.

The negative sign indicates that the restoring force is always in a direction opposite to the direction of motion of the mass.

From Newton's second law of motion we have

$$f_r = -ky = m\frac{d^2y}{dt^2} \tag{1.2}$$

From Equation 1.2, the following differential equation can be written:

$$m\ddot{y} + ky = 0, \tag{1.3}$$

where $\ddot{y}$ is the short form for the second derivative of time.

Dividing Equation 1.3 by m throughout, we get the following equation:

$$\ddot{y} + \frac{k}{m}y = 0 \tag{1.4}$$

This differential equation can be solved using the method of differential operators as follows.

1.1.2.1 Method of Differential Operators

Equation 1.4 can be written as

$$\left(D^2 + \frac{k}{m}\right)y = 0 \tag{1.5}$$

where the differential operator $D = d/dt$. The characteristic equation of the differential equation represented by Equation 1.5 in terms of the differential operator can therefore be written as

$$D^2 + \frac{k}{m} = 0 \tag{1.6}$$

or

$$D = \pm j\sqrt{\frac{k}{m}} \tag{1.7}$$

Since the differential operator has purely imaginary roots, the general solution for displacement is of the form

$$y(t) = A\sin\sqrt{\frac{k}{m}}t + B\cos\sqrt{\frac{k}{m}}t \tag{1.8}$$

Although Equation 1.8 contains sine and cosine terms, the periodic motion can only be either of them, because either of them can only represent the motion corresponding to a sine or a cosine term, depending on whether the initial condition is a displacement or velocity. The initial condition of displacement of the mass results in storing potential energy of the spring and velocity input due to the impulse applied to the mass results in the mass acquiring kinetic energy. Once the system acquires kinetic or potential energy, depending upon the initial condition imposed on it, it begins to vibrate as per the energy exchange mechanism explained earlier.

From Equation 1.8 it is clear that the term $\sqrt{\dfrac{k}{m}}$ represents the natural frequency of periodic motion of the system. Therefore, the undamped natural frequency is given by the following equation:

$$\omega_n = \sqrt{\frac{k}{m}} \text{ rad/s}$$

$$f_n = \frac{1}{2\pi}\sqrt{\frac{k}{m}} \text{ cycles/s(Hz)}$$

(1.9)

Frequency expressed in hertz (Hz) is the choice for experimental results and all frequencies used in mathematics must be in radians per second (rad/s). Therefore, we need to switch between these units of frequency depending on the application.

In terms of the undamped natural frequency, Equation 1.8 can be written as

$$y(t) = A\sin\omega_n t + B\cos\omega_n t$$

(1.10)

which can also be written as

$$y(t) = C\cos(\omega_n t + \theta)$$

(1.11)

A and B in Equation 1.10 and C and θ in Equation 1.11 are constants, which can be obtained from the initial conditions of displacement or velocity that were provided to initiate motion. For example, let us consider the case of an initial displacement, y_0, given to an undamped SDOF system. The initial conditions for giving an initial displacement can be written in the following form:

$$y(0) = y_0$$

(1.12)

and

$$\dot{y}(0) = 0$$

(1.13)

The initial condition for displacement is used in Equation 1.10 and the initial condition for velocity is applied after obtaining the velocity equation by differentiating

Equation 1.10 with respect to time. From this we get the following values of the constants A and B:

$$A = 0, \ B = y_0 \tag{1.14}$$

Substituting the values of A and B in Equation 1.10 we get

$$y(t) = y_0 \cos \omega_n t \tag{1.15}$$

Therefore, the given initial condition results in the system performing a cosine periodic motion at its undamped natural frequency, with peak amplitude equal to the initial displacement of the mass.

Similarly, the constants in Equation 1.11, for the initial conditions given in Equations 1.12 and 1.13, can be obtained as

$$C = y_0, \ \theta = 0. \tag{1.16}$$

The constants obtained in Equation 1.16 will give the same equation given in Equation 1.15.

An SDOF can be studied by solving the differential equation obtained by applying Newton's law by using conventional techniques, which might have been studied in earlier courses. But in this chapter, we shall borrow many concepts from the circuit theory that can better explain the dynamic characteristics of an SDOF that will be useful in later chapters. Circuit theory also helps in the definition of impedance, and impedance is a concept that can, not only clearly explain many dynamic characteristics of an SDOF but can also simplify the dynamic analysis of many complex dynamic systems related to vibration. In order to define the impedance of a SDOF, it is necessary to use the concept of a complex exponential in the solution of the differential equation. Before using this complex exponential approach, sufficient background material is covered beginning from a single complex number to a rotating complex number. A complex exponential is a very compact representation of a dynamic variable and thus simplifies the study of dynamic systems. It is important to understand that there is nothing complex about it and that it is only used for compactness. Once this approach is clearly understood, one will be able to appreciate the rest of the book.

1.1.2.2 Complex Exponential Method

Before we discuss the complex exponential method, let us discuss the basics of a complex number. Then, the complex number is rotated at a certain frequency that will lead to the solution differential equations using the complex exponential method. A complex number is an elegant and compact method of representing information and should not be mistaken as an abstract method.

A complex number is represented as shown in Figure 1.3a. It is basically a vector of magnitude A and is inclined at an angle θ with respect to the horizontal axis. The horizontal axis is referred to as the real axis and the vertical axis perpendicular to it is named as the imaginary axis. There is nothing imaginary about this axis except that it corresponds to the coordinate corresponding to $j = \sqrt{-1}$. The complex number is represented as

$$\bar{A} = A(\cos \theta + j \sin \theta) \tag{1.17}$$

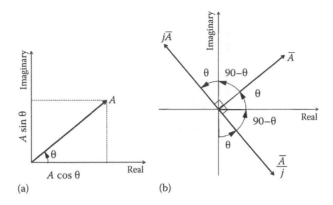

FIGURE 1.3
(a) Complex number. (b) Rotation of a complex number by using j.

Using Euler's formula, Equation 1.17 can also be represented as

$$\bar{A} = Ae^{j\theta} \tag{1.18}$$

The bar on A shows that it is a complex number having magnitude A and components along the real and imaginary axis: $A \cos \theta$ and $A \sin \theta$, respectively. The real and imaginary components also have a relationship in terms of their derivatives—the real component is the derivative of the imaginary component, which gives an additional property to j of representing derivative or integral action, depending on how it is used; this property is very useful in obtaining the derivative or integral of complex numbers.

Multiplying any complex number with j gives another complex number rotated 90° counterclockwise; this will be a derivative of the original complex number. Multiplying Equation 1.17 by j thus gives

$$j\bar{A} = A(-\sin\theta + j\cos\theta) \tag{1.19}$$

On the other hand, dividing a complex number by j gives another complex number that is rotated clockwise by 90°; this will be an integral of the original complex number. Dividing Equation 1.17 by j thus gives

$$\frac{\bar{A}}{j} = A(\sin\theta - j\cos\theta) \tag{1.20}$$

The rotation of a complex number by multiplying or dividing by j is shown in Figure 1.3b.

The complex number shown in Figure 1.4 can be continuously rotated at a frequency ω. So, the angle of this vector with respect to the real axis changes with respect to time, and therefore the projection of this rotating vector on the respective axes also changes with respect to time. The complex number now becomes a time-dependent vector and is given by

$$\bar{A}(t) = A(\cos\omega t + j\sin\omega t) \tag{1.21}$$

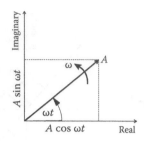

FIGURE 1.4
Rotating complex number.

The complex number in Equation 1.21 coincides with the real axis at $t = 0$. But there are many situations in which that may not be the case. Therefore, a general approach is to assume that the vector makes an angle ϕ with respect to the real axis at $t = 0$. Now, the rotating complex number, also known as phasor, becomes

$$\overline{A}(t) = A[\cos(\omega t + \phi) + j\sin(\omega t + \phi)] = Ae^{j\phi}e^{j\omega t} = \overline{A}e^{j\omega t} \tag{1.22}$$

Equation 1.22 is shown in Figure 1.5. The real and imaginary parts of the complex number are, respectively, a projection on the real and imaginary axis as shown. And they represent the equation of periodic motion at a frequency ω.

The preceding discussion regarding rotating complex numbers or phasors can be used for solving differential equations, wherein we are specifically looking for frequency response. Whenever a dynamic system is responding at a frequency, it can be represented either in terms of a sine or cosine. Since a complex number has both sine and cosine components, we can therefore assume the response be represented in terms of a rotating complex number similar to Figure 1.5. The actual solution can then be taken as a projection of the real or imaginary component of this rotating complex number, and both of them represent the response of the dynamic system.

The first step in using a complex exponential to solve differential equations is to assume that the dependent variable of the response is also complex. This is possible because we are combining two possible solutions through a complex number and both of them satisfy the original differential equation. Hence a differential equation with a real dependent variable can be converted into a differential equation with a complex dependent variable. Since physical variables such as displacement and velocity are real numbers, the final solution

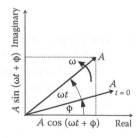

FIGURE 1.5
Rotating complex number with an initial phase.

for these variables is taken either as the real or complex part of the complex form. The complex exponential form of the solution is retained as long as it provides mathematical simplicity to use the complex solution for further analysis. For example, if a reference was made to complex displacement, it would only refer to the displacement in the complex form used in the process of simplifying the mathematics and should not lead to a misconception that a complex displacement exists in an abstract form; the complex form is used only for compactness. The biggest advantage of using a complex exponential is that it eliminates the need for keeping track of sines and cosines. In addition, the complex exponential form converts differential equations into algebraic equations.

The complex exponential method of solving linear differential equations arising in vibration and acoustics has many advantages. The most important advantage is its mathematical simplicity. This method of solution is useful in studying the phase relationships between mechanical and acoustic variables. In addition, problems arising in acoustics are similar to those in alternate current circuit theory. It is well known that the complex exponential method is extensively used in solving alternating current (ac) circuits. Measurement of vibration and noise signals invariably consists of converting the acceleration and dynamic pressure signals respectively to an electrical signal, which can be easily quantified in the complex form. Therefore, the use of the complex exponential method in studying differential equations arising out of vibration or noise problems is highly justified and very useful.

Coming back to the solution of the homogeneous linear differential equation using the complex exponential method, Equation 1.4 is now expressed in terms of a complex dependent variable as

$$\ddot{\tilde{y}} + \frac{k}{m}\tilde{y} = 0 \tag{1.23}$$

The solution to Equation 1.23 can be assumed as a phasor $\bar{A}$ that rotates at ω

$$\tilde{y} = \bar{A}e^{j\omega t} \tag{1.24}$$

In assuming this general form of the equation, we have not yet related the undamped natural frequency of the system to mass and stiffness. However, we will prove very soon that $\omega = \omega_n$, the undamped natural frequency of the system.

Expressing in terms of the complex form of the solution $\tilde{y}(t) = \bar{A}e^{j\omega t}$, Equation 1.23 can be written as follows:

$$\left(-\omega^2 + \frac{k}{m}\right)\bar{A}e^{j\omega t} = 0 \tag{1.25}$$

From Equation 1.25, the undamped natural frequency of vibration is given by

$$\omega = \omega_n = \sqrt{\frac{k}{m}} \tag{1.26}$$

Therefore, the frequency of vibration is the natural frequency of the system as obtained in the earlier section. The complex solution then will be of the form $\tilde{y}(t) = \bar{A}e^{j\omega_n t}$; a tilde is used to denote a phasor throughout. Now the actual solution is given by either the real part (denoted by $\Re$) or imaginary part of the complex solution. We shall, however, use the real part given by

$$y(t) = \Re(\bar{A}e^{j\omega_n t}) \tag{1.27}$$

This solution is explained by means of a phasor diagram as shown in Figure 1.6.

The magnitude and phase angle of the phasor $\tilde{y}$ are dependent on the initial conditions. The initial position of the phasor is as shown in the figure. It has a magnitude A and makes angle ϕ with the real axis at $t = 0$. For $t > 0$, the phasor advances in the counterclockwise direction to its new position making an angle $\omega_n t$ with respect to the initial position of the phasor and a total angle of $\omega_n t + \phi$ with respect to the real axis. The real part of the phasor is given by

$$y(t) = A\cos(\omega_n t + \phi) \tag{1.28}$$

which is the solution for the equation of motion of the SDOF without damping. This is same as Equation 1.11 obtained in the earlier section. This method may appear to be rather trivial when we are trying to solve for free vibration of the simple undamped SDOF. The usefulness of the complex exponential method will be clearly shown in future sections when we introduce damping and then proceed to study the response due to external harmonic excitation. Now it must be clear that the complex exponential method is used for its mathematical simplicity, compactness, and other advantages as stated before at the beginning of this section.

In the complex form, the velocity is given by

$$\tilde{v} = j\omega_n \bar{A}e^{j\omega_n t} = j\omega_n \tilde{y} \tag{1.29}$$

and the acceleration of the mass is given by

$$\tilde{a} = -\omega_n^2 \bar{A}e^{j\omega_n t} = j\omega_n \tilde{v} = -\omega_n^2 \tilde{y} \tag{1.30}$$

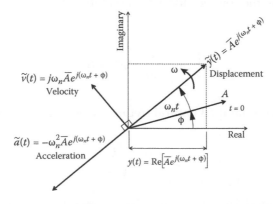

FIGURE 1.6
Phasor diagram of the response of an undamped system.

Therefore, velocity leads displacement by 90° and acceleration lags displacement by 180°, as shown in Figure 1.6. Now the similarity between Figure 1.3b and 1.6 should be clearly understood. Since the complex number is now a rotating vector at ω_n, multiplying the complex displacement by $j\omega_n$ gives the complex velocity, and multiplying the complex velocity by $j\omega_n$ gives complex acceleration. Similarly, one can get the complex velocity by dividing the complex acceleration by $j\omega_n$ and the complex displacement by dividing the complex velocity by $j\omega_n$.

1.2 Damped SDOF

Although undamped systems were considered in the initial analysis, energy dissipation by damping is a natural characteristic of most practical systems. The energy dissipating mechanism will decrease the amplitude of the free oscillations with time. Damping will also restrict the amplitude of response due to force excitation at the resonance frequency. In the present case, the damping force opposing the motion is assumed to be proportional to the velocity of the mass and this type of damping is known as viscous damping. The proportionality constant is defined as the damping coefficient R_m, expressed as N-s/m. R_m is used to represent damping since we shall prove later that it is equivalent to a resistor in the ac circuit. So, damping can be called mechanical resistance.

1.2.1 Differential Equation

The damping force is given by

$$f_d = -R_m v \tag{1.31}$$

Applying Newton's second law to the SDOF system shown in Figure 1.1, we have the following differential equation of motion:

$$m\ddot{y} + R_m \dot{y} + ky = 0 \tag{1.32}$$

Equation 1.32 can be written in the complex form as

$$m\ddot{\tilde{y}} + R_m \dot{\tilde{y}} + k\tilde{y} = 0 \tag{1.33}$$

Assuming a complex solution of the form

$$\tilde{y} = \overline{A}e^{j\omega t} \tag{1.34}$$

From Equation 1.34, Equation 1.33 becomes

$$(j\omega)^2 m + (j\omega)R_m + k = 0 \tag{1.35}$$

Solving the preceding quadratic equation for $j\omega$, we get the following equation:

$$j\omega = -\frac{R_m}{2m} \pm \sqrt{\left(\frac{R_m}{2m}\right)^2 - \frac{k}{m}} \qquad (1.36)$$

The nondimensional constant, *damping factor* ζ, is defined as follows:

$$\zeta = \frac{R_m}{2m\omega_n} = \frac{R_m}{2\sqrt{km}} \qquad (1.37)$$

In terms of the damping factor ζ, Equation 1.36 is written as follows:

$$j\omega = -\zeta\omega_n \pm \omega_n\sqrt{\zeta^2 - 1} \qquad (1.38)$$

The damping ratio can be defined as the ratio of the damping coefficient of the system to the maximum possible damping coefficient that will allow periodic motion of the same system. Therefore, the condition for periodic motion of the system is that the damping factor be less than one. Most engineering systems of practical interest have a damping ratio less than one and they are known as underdamped systems. Hence, in this chapter and throughout the book, only underdamped systems are generally considered.

For a damping ratio $\zeta < 1$, Equation 1.38 becomes

$$j\omega = -\zeta\omega_n \pm j\omega_n\sqrt{1 - \zeta^2} = -\zeta\omega_n \pm j\omega_d \qquad (1.39)$$

Substituting the value of $j\omega$ from Equation 1.39 and considering only one of the solutions with positive imaginary root, Equation 1.34 becomes

$$\tilde{y} = \bar{A}e^{-\zeta\omega_n t}e^{j\omega_d t} = Ae^{j\phi}e^{-\zeta\omega_n t}e^{j\omega_d t} \qquad (1.40)$$

This represents the equation of motion of a damped system, expressed in terms of the complex displacement. The complex number now has an exponentially decaying term and another complex exponential for oscillatory response. The exponentially decaying term physically represents the decay of amplitude of system response, and the complex exponential term represents the cyclic frequency of oscillation corresponding to the damped natural frequency. The actual motion of the system can be taken as the real part of $\tilde{y}$ and is given by

$$y = Ae^{-\zeta\omega_n t}\cos(\omega_d t + \phi) \qquad (1.41)$$

where A and ϕ are constants dependent on initial conditions.

1.2.2 Phasor of a Damped System

Equation 1.40, representing free vibration of a damped SDOF, can be represented on the complex plane for a better understanding of its physical significance. Figure 1.7 shows a

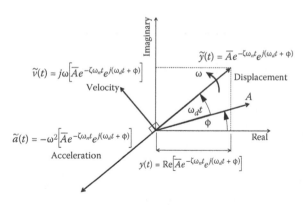

FIGURE 1.7
Phasor of a damped system.

phasor of shrinking magnitude rotating at a frequency ω_d corresponding to the damped natural frequency. The length of the phasor is $Ae^{-\zeta\omega_n t}$ and it gets shortened as it rotates in the complex plane. This is equivalent to the reduction of the amplitude of vibration in every cycle of oscillation of a damped system in free vibration. The extent of reduction of the amplitude of vibration in each cycle and the time it takes to stop vibrating after losing all the energy is primarily governed by the damping factor of the system. The motion of the mass is given by the real part of the phasor $\tilde{y}$ and is given by $Ae^{-\zeta\omega_n t}\cos(\omega_d t + \phi)$, which is the projection of the phasor on the real axis, as shown in Figure 1.7. Velocity and acceleration of damped vibration are also shown in the figure; the phase difference between displacement, velocity, and acceleration remain the same as the undamped system, although their magnitudes are now different.

1.3 Logarithmic Decrement

As mentioned earlier, the peak amplitudes of oscillation during free vibration reduce by a certain amount after each period of oscillation until the oscillations come to rest. This is shown in Figure 1.8, in which the decay of vibration of an SDOF of natural frequency 0.796 Hz and a damping factor of 0.1 is traced for six cycles after giving an initial displacement y_0 of 0.01 m.

Let the equation of motion of the damped system shown in Figure 1.8 be given by

$$y = y_0 e^{-\zeta\omega_n t}\cos(\omega_d t) \tag{1.42}$$

After giving an initial displacement of y_0, the second peak will come after the system executes one cycle corresponding to the damped natural frequency ω_d and the period of this cycle is given by $2\pi/\omega_d$. The second peak therefore appears at $t = 2\pi/\omega_d$ and is given by

$$y_1 = Ae^{-\zeta\omega_n(2\pi/\omega_d)}\cos 2\pi \tag{1.43}$$

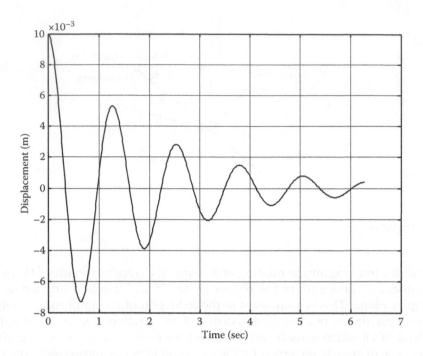

FIGURE 1.8
Decay of damped vibration.

Similarly, the third peak, y_2, can be obtained as

$$y_2 = Ae^{-\zeta\omega_n(4\pi/\omega_d)}\cos 4\pi \tag{1.44}$$

Natural logarithm of the ratio of successive peak amplitudes is known as *logarithmic decrement*, which remains constant as follows.

From Equations 1.42 to 1.44, the logarithmic decrement is given by

$$\Delta = \ln\frac{y_0}{y_1} = \ln\frac{y_1}{y_2} = \frac{2\pi\zeta\omega_n}{\omega_d} = \frac{2\pi\zeta}{\sqrt{1-\zeta^2}} \tag{1.45}$$

Determination of the successive peak amplitudes of an underdamped system performing free vibration is one of the methods of estimating the damping of a system using Equation 1.45. It can also be determined based on half-power frequencies, which will be discussed in later sections of this chapter.

Example 1.1

A single-degree-of-freedom spring–mass–damper system statically deflects by 1 mm due to a 100 N load. When it is given an initial displacement of 20 mm, it executes 7 cycles of oscillations in 1 sec and its amplitude decreases to 1 mm in the next 3 cycles. This is shown in Figure 1.9. Determine its undamped natural frequency, damping factor, and damped natural frequency.

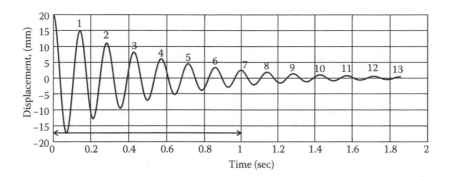

FIGURE 1.9
Displacement response of an SDOF system (Example 1.1).

From Equation 1.45, the logarithmic decrement for successive oscillations is given by

$$\Delta = \ln \frac{y_0}{y_1} = \frac{2\pi\zeta}{\sqrt{1-\zeta^2}} \tag{1}$$

From Equation 1, the logarithmic decrement between the initial and nth oscillation is given by

$$\ln \frac{y_0}{y_n} = \frac{2\pi n\zeta}{\sqrt{1-\zeta^2}} \tag{2}$$

From Equation 2, the damping factor is given by

$$\zeta = \sqrt{\frac{1}{\left(\dfrac{2\pi n}{\ln(y_0/y_n)}\right)^2 + 1}} \tag{3}$$

given $n = 10$, $y_0 = 20$ mm, and $y_n = 1$ mm.
From the aforementioned data and Equation 3, the damping factor is given by

$$\zeta = \sqrt{\frac{1}{\left(\dfrac{2\pi \times 10}{\ln(20/1)}\right)^2 + 1}} = 0.0476 \tag{4}$$

The time taken for seven damped oscillations is 1 s. Therefore, the time period of one damped oscillation is given by

$$\tau_d = \frac{2\pi}{\omega_d} = \frac{1}{7} \text{ s} \tag{5}$$

From Equation 5, the damped natural frequency is given by

$$\omega_d = 14\pi = 43.98 \text{ rad/s} \tag{6}$$

Using the data from Equations 4 and 6 and using Equation 1.39, the undamped natural frequency is given by

$$\omega_n = \frac{\omega_d}{\sqrt{1-\zeta^2}} = \frac{43.98}{\sqrt{1-0.0476^2}} = 44.03 \text{ rad/s} \tag{7}$$

The static deflection is 1 mm for a static load of 100 N. Therefore, the spring stiffness is given by

$$k = \frac{F_s}{\delta_{stat}} = \frac{100}{10^{-3}} = 10^5 \text{ N/m} \tag{8}$$

Using the data from Equations 7 and 8, the mass is given by

$$m = \frac{k}{\omega_n^2} = \frac{10^5}{44.03^2} = 51.58 \text{ kg} \tag{9}$$

1.4 Impulse Response Function

The impulse response function of an SDOF is defined as the response of the system defined by Equation 1.32 to a unit impulse. This can be solved as an initial value problem with initial conditions of zero displacement and unit impulse at $t = 0$. These initial conditions can be expressed as

$$y(0) = 0 \quad m\dot{y}(0) = 1 \tag{1.46}$$

The general equation for the free response of a damped SDOF is given by

$$y = Ae^{-\zeta\omega_n t} \cos(\omega_d t + \phi) \tag{1.47}$$

The first time derivative of Equation 1.47 is given by

$$\dot{y} = -Ae^{-\zeta\omega_n t} \left[\zeta\omega_n \cos(\omega_d t + \phi) + \omega_d \sin(\omega_d t + \phi) \right] \tag{1.48}$$

Applying the initial conditions of Equation 1.46 to Equations 1.47 and 1.48 gives

$$\phi = \frac{\pi}{2} \quad A = -\frac{1}{m\omega_d \sin\phi} \tag{1.49}$$

Using the constants obtained in Equation 1.49, Equation 1.47 thus becomes

$$y(t) = \frac{e^{-\zeta\omega_n t}\sin(\omega_d t)}{m\omega_d} \tag{1.50}$$

Equation 1.50 is the free response of a damped SDOF to unit impulse, which is known as the impulse response function. It is given a special symbol $h(t)$. Thus the impulse response function of a damped SDOF is given by

$$h(t) = \frac{e^{-\zeta\omega_n t}\sin(\omega_d t)}{m\omega_d} \tag{1.51}$$

The impulse response function and the system transfer function, which will be discussed in the next section, are related through Fourier transformation that will be discussed in Chapter 4.

We experience impulse excitation in everyday life, although we are unaware of it. When we knock on the door, drop a spoon, run on the floor, or ring a bell we are applying impulse excitation. An important aspect of impulse excitation is that it results in the body vibrating at its natural frequency; this may result in a large sound. Therefore, impulse excitation is an important source of vibration and noise generation in many machines. For an SDOF system, the response due to a nonunity impulse excitation, by impacting it, will be a scaled version of Equation 1.51 since we may not be able to apply unit impulse nor be able to measure the amplitude of impulse without a force transducer attached to the impacting element.

Since impulse excitation of any structure results in vibration at its natural frequencies, natural frequency of the structure can be measured by subjecting it to impulse excitation. Based on a similar principle, natural frequencies of teeth are measured to check for gum disease; if the natural frequencies are below a certain value, it means that they are weakened by gum disease or weak bone support. A commercially available instrument known as the Periotest works on this impulse principle that is used by dentists to determine the strength of tooth support by a combination of gums and bones. It will be proved later in Chapter 4 that the system transfer function and impulse response function of a dynamic system form a Fourier transform pair that will help visualize natural frequencies by converting impulse response into frequency domain using spectral analyzers.

1.5 Force Excitation

A single-degree-of-freedom system subjected to an arbitrary external force excitation, $F(t)$, is shown in Figure 1.10. By using conventional techniques it is possible to obtain the time-domain response. One such technique is the use of convolution or Duhamel's integral that can compute the response for any arbitrary excitation using the impulse response function. The main objective of this chapter, however, is not concerned with arbitrary excitation but only harmonic excitation. In any case, we shall first discuss the convolution integral for the sake of completeness and then move on to harmonic excitation. The concept of convolution integral is useful in determining the response of dynamic systems due to random excitation and is covered in Chapter 4.

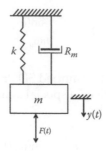

FIGURE 1.10
SDOF subjected to force excitation.

1.5.1 Convolution Integral

The convolution integral is useful in determining the time-domain response to any arbitrary force excitation. Although we are generally concerned with harmonic excitation in this book, knowledge of this integral is definitely useful in completing the dynamic study of SDOF systems.

Consider an arbitrary force excitation as a function of time, as shown in Figure 1.11. This force excitation can be divided into a number of small parts separated by a small time interval, $\Delta\tau$. Therefore, the force excitation is assumed to be a combination of many impulses as shown in the figure.

Consider a time instant τ and the impulse corresponding to this time is given by

$$F(\tau)\Delta\tau \tag{1.52}$$

Since we know the response of a system to a unit impulse from Equation 1.51, the response due to the aforementioned impulse is its product with the impulse response function. However, since we are applying the impulse at the time instant τ, the time should be counted from this instant onward; the response will start only after applying the force.

The response only due to the impulse force $F(\tau)\Delta\tau$ applied at the instant τ is given by

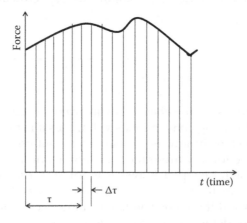

FIGURE 1.11
Arbitrary force excitation.

$$y(t)\Big|_{F(\tau)\Delta\tau} = \frac{F(\tau)\Delta\tau e^{-\zeta\omega_n(t-\tau)} \sin\left[\omega_d(t-\tau)\right]}{m\omega_d} \tag{1.53}$$

The response due to all the impulses will be an integral value of the preceding response, given by

$$y(t) = \int_0^t \frac{F(\tau)e^{-\zeta\omega_n(t-\tau)} \sin\left[\omega_d(t-\tau)\right]d\tau}{m\omega_d} \tag{1.54}$$

The short form of Equation 1.54 is

$$y(t) = \int_0^t F(\tau)h(t-\tau)d\tau \tag{1.55}$$

Equation 1.55 is known as Duhamel's integral or convolution integral. If the time history of any arbitrary force is known, Equation 1.55 can be used to compute the response. Since we are focusing on frequency response in this chapter, using the complex exponential approach to obtain the frequency response is much easier than using the preceding equation.

1.5.2 System Transfer Function

The characterization of the dynamic behavior of a system through a system transfer function is also a reason for studying systems excited by a single harmonic excitation. The *system transfer function* is defined as the ratio of the amplitude of displacement (or velocity or acceleration) to the amplitude of force, evaluated at various frequencies of the force excitation. The amplitude of the force is assumed constant for all frequencies of the harmonic excitation, which allows for fair comparison of system characteristics. To ensure this in an actual experiment, the amplitude of force excitation is controlled to remain constant at all frequencies of excitation by means of a feedback control system.

A system transfer function is defined as the ratio of the response and excitation, when both of them are expressed in the frequency domain. The dynamic system, to which the harmonic force excitation is applied, vibrates at the same frequency of the periodic excitation, whereas amplitude and phase of the response due to harmonic excitation depends upon the frequency of the force excitation, in addition to mass, stiffness, and damping. Therefore, by knowing the amplitude and phase of response, one can determine the transfer function that gives characteristics of the dynamic system. Since response to harmonic excitation defines the transfer function, frequency response plays an important role in dynamic analysis.

In the following sections, we shall study the response of an SDOF system to harmonic excitation.

1.5.3 Harmonic Excitation

An SDOF subjected to a harmonic force can be written as follows:

$$m\ddot{y} + R_m\dot{y} + ky = F\cos\omega t \tag{1.56}$$

or

$$m\ddot{y} + R_m\dot{y} + ky = F\sin\omega t \tag{1.57}$$

Equations 1.56 and 1.57 can be written in complex form as follows:

$$m\ddot{\tilde{y}} + R_m\dot{\tilde{y}} + k\tilde{y} = \tilde{F} \tag{1.58}$$

where

$$\tilde{F} = \bar{F}e^{j\omega t} \tag{1.59}$$

The amplitude of the force is complex in Equation 1.59 suggesting that the force in general need not start exactly at $t = 0$ from which the response is measured, and the phase difference can be accounted for in the complex number.

The steady-state solution to Equation 1.58 can be expressed in the form

$$\tilde{y} = \bar{A}e^{j\omega t} \tag{1.60}$$

Equation 1.58 representing a complex differential equation can be used to compute the response of the SDOF system to both sine and cosine excitations. We can take the real and imaginary part of the response to obtain the time-domain response due to cosine or sine excitation. However, the time-domain response due to harmonic excitation is of little interest to us; the objective is to know the amplitude and phase of the response that brings out characteristic features of the dynamic system, rather than finding a specific time-domain response for sine or cosine excitation. Therefore, we retain the solution in the complex form and then try to obtain information about amplitude and phase of response that are more important in relating the response to the characteristic features of an SDOF.

The real part of the solution of Equation 1.60 represents a solution corresponding to the force excitation $F\cos\omega t$, and the imaginary part of the solution of Equation 1.60 represents a solution corresponding to the force excitation $F\sin\omega t$. As mentioned earlier, since the objective is to determine the amplitude and phase of frequency response, we are much less concerned about the real or imaginary part of the solution in time domain; the frequency response gives characteristic features of the dynamic system that is more important.

By using a solution of the form $\tilde{y} = \bar{A}e^{j\omega t}$ of Equation 1.60, Equation 1.58 becomes

$$(-\omega^2 m + j\omega R_m + k)\bar{A}e^{j\omega t} = \bar{F}e^{j\omega t} \tag{1.61}$$

which yields the amplitude of displacement response

$$\bar{A} = \frac{\bar{F}}{k - m\omega^2 + j\omega R_m} \tag{1.62}$$

Therefore, the complex displacement is given by

$$\tilde{y} = \frac{\bar{F}e^{j\omega t}}{k - m\omega^2 + j\omega R_m} = \frac{\tilde{F}}{k - m\omega^2 + j\omega R_m} \tag{1.63}$$

which can be written as

$$\tilde{y} = \frac{-j\tilde{F}}{\omega[R_m + j(\omega m - k/\omega)]} \tag{1.64}$$

and the velocity is given by $j\omega\tilde{y}$, which can be expressed as

$$\tilde{v} = \frac{\tilde{F}}{R_m + j(\omega m - k/\omega)} \tag{1.65}$$

The denominator of Equation 1.65 defines the mechanical impedance that is the ratio of complex force to complex velocity.

1.5.4 Complex Mechanical Impedance

Complex mechanical impedance is defined as the ratio of complex harmonic force to complex velocity response, which can be obtained from Equation 1.65 as follows

$$\bar{Z}_m = \frac{\tilde{F}}{\tilde{v}} = R_m + j(\omega m - k/\omega) \tag{1.66}$$

In the complex exponential form,

$$\bar{Z}_m = Z_m e^{j\phi_m} \tag{1.67}$$

where Z_m is the magnitude of the complex mechanical impedance given by

$$Z_m = \sqrt{R_m^2 + \left[(\omega m) - \left(\frac{k}{\omega}\right)\right]^2} \tag{1.68}$$

And the phase of the mechanical impedance, denoted as the mechanical phase angle, is given by

$$\phi_m = \tan^{-1}\left(\frac{\omega m - k/\omega}{R_m}\right) \tag{1.69}$$

where ϕ_m is the phase angle of the complex mechanical impedance $\bar{Z}_m$. The unit of mechanical impedance is the *rayl*, which is equivalent to one newton second per meter (N-s/m).

Using this phase angle, the *mechanical power factor*, cos ϕ_m, can be defined, analogous to the power factor used in ac circuits. The reciprocal of mechanical impedance is known as *mobility*, which is a useful parameter in determining the response of dynamic systems.

The magnitude and phase angle of mechanical impedance, respectively from Equations 1.68 and 1.69), as a function of frequency are shown in Figure 1.12a and b. The magnitude of the impedance represents the resisting capacity of the SDOF to various frequencies of harmonic excitation; it has a large magnitude at very low and very large frequencies, but the magnitude is the lowest at a forcing frequency equal to the undamped natural frequency. This is a very important frequency dependent property of the SDOF that is responsible for large values of response at the excitation frequency, the same as the undamped natural frequency. This physical reason, based on the lowest magnitude of impedance at resonance, is a very convincing argument that forms the basis for resonance in complex structures. We shall use the property of impedance to determine the frequency response for displacement, velocity, and acceleration in later sections.

The Nyquist plot of mechanical impedance in Figure 1.12b is equally interesting. This figure is based on Equation 1.66 in which the mechanical impedance has a constant real part of the mechanical resistance and an imaginary part that varies with frequency. Hence, the impedance vector is a vertical line parallel to the imaginary axis. The mechanical phase angle has a value of $-90°$ at very low frequencies of harmonic excitation and $90°$ at very large frequencies of harmonic excitation. But when the harmonic excitation is the same as the undamped natural frequency at resonance, the mechanical phase angle is zero. Since impedance is derived from velocity response, velocity and force are in phase at resonance. Therefore, the mechanical phase angle has to be close to zero for large response, and it is a result of the action of mass and stiffness canceling with each other.

The phase between displacement and force can be written as

$$\tilde{y} = \frac{-j\tilde{F}}{\omega \bar{Z}} = \frac{\tilde{F}e^{-j(\pi/2+\phi_m)}}{\omega Z_m} \tag{1.70}$$

The phase between velocity and force can be written as

$$\tilde{v} = \frac{\tilde{F}e^{-j\phi_m}}{Z_m} \tag{1.71}$$

The phase between acceleration and force can be written as

$$\tilde{a} = \frac{\tilde{F}\omega e^{-j(\phi_m - \pi/2)}}{Z_m} \tag{1.72}$$

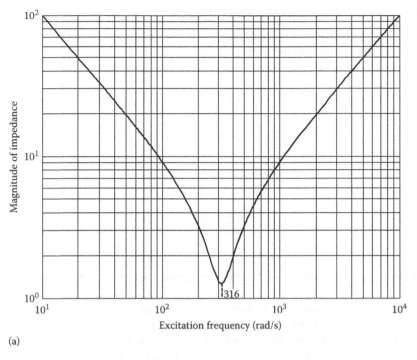

(a)

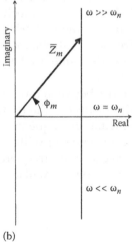

(b)

FIGURE 1.12
(a) Magnitude of impedance as a function of frequency (k = 1000 N/m, m = 0.01 kg, and R_m = 1.27 N-s/m).
(b) Nyquist plot of the impedance function.

1.5.5 Phasor Representation of Harmonic Response

The displacement, velocity, and acceleration can be written, respectively, in terms of complex mechanical impedance as follows:

$$\tilde{y} = \frac{-j\tilde{F}}{\omega \bar{Z}_m} \tag{1.73}$$

$$\tilde{v} = \frac{\tilde{F}}{\bar{Z}_m} \tag{1.74}$$

$$\tilde{a} = \frac{j\omega \tilde{F}}{\bar{Z}_m} \tag{1.75}$$

These equations are presented as phasors for various excitation frequencies in the next section.

The response of a single-degree-of-freedom system to harmonic excitation can be represented on a phasor diagram, as shown in Figure 1.13. The displacement, velocity, and acceleration responses will have the same frequency of excitation (ω) as the excitation force, and their amplitude will depend on the magnitude of impedance at the frequency; their phase with respect to the force will depend on the mechanical phase angle. As in the case of free response, velocity will lead displacement by 90° and acceleration will lead velocity by 90°. For a particular excitation frequency, all the four phasors—displacement, velocity, acceleration, and force—will rotate at the same excitation frequency and will keep their phase relationship the same throughout.

Figure 1.13 shows the phase difference between force and other phasors for various possibilities of excitation frequency with respect to the undamped natural frequency. Figure 1.13a shows the case in which the excitation frequency is much less than the undamped natural frequency of the system; the force phasor will be close to the displacement phasor. Figure 1.13b shows the case when the excitation frequency is very close to the undamped natural frequency of the system; the force phasor will be very close to the velocity phasor of

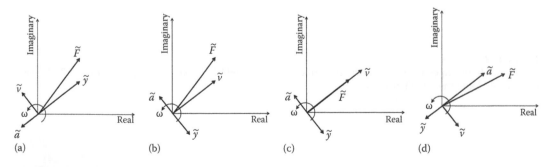

FIGURE 1.13

Phasor representation of harmonic force excitation. (a) Force very close to displacement in phase ($\omega \ll \omega_n$). (b) Force very close to velocity in phase $\omega \approx \omega_n$. (c) Force in phase with velocity, resonance ($\omega = \omega_n$). (d) Force very close to acceleration in phase $\omega \gg \omega_n$.

the system. If the excitation frequency is exactly equal to the undamped natural frequency of the system as shown in Figure 1.13c, the force and the velocity phasor are in phase with each other causing resonance resulting in large amplitude of velocity. In Figure 1.13d the excitation frequency is much higher than the undamped natural frequency of the system, and the force and acceleration phasors are close to each other. The mechanical phase angle also will be distinctly different in the above three regions. It is clear that the system behaves basically in three different ways based on the parameter with which the phasor is in phase with—displacement, velocity, and acceleration. Based on this discussion, we can see the possibility of three frequency regions based on the phase difference between force with displacement, velocity, and acceleration. However, an exact demarcation is only possible based on the power input to the system. This will be further discussed in the next section.

Example 1.2

A mass of 0.1 kg hangs on a spring of negligible mass. The stiffness constant of the spring is 1000 N/m and the damping coefficient is 0.15 N-m/s. The forces driving the system are represented by $F_1 = 2\cos\left(85t + \dfrac{\pi}{6}\right)$ N and $F_2 = 2\sin\left(80t + \dfrac{\pi}{9}\right)$ N. Using the method of complex exponential, determine the velocity response of the system.

From Equation 1.66, the complex impedance of the SDOF at 85 rad/s is given by

$$\overline{Z}_m(85) = 0.15 + j\left(85 \times 0.1 - \frac{1000}{85}\right)$$
$$= 0.15 - 3.26j = 3.26\angle -87° \text{ rayl} \tag{1}$$

From Equation 1.66, the complex impedance of the SDOF at 80 rad/s is given by

$$\overline{Z}_m(80) = 0.15 + j\left(80 \times 0.1 - \frac{1000}{80}\right)$$
$$= 0.15 - 4.5j = 4.5\angle -88° \text{ rayl} \tag{2}$$

The complex excitation force for the 85 rad/s excitation is given by

$$\tilde{F}_1 = 2e^{j\pi/6}e^{j85t} \text{ N} \tag{3}$$

The complex excitation force for the 80 rad/s excitation is given by

$$\tilde{F}_2 = 2e^{j\pi/9}e^{j80t} \text{ N} \tag{4}$$

From Equation 1.66, the complex velocity response due to the first excitation frequency is given by

$$\tilde{v}_1 = \frac{\tilde{F}_1}{Z_m(85)} = \frac{2e^{j\pi/6}e^{j85t}}{3.26\angle -87°} = 0.61\angle 107°e^{j85t} \text{ m/s} \tag{5}$$

From Equation 1.66, the complex velocity response due to the second excitation frequency is given by

$$\tilde{v}_2 = \frac{\tilde{F}_2}{Z_m(80)} = \frac{2e^{j\pi/9}e^{j80t}}{4.5\angle -88°} = 0.44\angle 108°e^{j80t} \text{ m/s} \tag{6}$$

From Equations 5 and 6, the time-domain velocity response for the given force excitation is given by

$$\begin{aligned} v(t) &= \text{Re}(\tilde{v}_1) + \text{Im}(\tilde{v}_2) \\ &= 0.61\cos(85t + 107°) + 0.44\sin(80t + 108°) \text{ m/s} \end{aligned} \tag{7}$$

Example 1.3

An SDOF has the following parameters: $m = 0.05$ kg, $k = 20$ kN/m, $R_m = 20$ N-s/m. If a harmonic force of root mean square (rms)* amplitude 0.1 N and frequency 90 Hz is applied, determine the rms magnitude of velocity and the phase difference between velocity and force. Draw a phase diagram showing the given values.

From Equation 1.68, the magnitude of impedance is given by

$$Z_m(f = 90 \text{ Hz}) = \sqrt{20^2 + \left(2\times\pi\times90\times0.05 - \frac{20\times10^3}{2\times\pi\times90}\right)^2} = 21.22 \text{ rayl} \tag{1}$$

From Equation 1.69, the mechanical phase angle is given by

$$\phi_m(f = 90 \text{ Hz}) = \tan^{-1}\left[\frac{\left(2\times90\times\pi\times0.05 - \dfrac{20\times10^3}{2\times90\times\pi}\right)}{20}\right] = -19.53° \tag{2}$$

The rms magnitude of velocity response is given by

$$V = \frac{F}{Z_m} = \frac{0.1}{21.22} = 4.7 \text{ mm/s} \tag{3}$$

From Equation 1.71, the phase difference between velocity and force is given by

$$\begin{aligned} \phi^{V/F} &= -\phi_m = -(-19.53) \\ &= 19.53° \end{aligned} \tag{4}$$

From Equations 3 and 4, the velocity phasor is given by

$$\tilde{v} = 4.7 \text{ mm/s} \angle 19.53°e^{j\omega t} \tag{5}$$

The force phasor is given by

$$\tilde{f} = 0.1 \text{ N} \angle 0°e^{j\omega t} \tag{6}$$

---

* If there is a signal of single frequency, there should be a clear distinction between peak value and rms value. But, for a multi-frequency signal, or for a space-, frequency-, or time-averaged signal, rms value is used.

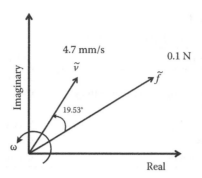

FIGURE 1.14
Phasor representation of velocity and force ($\omega = 180\pi$ rad/s (90 Hz)) (Example 1.3).

Both the force and velocity phasors will be rotating at a common excitation frequency of ω (180π rad/s in this case), but with a phase difference of 19.53°, the velocity will therefore lead the force. They are shown in Figure 1.14.

1.6 Electromechanical Analogy

Before we proceed further, it would be useful to discuss the analogy that exists between electrical circuits and mechanical systems. To begin with, we may try to establish the analogy between a series LRC circuit and an SDOF mechanical system.

Equation 1.66 for complex mechanical impedance can be rewritten as

$$\bar{Z}_m = R_m + j(\omega m - k/\omega) \tag{1.76}$$

The complex impedance of a series LRC circuit is given by

$$\bar{Z} = R + j(\omega L - 1/\omega C) \tag{1.77}$$

where R, L, C are, respectively, resistance, inductance, and capacitance. Figure 1.15 shows such an arrangement of these elements in an ac circuit.

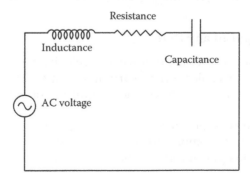

FIGURE 1.15
LRC circuit.

TABLE 1.1

Electric Analog of Mechanical Systems

Mechanical	Electrical
Mass	Inductance
Compliance (reciprocal of stiffness)	Capacitance
Damping	Resistance
Force	Voltage
Velocity	Current
Displacement	Charge

By comparing the complex impedance of the mechanical and electric systems given in the above equations, analogy between its elements can be established. Furthermore, by examining the relation between force, velocity, and mechanical impedance, analogy of other parameters related to input and response can be obtained. They are as shown in Table 1.1.

The earlier analogy, where force and voltage are analogous, is known as *force–voltage analogy* or *direct analogy*. Similarly, a force–current analogy can be obtained for a parallel LRC circuit driven by a current source. Parallel LRC and the series LRC are dual circuits. Therefore, depending upon the mechanical system that is being modeled, either type of analogy can be used. At the moment the force–voltage analogy is sufficient and it will be very useful to understand the concept of power input to a SDOF system, which is discussed in the next section.

1.7 Power Input

Unless a dynamic system is willing to accept power, any amplitude of external excitation cannot cause significant response. In addition, since impedance is a frequency-dependent property, the power input will also be frequency dependent. Therefore, the power that is input to a dynamic system at various frequencies is a very useful measure of its frequency characteristics. Hence, determination of power input to a dynamic system as a function of excitation frequency can give better insight into its characteristics.

1.7.1 Average Power Input to an SDOF

Power is defined as the scalar product of force and velocity. Since force and velocity vary with respect to time, power is also a time-varying quantity. Therefore, at any instant of time, the product of the magnitude force and magnitude of velocity gives the instantaneous power.

Since the real part of the complex force results in the real part of the complex response and the imaginary part of the complex force results in the imaginary part of the complex response, the instantaneous power can thus be written as

$$W_i = \Re(\tilde{F})\Re(\tilde{v}) = \mathrm{Im}(\tilde{F})\,\mathrm{Im}(\tilde{v}) \tag{1.78}$$

We shall, however, use the product of the real part of the complex force and the real part of the complex velocity.

Equation 1.65 for complex velocity can be written as

$$\tilde{v} = \frac{\tilde{F}}{Z_m}(\cos\phi_m - j\sin\phi_m) \tag{1.79}$$

where ϕ_m is the phase angle of the complex mechanical impedance defined by Equation 1.69.

For a complex harmonic force $\tilde{F} = Fe^{j\omega t}$, the real part of velocity from Equation 1.79 is given by

$$\Re(\tilde{v}) = \left(\frac{F\cos\omega t\cos\phi_m + F\sin\omega t\sin\phi_m}{Z_m}\right) = \frac{F\cos(\omega t - \phi_m)}{Z_m} \tag{1.80}$$

and the real part of the force is given by

$$\Re(\tilde{F}) = F\cos\omega t \tag{1.81}$$

To keep the derivation simple, it is assumed that the amplitude of the force is real (F instead of $\bar{F}$). That means the force has started at $t = 0$. This, however, does not result in any loss of generality.

The instantaneous power input to the SDOF system from Equations 1.78, 1.80, and 1.81 is given by

$$W_i = \frac{F^2\cos\omega t\cos(\omega t - \phi_m)}{Z_m} \tag{1.82}$$

The average power supplied during a cycle can be obtained by integrating the instantaneous power input over a period T:

$$\langle W \rangle_t = \frac{1}{T}\int_0^T W_i \, dt \tag{1.83}$$

(angle brackets, also known as chevrons, are used to indicate average, and the subscript next to the closing bracket indicates the parameter that is used for averaging) which yields

$$\langle W \rangle_t = \frac{F^2}{2Z_m}\cos\phi_m = \frac{F_{rms}^2}{Z_m}\cos\phi_m = F_{rms}V_{rms}\cos\phi_m \tag{1.84}$$

where the rms value of force amplitude is given by

$$F_{rms} = \frac{F}{\sqrt{2}} \tag{1.85}$$

The maximum value of the average power occurs when the mechanical power factor, $\cos \phi_m$ is unity, that is when the reactance is zero and its value is given by

$$W_{\max} = \frac{F_{rms}^2}{R_m} \tag{1.86}$$

The average power input to an LRC circuit is given by

$$\langle W \rangle_t = E_{rms} I_{rms} \cos \phi \tag{1.87}$$

By using the mechanical equivalent of Equation 1.87 and the force–voltage analogy, Equation 1.87 can be directly derived. A general procedure for directly computing the input power using complex force and velocity is presented in the next section, which can be used to obtain Equation 1.84 more easily. In addition, the general equation will be useful in later chapters also for computing sound intensity using dynamic pressure and particle velocity.

1.7.2 General Power Input Equation

A general power input equation can be derived based on the SDOF model, which can as well be applied to any complex dynamic system. It is obtained as follows:

Let $\tilde{F} = \bar{F} e^{j\omega t}$ be a harmonic force applied to an SDOF of impedance $\bar{Z}_m$; $\tilde{v}(t)$ be the velocity response and $\tilde{v}^*$ be the complex conjugate of $\tilde{v}$.

The instantaneous power input is given by

$$W_i(t) = \operatorname{Re}(\tilde{F})\operatorname{Re}(\tilde{v}) = \operatorname{Im}(\tilde{F})\operatorname{Im}(\tilde{v}) \tag{1.88}$$

We will, however, not directly use Equation 1.88, which would require integration with respect to time. Instead, we will try to eliminate the time component by taking the product of force and velocity conjugate. The product of complex force and conjugate of velocity is given by

$$\tilde{F}\tilde{v}^* = \bar{F} e^{j\omega t} \bar{V}^* e^{-j\omega t} = \bar{F}\bar{V}^* \tag{1.89}$$

Let the complex amplitude of force be given by

$$\bar{F} = F e^{j\theta} \tag{1.90}$$

And the complex amplitude of velocity response be given by

$$\bar{V} = V e^{j\phi} \tag{1.91}$$

where the angles θ and ϕ are measured with respect to the same frame of reference.

Equations 1.90 and 1.91 are substituted in Equation 1.89 as

$$\bar{F}\bar{V}^* = Fe^{j\theta}Ve^{-j\phi} = FVe^{j(\theta-\phi)} \tag{1.92}$$

From Equation 1.92, the real part of the product of complex amplitude of force and the conjugate of the complex amplitude of velocity is given by

$$\text{Re}(\bar{F}\bar{V}^*) = FV\cos(\theta-\phi) \tag{1.93}$$

Equation 1.93 can be written as

$$\frac{1}{2}\text{Re}(\bar{F}\bar{V}^*) = \frac{1}{2}FV\cos(\theta-\phi) \tag{1.94}$$

Expressing the peak amplitudes of force and velocity in terms of their rms values, Equation 1.94 becomes

$$\frac{1}{2}\text{Re}(\bar{F}\bar{V}^*) = F_{rms}V_{rms}\cos(\theta-\phi) = F_{rms}V_{rms}\cos\phi_m \tag{1.95}$$

where $\theta - \phi = \phi_m$ is the mechanical phase angle. The right-hand side of Equation 1.95 represents the average power input, which was obtained in the previous section using Equation 1.88 or 1.78.

Therefore, the time-averaged power input in terms of complex force and velocity of any dynamic system is given by

$$\langle W \rangle_t = \frac{1}{2}\text{Re}\{\tilde{F}\tilde{v}^*\} = \frac{1}{2}|\tilde{v}|^2 \text{Re}\{\bar{Z}\} \tag{1.96}$$

Equation 1.96 is a very important equation that is used to compute the power input to a structure of any complexity using the driving point force and its velocity response, and is, therefore, one of the widely used equations. An extension of the same equation can be used to compute sound intensity measured by two closely spaced microphones.

1.7.3 Power Input and Resonance

At the condition of maximum power input, impedance due to the mass and stiffness cancel out each other and the only resistance to the motion is due to damping. This condition refers to the resonance condition at which the system can draw maximum power from the excitation system. If this excess power is not dissipated by a suitable damping mechanism, it may result in large amplitudes of vibration resulting in the possible failure of the structure. In case of sound radiation, when the vibration of the body couples well with the surrounding fluid medium, the resonance condition will result in high noise levels. Due to these reasons, conditions around the resonance frequency and damping mechanisms are a focal point of discussion in dealing with vibration and noise problems.

1.7.4 Power Input Frequency Characteristics

The average power input to an SDOF system can be nondimensionalized with respect to the maximum power that can be drawn from the source at the resonance frequency. From Equations 1.84 and 1.86 this nondimensionalized value, denoted as power ratio W_α, and is given by

$$W_\alpha = \frac{W_{\text{avg}}}{W_{\text{max}}} = \frac{R_m^2}{Z_m^2} = \frac{(2\zeta r)^2}{(2\zeta r)^2 + (1 - r^2)^2} = (\cos\phi_m)^2 \tag{1.97}$$

where $r = \omega/\omega_n$ is the frequency ratio. Equation 1.97 can be plotted for values of damping ratio and frequency, as shown in Figure 1.16.

The abscissa represents the ratio of the forcing frequency and natural frequency, commonly called frequency ratio and the ordinate represents the power ratio W_α.

The curves shown in Figure 1.16 have a peak value at the resonance frequency and fall off at lower and higher frequencies. For small values of damping, the curves are symmetric about the resonance frequency and have a very sharp peak near the resonance. For large values of damping, the curves are asymmetric about the resonance frequency and have a wider frequency band around resonance. It is interesting to note that the power ratio is directly related to the square of the mechanical power factor. For large values of damping, the power ratio has a significant value even away from the resonance frequency. This can be explained by revisiting Figure 1.12b. For small values of damping, the locus of the magnitude of impedance is closer to the imaginary axis. Hence, the mechanical phase angle is close to ±90° at excitation frequencies far away from resonance. Whereas for large values of damping, the

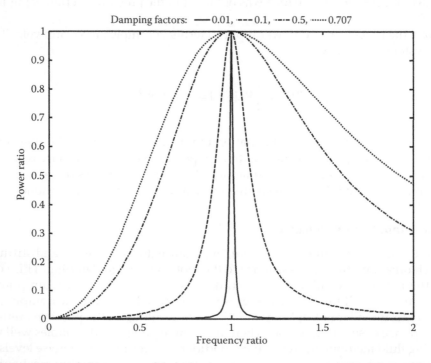

FIGURE 1.16
Power input as a function of frequency.

locus of the magnitude of mechanical impedance is far away from the imaginary axis and the mechanical phase angle is much less than ±90° at excitation frequencies far away from resonant frequency. Therefore, even at nonresonant frequencies, large values of damping allow the system to draw some power from the source and dissipate it in the from of heat. This aspect is useful in using damping materials for vibration and noise reduction.

In Figure 1.16, the power ratio W_α of an SDOF system is presented for different values of damping. In terms of the absolute values of power input as a function of excitation frequency, the shape of the curves would look slightly different. This is obvious due to the fact that systems with less damping would draw more power than those with higher damping.

1.7.5 Half-Power Frequencies

Half-power frequencies of an SDOF system are those frequencies at which the power input to the system is half the power that is input at the resonance. With reference to Figure 1.16, half-power frequencies are those frequencies at which the power ratio is equal to 0.5. There are two such frequencies, on either side of the resonance, which meet this criterion. Half-power frequencies play a very important role in defining the bandwidth of an SDOF system and to determine damping based on experimental measurement of the response. From Equation 1.97, the condition for half-power frequencies can be expressed as

$$\cos^2 \phi_m = \frac{R_m^2}{Z_m^2} = 0.5 \tag{1.98}$$

Equation 1.98 results in

$$\cos \phi_m^{hp} = \pm \frac{1}{\sqrt{2}} \tag{1.99}$$

Therefore, the mechanical phase angle corresponding to the half-power frequencies are ±45° and hence it is the frequency at which the resistive impedance due to damping is exactly equal to the reactive impedance due to the mass and spring. Hence, in the half-power bandwidth, the action of damping starts dominating in comparison to the actions of mass and spring.

From Equations 1.69 and 1.99

$$\tan \phi_m^{hp} = \frac{\omega m - \dfrac{k}{\omega}}{R_m} = \pm 1 \tag{1.100}$$

$$\left(\frac{\omega m}{R_m} - \frac{k}{\omega R_m} \right)^2 = 1 \tag{1.101}$$

In terms of the frequency ratio, r, and damping factor, ζ, Equation 1.101 can be expressed as

$$\left\{ \frac{r}{2\zeta} - \frac{1}{2\zeta r} \right\} = \pm 1 \tag{1.102}$$

which results in two quadratic equations

$$r^2 + 2\zeta r - 1 = 0 \qquad r^2 - 2\zeta r - 1 = 0 \tag{1.103}$$

Solving the preceding quadratic equations and eliminating the roots that are less than zero, the half-power frequency ratios can be obtained as

$$r_1 = \sqrt{1+\zeta^2} - \zeta \qquad r_2 = \sqrt{1+\zeta^2} + \zeta \tag{1.104}$$

When the natural frequency of the system is known, its corresponding half-power frequencies can be computed from the preceding equations. The frequency band between the half-power frequencies is known as the *half-power bandwidth*.

1.7.6 Quality Factor and Damping Ratio

A precise definition of sharpness of resonance can be defined in terms of quality factor Q. Quality factor is defined as the ratio of the resonance frequency to half-power bandwidth. Quality factor Q is given by

$$Q = \frac{\omega_n}{\omega_2 - \omega_1} = \frac{1}{r_2 - r_1} \tag{1.105}$$

where ω_2 is the half-power frequency above the resonance frequency and ω_1 is the half-power frequency below the resonance frequency; r_2 and r_1 are the corresponding frequency ratios given by Equation 1.104.

From Equations 1.104 and 1.105, it can be proven that, for lightly damped systems, the quality factor is related to the damping factor by the following equation:

$$r_2 - r_1 = 2\zeta; \quad Q = \frac{1}{2\zeta} \tag{1.106}$$

Therefore, quality factor obtained from the half-power points can be used to determine the damping coefficient of an SDOF. This was the earlier method used for determining damping. However, due to sharpness of peaks for low damping values, the half-power points could not be located with consistency. But if the half-power points can be located using the phase information, the damping value can be more accurately obtained.

1.8 Frequency Regions

An important property of an SDOF is its frequency dependence that makes it behave differently at various excitation frequencies; in fact, most dynamic systems have frequency-dependent properties. Hence they react according to the frequency at which they are excited. In addition, these dynamic systems have frequency-dependent properties that are unique to them and hence follow a definite pattern in their response. Hence, the input power the system can draw at various excitation frequencies is frequency dependent. Even living beings have dynamic characteristics of many sensory organs related to frequencies. For example, our hearing mechanism responds to frequencies in the audible range. Therefore, the frequency-centered study is a natural consequence of the study of how living organisms evolved. Since the impedance of an SDOF is also frequency dependent, most of the discussion is centered on the frequency response of an SDOF, which is simplified by the use of a complex exponential.

1.8.1 Input Power versus Frequency

Let us consider the power curve of Figure 1.16 corresponding to a damping factor of 0.1 and show it as Figure 1.17. Three distinct regions are marked on this figure. The region between the half-power points is the damping-controlled region. The region to the left of the damping-controlled region, below the half-power band is the stiffness-controlled region and the region to the right of the damping-controlled region, above the half-power band, is the mass-controlled region.

As a first step, let us demonstrate the existence of these regions by examining the complex impedance given by Equation 1.66. The following observations can be made:

1. In the stiffness-controlled region, the forcing frequency is very low in comparison to the natural frequency, and therefore the term k/ω dominates the impedance equation that is approximately equal to $\bar{Z}_m \approx -jk/\omega$. Therefore impedance is determined by the stiffness of the system and hence the name *stiffness-controlled region*.

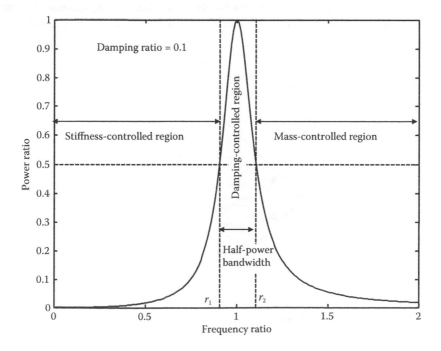

FIGURE 1.17
Frequency regions of a SDOF.

TABLE 1.2

Frequency Responses in Various Regions

	Stiffness-Controlled Region	Damping-Controlled Region	Mass-Controlled Region
Displacement	$\tilde{F}/k$	$-j\tilde{F}/\omega_n R_m$	$-\tilde{F}/\omega^2 m$
Velocity	$j\omega\tilde{F}/k$	$\tilde{F}/R_m$	$\tilde{F}/j\omega m$
Acceleration	$-\tilde{F}\omega^2/k$	$j\omega_n\tilde{F}/R_m$	$\tilde{F}/m$

2. Around the resonance region, the action of the stiffness and mass cancel each other resulting in the domination of the mechanical resistance in this region and hence the name *resistance-controlled region*. This region exists between the half-power points where $Z_m \approx R_m$.

3. At frequencies higher than the resonance frequency, beyond the higher half-power point, the term ωm dominates the complex impedance, that is $\bar{Z}_m \approx j\omega m$. This region is called the *mass-controlled region* since changing the mass can change the impedance.

Displacement, velocity, and acceleration in different frequency regions are as shown in Table 1.2.

1.8.2 Mechanical Phase Angle versus Frequency

Since the power ratio is directly related to the mechanical phase angle through Equation 1.98, frequency regions as shown in Figure 1.18 can also be marked on the magnitude and phase

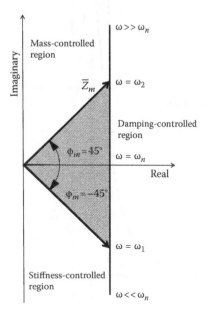

FIGURE 1.18
Nyquist plot of impedance indicating half-power points.

plot of impedance. Since half-power frequencies correspond to a power ratio of 0.5, the corresponding mechanical phase angle of impedance is ±45°. The region within this range of phase angle is marked as the damping-controlled region. The mechanical phase angle is in the fourth quadrant for the stiffness-controlled region and in the first quadrant for the mass-controlled region, as shown in Figure 1.18. It should also be noted that corresponding to the half-power frequencies, the resistive and reactive impedances are of the same order of magnitude, and within the damping-controlled region, resistive impedance becomes dominant.

Example 1.4

For the system described in Example 1.3 ($\zeta = 0.32, f_n = 100.66$ Hz):

    a. Determine the half-power frequencies.
    b. If a harmonic force of peak amplitude 0.1 N and the following frequencies is applied, determine the power input in each case:
       i.   20 Hz
      ii. 90 Hz
    iii. 120 Hz
    iv. 200 Hz

From Equation 1.104, the half-power frequency ratios are given by

$$
\begin{aligned}
r_{1,2} &= \sqrt{1+\zeta^2} \pm \zeta \\
&= \sqrt{1+0.32^2} \pm 0.32 \\
&= 0.73, 1.37
\end{aligned}
\tag{1}
$$

From Equation 1, the half-power frequencies are given by

$$
\begin{aligned}
f_{1,2} &= f_n \times r_{1,2} \\
&= 100.66 \times 0.73 = 73.48 \text{ Hz} \\
&= 100.66 \times 1.37 = 137.90 \text{ Hz}
\end{aligned}
\tag{2}
$$

The rms magnitude of velocity is given by

$$
v_{rms} = \frac{F}{\sqrt{2}Z_m}
\tag{3}
$$

From Equation 1.84, the average power input is given by

$$
\langle W \rangle_t = F_{rms} v_{rms} \cos\phi_m
\tag{4}
$$

For various excitation frequencies, the corresponding value of rms velocity response, magnitude and phase of the complex mechanical impedance, power factor, and power input are shown in Table 1.3. The same results are plotted in Figure 1.19.

Since the average power consumed at resonance is 0.25 mW, 0.125 mW corresponds to the average power consumed at half-power points. Therefore, excitation frequencies of 90 and 120 Hz, which consume more than 0.125 mW, are within the half-power bandwidth.

TABLE 1.3

Excitation Frequency, rms Velocity, Phase, and Power Input (Example 1.4)

No.	Excitation Frequency (Hz)	Excitation Frequency (rad/s)	Z_m (rayl)	ϕ_m (deg)	v_{rms} (mm/s)	$\cos \phi_m$	<W> Watt
1	20	40 π	154.17	−82.50	0.469	0.13	4.20 μW
2	90	180 π	21.22	−19.53	3.300	0.94	0.22 mW
3	100.66	201.32 π	20.00	0	3.53	1	0.25 mW
4	120	240 π	22.91	29.19	3.100	0.87	0.19 mW
5	200	400 π	51.00	66.90	1.400	0.39	38.4 μW

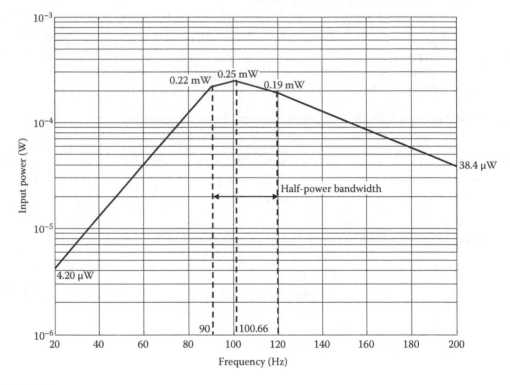

FIGURE 1.19
Input power versus frequency of the SDOF system for $F = 0.1$ N (Example 1.4).

Example 1.5

Consider a single-degree-of-freedom system with spring, mass, and viscous damper. The coefficient of viscous damping is 10 N-s/m. When it is subjected to a force of rms value 0.25 N, through an exciter at a single excitation frequency of 100 rad/s in the stiffness-controlled region, it consumes one-fourth of the power it would have consumed when excited at its resonance frequency.

a. Determine the actual power consumed at this excitation frequency (100 rad/s).
b. What is the power consumed and the rms velocity when excited at its resonance frequency?

The following is given:

$R_m = 10$ N-s/m
$F_{rms} = 0.25$ N
$\omega_1 = 100$ rad/s (stiffness-controlled region)
$W_\alpha =$ power ratio $= 1/4 = \cos^2 \phi_m$

ANSWERS

a. Since $\cos^2 \phi_m = 1/4$, $\phi_m = -60°$, the excitation frequency is in the stiffness-controlled region.

The impedance at the given excitation frequency is given by

$$Z_m(100) = \frac{R_m}{\cos \phi_m} = \frac{10}{0.5} = 20 \text{ rayl} \tag{1}$$

Using the preceding value of impedance, the rms velocity of response at the excitation frequency of 100 rad/s is given by

$$v_{rms}(100) = \frac{F_{rms}}{Z_m(100)} = \frac{0.25}{20} = 12.5 \text{ mm/s} \tag{2}$$

Power consumed by this system at an excitation frequency of 100 rad/s is given by

$$W_{avg}(100) = F_{rms} v_{rms} \cos \phi_m = 0.25 \times 0.0125 \times 0.5 = 1.56 \text{ mW} \tag{3}$$

b. Power consumed at an excitation frequency equal to the resonance frequency is given by 4 times the power consumed at an excitation frequency of 100 rad/s, that 1.56 mW × 4 = 6.2 mW.

The velocity response at an excitation frequency equal to resonance frequency is given by $W_{avg}(\text{res})/F_{rms} = 0.0062/0.25 = 25$ mm/s.

Example 1.6

A single-degree-of-freedom system consists of a spring ($k = 1000$ N/m), mass ($m = 0.1$ kg), and dashpot ($R_m = 0.71$ N-s/m) connected in the usual arrangement. The phase angle between velocity and force at an excitation frequency is 25°.

a. What is the value of the excitation frequency?
b. What is the magnitude of impedance at this excitation frequency?
c. What is the rms value of velocity at this excitation frequency, if the rms value of force is kept at 0.05 N at all excitation frequencies?
d. What is the power input to the system at this excitation frequency?
e. What is the maximum possible value of power input to this system from the given excitation force?

ANSWERS

a. The mechanical phase angle at the given excitation frequency is given by

$$\phi_m = -\phi_{v/F} = -25° \tag{1}$$

The tangent of the mechanical phase angle is given by

$$\tan\phi_m = \frac{\omega m - \dfrac{k}{\omega}}{R_m} \tag{2}$$

In terms of excitation frequency, Equation 2 can be expressed in the quadratic form as

$$\omega^2 m - \omega\tan\phi_m R_m - k = 0 \tag{3}$$

The positive root of Equation 3 gives the excitation frequency, given by

$$\omega = \frac{\tan\phi_m R_m \pm \sqrt{(\tan\phi_m R_m)^2 + 4km}}{2m}$$

$$= \frac{\tan(-25)\times 0.71 \pm \sqrt{(\tan(-25)\times 0.71)^2 + 4\times 1000\times 0.1}}{2\times 0.1} = 98.35 \text{ rad/s} \tag{4}$$

b. From Equation 1.68, the magnitude of mechanical impedance at the excitation frequency of 98.35 rad/s is given by

$$Z_m = \sqrt{R_m^2 + \left(\omega m - \frac{k}{\omega}\right)^2}$$

$$= \sqrt{0.71^2 + \left(98.35\times 0.1 - \frac{1000}{98.35}\right)^2} \tag{5}$$

$$= 0.784 \text{ rayl}$$

c. The rms velocity at the given excitation frequency is given by

$$v_{rms} = \frac{F_{rms}}{Z_m} = \frac{0.05}{0.784} = 63.8 \text{ mm/s} \tag{6}$$

The average power consumed at the given excitation frequency is given by

$$\langle W\rangle_t = F_{rms} v_{rms} \cos\phi_m$$

$$= 0.05\times 0.00638\times \cos 25 \tag{7}$$

$$= 2.9 \text{ mW}$$

d. The maximum rms velocity response at resonance is given by

$$v_{rms}(\text{max}) = \frac{F_{rms}}{R_m} = \frac{0.05}{0.71} = 70.4 \text{ mm/s} \tag{8}$$

e. The maximum average power input at resonance is given by

$$W_{avg}(\text{max}) = F_{rms}v_{rms}(\text{max}) = 0.05 \times 0.0704 = 3.5 \text{ mW} \tag{9}$$

1.9 Phase Relations

Phase relations play a very important role in the analysis of many vibration and acoustic problems. Although their definition has been known for a long time, only recently have they been used for important practical applications. Simply stated, phase difference is the time difference measured from a common reference that exists between two signals of the same frequency. If the time signals are represented as phasors rotating in a complex plane, the phase difference is the angle measured between the phasors. Due to the simplicity of representing phase angle through phasors, phase relations are always represented in terms of angular measurement. It will be seen in this section how the use of complex analysis can be used for determining the phase relationship of two time-varying quantities.

The equations for displacement, velocity, and acceleration are rewritten as follows, in order to examine the phase relations between them and the force in greater detail.

1.9.1 Phase Relations in Terms of ϕ_m

The phase angle ϕ_m is related to the complex mechanical impedance and is given by

$$\tan\phi_m = \frac{\omega m - k/\omega}{R_m} = \frac{r^2 - 1}{2\zeta r} \tag{1.107}$$

Equation 1.107 can be used to compute the phase angle ϕ_m of the impedance in the stiffness-, resistance-, and mass-controlled regions. For low values of damping, ϕ_m is approximately $-90°$, $0°$, and $90°$ in the respective regions.

1.9.2 Phase between Displacement and Force

The phase diagram for displacement and force is shown in Figure 1.20. For small values of damping, the displacement is in phase with force in the stiffness-controlled region and out of phase in other regions.

The in-phase relationship between the displacement and force basically means that force and displacement attain maximum, minimum, and zero values at the same time, although they may have different values of maxima or minima and are different quantities in the

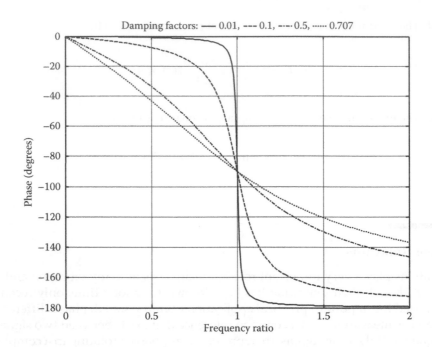

FIGURE 1.20
Phase difference between displacement and force.

physical sense. Since displacement follows the force in this region it is more like a static load compressing a spring to store potential energy. Energy is stored by the system during half the cycle and returned back to the input excitation system during the other half cycle. Therefore, the displacement of the mass is entirely controlled by the stiffness of the spring at all frequencies in this region. The average power input due to the system is negligible and is approximately $F_{rms}^2 R_m/(k/\omega)^2$. The power factor of the impedance is close to zero in this region and therefore the system can draw very little power from the source. At higher values of damping, the phase relationship significantly deviates from that described earlier. These phase relationships will have to be considered while increasing the damping of a system through external means.

1.9.3 Phase between Velocity and Force

In the damping-controlled region, at the resonance frequency, for all values of damping, velocity, and force are in phase with each other as shown in Figure 1.21. Since the system is vibrating near the resonance frequency, exchange of kinetic, and potential energy takes place. We have earlier seen that at the resonance frequency, the magnitude of the power input is independent of the frequency of vibration. The mass and stiffness nullify each other's action due to which the system cannot offer resistance to the input power, except due to damping. Therefore, the input energy is only dissipated through damping, and hence displacement, velocity and acceleration are much larger than in the other regions. Due to the above reasons, the system draws maximum power from the source and the mechanical power factor is unity in this region. In the absence of sufficient damping, the velocity and displacement amplitudes reach very high values resulting in the failure of the system. For low values of damping ($\zeta = 0.01$), velocity and force are in quadrature with each other

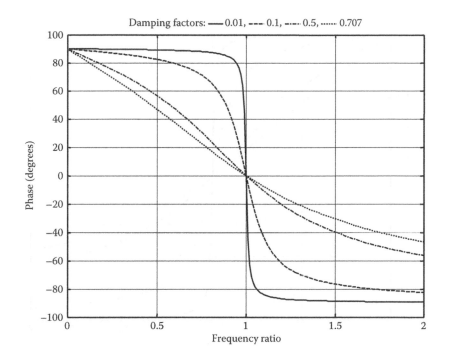

FIGURE 1.21
Phase difference between velocity and force.

outside the resonance frequency and therefore cannot draw any power from the source. For higher values of damping, the phase difference between velocity and force is almost linear, varying from 90° to –90°. This results in a nonzero mechanical power factor outside the damping-controlled region, which helps the system to draw power from the source and dissipate it in the form of heat.

1.9.4 Phase between Acceleration and Force

The phase between acceleration and force is shown in Figure 1.22. In the mass-controlled region, the acceleration amplitude is greater than displacement and velocity amplitudes. First consider the curve corresponding to low damping, $\zeta = 0.01$. Since force and acceleration are directly related, acceleration is in phase with the force. Velocity lags behind the force by 90° and the displacement by 180°. Similar to the stiffness-controlled region, the power factor of the impedance is close to zero due to which the system can draw very little power from the system. The displacement amplitude is very less in this region. Velocity amplitudes are also less than the acceleration amplitudes, but much higher than the displacement amplitudes. For higher values of damping, the phase relationships show a significant change in their behavior as shown in Figure 1.22.

The results shown in Figures 1.20, 1.21, and 1.22 can be summarized as follows. The phase relationship is flat in the stiffness- and mass-controlled regions. For higher values of damping, the phase relations tend to be linear. These characteristics of phase behavior are important in designing sensing elements of transducers. It is important that transducers have a flat response for amplitude so they can be calibrated at one frequency and used in a wider frequency band. Therefore the sensing elements must have the right amount of damping

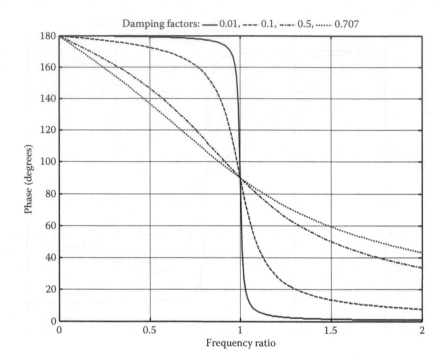

FIGURE 1.22
Phase difference between acceleration and force.

TABLE 1.4

Phase (Degrees) Relationship in Various Regions

	Stiffness-Controlled Region	Damping-Controlled Region	Mass-Controlled Region
Displacement and force	0	−90	−180
Velocity and force	90	0	−90
Acceleration and force	180	90	0

and natural frequency to have a flat response for amplitude within the frequency band of interest. The phase should be either zero or linearly change with frequency. For small values of damping, the phase relationships are summarized in Table 1.4 for various regions.

1.10 Frequency Response

Based on the discussion thus far, the frequency response for displacement, velocity, and acceleration can be presented as follows.

1.10.1 Displacement Frequency Response

A plot of displacement response of an SDOF system as a function of frequency is shown in Figure 1.23. The response is due to a complex exponential force of constant amplitude and

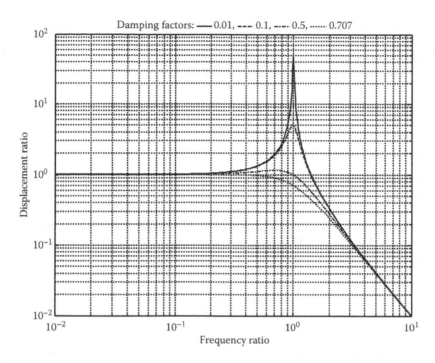

FIGURE 1.23
Displacement ratio versus frequency.

varying forcing frequency. The displacement response is expressed in terms of a nondimensionalized value known as the displacement ratio. Displacement ratio is defined as the ratio of displacement response at the excitation frequency to the displacement response at very low frequencies, known as *equivalent static deflection*. Equivalent static deflection is the deflection of the mass of an SDOF due to a static force of magnitude equal to that of the amplitude of the complex harmonic force. This is given by the ratio of amplitude of the force and stiffness. If X is the amplitude of the displacement response due to a harmonic force, the displacement ratio, x_α is given by

$$x_\alpha = \frac{X}{F/k} = \frac{1}{\sqrt{(2\zeta r)^2 + (1 - r^2)^2}} \qquad (1.108)$$

The term equivalent static deflection is used for the reason that at very low frequencies, the dynamic response equals this value. This is clearly seen in Figure 1.23 where the displacement ratio, which is the amplitude of the displacement response divided by the static deflection, is equal to one at frequencies up to 20% of the natural frequency for all values of damping. Therefore, the displacement ratio is approximately equal to one in the stiffness-controlled region. This flat response of the SDOF in this region is useful in the construction of accelerometers, which are the universal transducers for vibration measurement. An SDOF is used as a basic sensing element or a first stage element in an accelerometer. Other applications in which the flat response is useful include loudspeaker cones and microphone diaphragms. At frequencies higher than the resonance, the displacement ratio reduces as the square of the frequency. This would mean that at frequencies higher than the natural frequency, the displacement is significantly reduced.

The displacement ratio has peak values in the damping-controlled region. The forcing frequency at which the displacement response is a maximum can be obtained by minimizing the denominator of Equation 1.108, that is $\omega \bar{Z}_m$. Its value is slightly less than the undamped resonance frequency and is given by

$$\omega_r^2 = \omega_n^2 (1 - 2\zeta^2) \tag{1.109}$$

1.10.2 Velocity Frequency Response

The ratio of complex velocity to complex force is known as mobility. Unfortunately, however, the velocity response expressed in terms of mobility cannot be entirely expressed in terms of the nondimensionalized parameters of the SDOF system and the excitation, namely, the frequency ratio and damping factor. Therefore, the amplitude of mobility can only be determined for an SDOF system of particular values of mass, stiffness, and damping. The amplitude of the mobility is given by

$$\frac{V}{F} = \frac{r}{\left(\sqrt{km}\right)\sqrt{(2\zeta r)^2 + (1 - r^2)^2}} \tag{1.110}$$

The amplitude of mobility for an SDOF system of mass 1 kg and stiffness 1000 N/m is plotted for various values of damping as shown in Figure 1.24. Unlike the displacement response, the velocity response is symmetrical about the resonance frequency. Because of

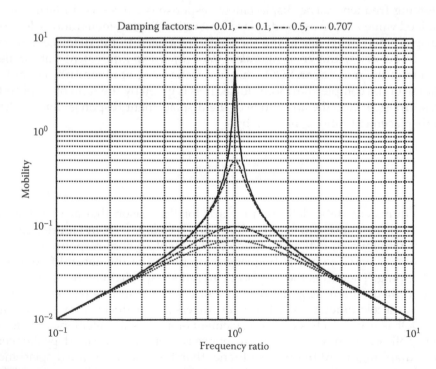

FIGURE 1.24
Mobility as a function of frequency.

this reason, the velocity response is more convenient than the displacement response and therefore preferred to be used in the analysis. In addition, velocity can be directly related to sound radiated by a vibrating structure. The mobility diagram (corresponding to the given damping factors) presented in Figure 1.24 can still be used to determine the mobility of any arbitrary SDOF system of mass m and stiffness k by multiplying by a scaling factor of $10/\sqrt{km}$.

1.10.3 Acceleration Frequency Response

From Equations 1.68 and 1.72, the following equation can be obtained for the magnitude of acceleration frequency response:

$$\frac{Am}{F} = \frac{1}{\sqrt{\left(\frac{2\zeta}{r}\right)^2 + \left(1 - \frac{1}{r^2}\right)^2}} \tag{1.111}$$

Equation 1.98 is the ratio of acceleration and force magnitude ratio, defined as accelerance, normalized by multiplying with the mass. Equation 1.111 is plotted for a damping factor of 0.1 for various frequencies and is shown in Figure 1.25. The normalized accelerance is very low at frequencies lower than the undamped frequency and reaches a peak at a frequency slightly more than undamped frequency; at higher frequencies, its value tends to unity. Since acceleration and force are in phase with each other at higher frequencies, the ratio of the magnitude of acceleration and force will be equal to the mass of SDOF.

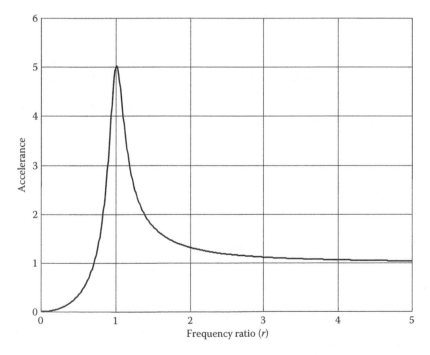

FIGURE 1.25
Magnitude of accelerance versus frequency.

1.11 Admittance

Although we have used the concept of mechanical impedance throughout most of the chapter, the concept of mechanical admittance is also useful in many applications. Admittance, defined as the ratio of velocity response to force input, is basically the reciprocal of impedance and for an SDOF is given by

$$\bar{Y}_m = \frac{\tilde{v}}{\tilde{F}} = \frac{1}{\bar{Z}_m} = \frac{1}{R_m + j(\omega m - k/\omega)} = \frac{e^{-j\phi_m}}{Z_m} \tag{1.112}$$

Therefore, the admittance of a dashpot is $1/R_m$, of a mass is $1/j\omega m$, and of a spring is $j\omega/k$. Since admittance is directly proportional to the velocity response, it is a direct measure of the velocity frequency response. Hence, it is much easier to comprehend the frequency response characteristics using admittance.

Admittance can be expressed in terms of mechanical resistance and mechanical phase angle as follows:

$$\begin{aligned}\bar{Y}_m &= \frac{1}{\bar{Z}_m} = \frac{1}{\bar{Z}_m} \times \frac{R_m}{R_m} = \frac{R_m}{Z_m} \times \frac{1}{R_m} \times e^{-j\phi_m} \\ &= \frac{\cos\phi_m e^{-j\phi_m}}{R_m}\end{aligned} \tag{1.113}$$

Example 1.7

Consider an SDOF of the following parameters: $k = 1000$ N/m, $m = 0.01$ kg, and $R_m = 1$ N-s/m. Plot the magnitude frequency response of admittance and its Nyquist plot showing half-power frequencies.

The frequency response of magnitude admittance is the reciprocal of the frequency response of magnitude impedance. It can therefore be plotted using

$$\frac{1}{\sqrt{R_m^2 + \left[(\omega m)^2 - \left(\frac{k}{\omega}\right)^2\right]}}.$$ It is shown in Figure 1.26.

The real and imaginary parts of admittance for mechanical phase angle values of $-90°$ to $+90°$ can be obtained using Equation 1.113 that corresponds to very small values of frequency to large values of frequency. The locus of this Nyquist plot is a circle of diameter $1/R_m$, as shown in Figure 1.27. The half-power points that correspond the mechanical phase angle values of $\pm45°$ are also shown in this figure.

As shown in Figure 1.27, the locus of admittance follows a circle of diameter equal to the reciprocal of mechanical resistance. The circle is divided into four symmetrical sections and represents the locus for very small to very large excitation frequency, and the locus starts from the origin for low frequencies and moves in the clockwise direction for higher frequencies. The half-power frequencies are represented by ω_1 and ω_2. The region from 0 to ω_1 is the stiffness-controlled region and the region from ω_1 to ω_n and from ω_n to ω_2 is the damping-controlled region; the region between ω_2 and ∞ is the mass-controlled region.

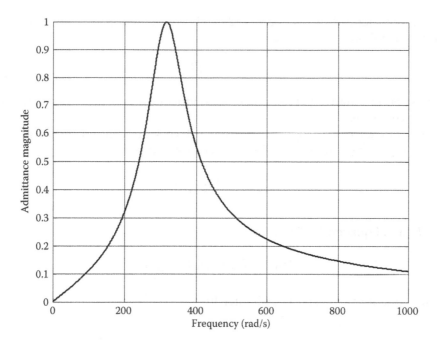

FIGURE 1.26
Frequency response magnitude of admittance (Example 1.7).

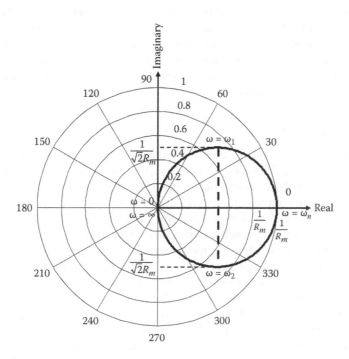

FIGURE 1.27
Nyquist plot of admittance (Example 1.7).

The admittance defined in Equation 1.112 can be expressed in terms of a real and complex numbers as follows:

$$\bar{Y}_m = G + jB \tag{1.114}$$

The real part of admittance (G) is known as conductance and the imaginary part (B) is known as susceptance.

The time-averaged power input expressed in terms of admittance has many useful applications. Recalling Equation 1.96 for average power input

$$\langle W_i(t) \rangle = \frac{1}{2} \text{Re}\{\tilde{F}\tilde{v}^*\} \tag{1.115}$$

Since $\tilde{v}^* = \tilde{F}^* \bar{Y}^*$, Equation 1.113 becomes

$$\langle W_i(t) \rangle = F_{rms}^2 \text{Re}\{\bar{Y}_m\} = F_{rms}^2 G \tag{1.116}$$

The above equations are useful in Chapter 10.

1.12 Vibration Transducers

A good choice of measuring instruments is extremely important in vibration analysis. The ability to obtain precise measurements will be very useful in trouble shooting and improving the design that has minimum vibration. The transducers used in vibration measurement are vibrometers, piezoelectric accelerometers, and piezoelectric impedance heads. In addition, vibration exciters and hammer excitation systems are required to induce dynamic motion to a system. In this section, transducers used for vibration measurement devices will be discussed.

1.12.1 Seismic Instrument

Vibration measurement requires a basic detection device that can be mounted on the vibrating part of a machine. This detection device must vibrate, picking up vibration of the machine, which can then be conveniently converted into an electric signal. The dynamic sensing systems of vibration pickups and accelerometers are usually of the "seismic mass" form, as shown in Figure 1.28. Since they are similar to those that were used in measuring earthquakes, they are known as seismic devices; in fact, earthquake measurement was done much before machinery vibration measurements were made.

The system consists of a spring-supported weight that is mounted in a suitable housing, with a sensing element provided to detect the relative motion between the mass and the housing. Suitable damping will also be provided, if necessary, and is represented by the dashpot. The output of a seismic device depends on the relative motion between the mass and the housing, which in turn is dependent on the natural frequency and damping of the system. Therefore, by appropriately designing natural frequency and damping, a seismic device can be made to measure motion (vibrometer) or acceleration (accelerometer). This is clear from the general theory described hereunder.

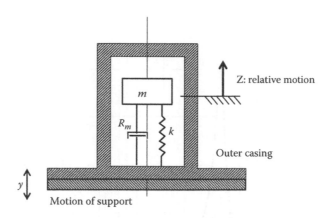

FIGURE 1.28
Seismic type of motion measuring instrument.

Let

k = spring stiffness

R_m = damping coefficient

Y = displacement amplitude of the supporting member

$y(t)$ = absolute displacement of the supporting member, $Y \cos \omega t$

Z = amplitude of relative displacement between mass and support

z = relative displacement between mass and support

ω = excitation frequency

ω_n = undamped natural frequency of the system

ζ = damping factor = $\dfrac{R_m}{2\sqrt{km}}$

r = frequency ratio = ω/ω_n

Applying Newton's law to the seismic system, the following equations for steady-state response can be obtained, either for measuring displacement or acceleration.

The peak amplitude of relative motion between the seismic and outer casing, Z, and the peak amplitude of input displacement, Y, can be obtained as

$$\frac{Z}{Y} = \frac{r^2}{\sqrt{(1-r^2)^2 + (2\zeta r)^2}} \tag{1.117}$$

Equation 1.117 represents a normalized response as a function of frequency ratio, r, and is shown in Figure 1.29.

In Equation 1.117, the output signal is proportional to the displacement of the input signal and hence it is used for designing a vibrometer for measuring the displacement amplitude of vibration. The frequency response of a vibrometer is shown in Figure 1.29. The useful range of frequencies is determined by the flat response region of the frequency response. Since the detection has to account for all the frequencies in a wider range, the

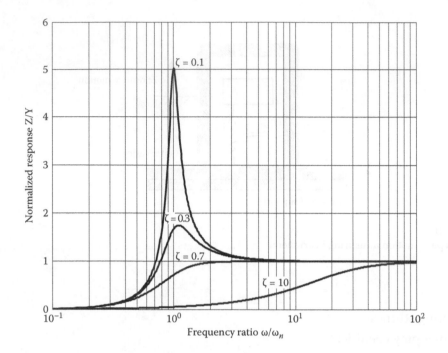

FIGURE 1.29
Response of a seismic instrument to harmonic displacement.

flat range ensures that the calibration of such a dynamic sensor needs to be done only at one of the frequencies. Therefore, the flat response range is the operation frequency range of any transducer. From the figure it is clear that the operational frequency for vibrometers is higher than the natural frequency of the detection system. As a result, depending on the lower limit of frequency, the vibrometer has a softer spring and a heavier mass. This results in serious limitations on the lower limit of frequency measurement. In addition, due to the large mass of the detection system, very high frequencies also cannot be accurately measured. Therefore, vibrometers have now become obsolete for vibration measurement. In the system shown in Figure 1.29, it is assumed that the next stage element converts relative displacement Z into suitable voltage. The same dynamic system can also be used to measure the amplitude of velocity, if the second stage element is sensitive to the relative velocity. However, both measurement systems will have the same disadvantages noted earlier.

If the detection system were to be made sensitive to acceleration, Equation 1.117 can be modified to obtain the following equation:

$$\frac{Z\omega_n^2}{\ddot{Y}} = \frac{1}{\sqrt{(1-r^2)^2 + (2\zeta r)^2}} \tag{1.118}$$

In Equation 1.118, the relative displacement is proportional to the input acceleration. The natural frequency of the detection system is only a scaling factor and hence moved to the numerator on the left-hand side. The right-hand side of Equation 1.118 therefore represents the normalized frequency response of an acceleration detection system. Equation 1.118 is plotted in Figure 1.30 as a function of frequency ratio.

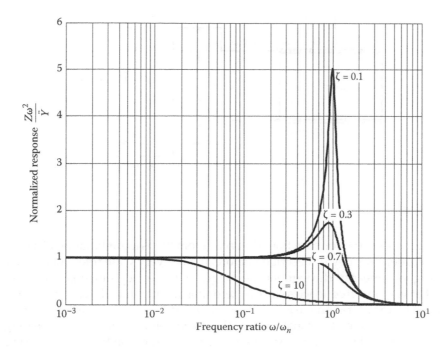

FIGURE 1.30
Response of a seismic instrument to sinusoidal acceleration.

From Figure 1.30, it is clear that the operational frequency of accelerometers is below the natural frequency of the system; the natural frequency is generally of the order of 50 kHz that would give the upper limit vibration measurement as 10 kHz that is sufficient for most applications. As a result, the accelerometer has a stiff spring and a lighter mass. Therefore, accelerometers are lighter and more rugged than vibrometers and hence are widely used for vibration measurement.

1.12.2 Piezoelectric Transducer

The piezoelectric transducer employs the property of certain types of materials that when deformed produce an electric charge. Alternately, when a voltage is applied across them they undergo deformation. This effect, called the piezoelectric effect, is exhibited by quartz and Rochelle salt crystals found in nature. Lithium sulfate, ammonium dihydrogen phosphate crystals, and polarized barium titanate, which can be synthetically prepared, also exhibit similar characteristics. In the piezoelectric transducer, the vibratory motion is made to cause a vibratory deformation of the crystal from which a proportionate voltage is obtained. Figure 1.31 shows the schematic of a piezoelectric transducer. The piezoelectric transducer is used as a second stage device in conjunction with a detection system that may be a vibrometer or an accelerometer, described earlier. Its input is the relative displacement output of the detection device.

1.12.3 Piezoelectric Accelerometer

The transducer, which is almost universally used for vibration measurement, is the piezoelectric accelerometer. It exhibits better all-round characteristics than any other type of

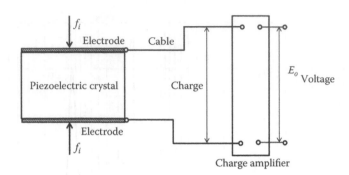

FIGURE 1.31
Piezoelectric transducer.

vibration transducer. The construction of an accelerometer is a combination of a seismic system and a piezoelectric device described before. It has a very wide frequency and dynamic range with good linearity throughout the range. Its frequency range is similar to Figure 1.30, except at very low frequencies close to zero that are attenuated due to the piezo device. Therefore, accelerometers can measure dynamic signals above a certain frequency only.

In an actual accelerometer construction, the piezoelectric element is so arranged that when the assembly vibrates, the mass applies an inertial force on the piezoelectric element, which is proportional to input acceleration. For frequencies well below the resonant frequency of the equivalent spring–mass system, the acceleration of the mass will be same as the acceleration of the base and the output signal magnitude will therefore be proportional to the acceleration to which the pickup is subjected. Two accelerometer combinations are in common use. In the compression type, the mass exerts a compressive force on the piezoelectric element and in the shear type the mass exerts a shear force. Figure 1.32 shows a compression type of accelerometer design.

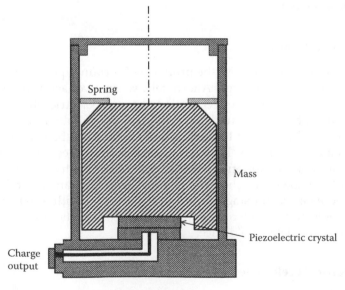

FIGURE 1.32
Piezoelectric accelerometer.

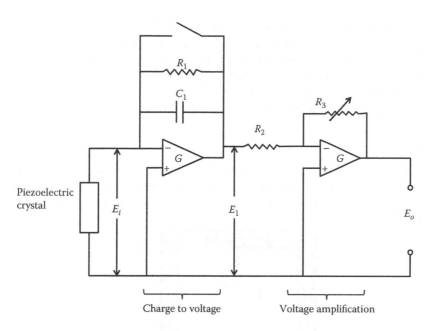

FIGURE 1.33
Charge amplifier.

1.12.4 Charge Amplifier

The output signal of a piezoelectric crystal is a charge at high impedance, which has to be converted into voltage at low impedance and amplified. A device, which can perform this task, is the charge amplifier as shown schematically in Figure 1.33. Since spectrum analyzers have a low impedance input, a charge amplifier is an essential part of any vibration measurement system. The first component in the system is a high impedance operational amplifier with capacitive feedback. This component serves as an integrator and converts the charge to voltage. The second operational amplifier with an adjustable feedback resistor is used to amplify the voltage by specified amounts.

Charge amplifiers are commercially available with input impedances of 1014 Ω and maximum output voltages of 10 V. Charge ranges can vary from 10 to 500000 pC. In recent times, charge amplifiers have been miniaturized and in-built into the piezoelectric transducers. These are known as ICP accelerometers manufactured by PCB or Deltatron manufactured by B&K. These integrated charge amplifiers inside accelerometers have become extremely popular in recent times and have thus made the use of external charge amplifiers redundant. However, only in applications where accelerometers are used at higher temperatures, are accelerometers with external charge amplifiers used.

1.13 Conclusions

An SDOF system was thoroughly studied in this chapter, focusing on important aspects of vibration: free and forced vibration. In addition, the complex exponential method of obtaining the response of the SDOF to harmonic excitation was presented that underscored the importance of frequency response, which is fundamental to the later chapters.

The concept of resonance was explained by using the definition of input power. In addition, input power was used to define distinct frequency regions of mechanical impedance: both magnitude and phase. Many example problems were presented that help understand the basic concepts that are required for studying vibration and noise. This chapter is therefore fundamental to all the material presented in this book. Hence, it may be useful to return to this chapter many times while reading the later chapters.

PROBLEMS

1.1. A spring–mass system shown in the following figure, formed by m and k, is mounted within a box of mass, M, and is dropped from a height, h. Determine the maximum displacement assuming the box does not leave the ground and that $M \gg m$.

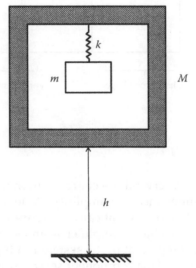

1.2. Determine the unit step-response of an SDOF system.

1.3. A single-degree-of-freedom spring–mass–damper system has the following data: spring stiffness, 10 kN/m; mass, 0.05 kg; and damping coefficient, 10 N-s/m. If it is subjected to a force excitation of $10e^{-5t} \cos 400t$ N, determine the time-domain response of this system using complex analysis.

1.4. A single-degree-of-freedom spring–mass–damper system has the following data: spring stiffness, 10 kN/m; mass, 0.05 kg; and damping coefficient, 10 N-s/m. If this system is subjected to two forces—5 sin $(50t + \pi/6)$ **N** and 10 cos$(250t + \pi/3)$ **N**—determine the time-domain acceleration response of this system and in addition, plot the phasors of acceleration response of the corresponding forces, clearly indicating the phase angles and amplitudes. Locate the phasors of the forces in their positions at $t = 0$.

1.5. A machine weighing 50 kg is mounted on an isolator of unknown stiffness and damping. When the machine excited by 50 N (rms) force at 50 Hz, the machine responds with a steady-state response of 100 μm (rms) lags the force by 30°. Determine the stiffness and damping of this system.

1.6. A machine weighing 50 kg is mounted on an isolator of unknown stiffness and damping. When the machine excited by a 50 N (rms) force at a number of frequencies. The machine responds with a maximum steady-state response of 100 μm (rms) occurs at 50 Hz. Determine the stiffness and damping.

1.7. Consider an SDOF that has a 30 kg mass and unknown stiffness and damping. There is an unbalance from a machine that is part of the mass. When the operating speed is 1000 rpm, the steady-state displacement response is 100 μm (rms), and at an operating speed of 2000 rpm the steady-state displacement response is 250 μm (rms). At very high speeds, the steady-state displacement response is 500 μm (rms). Determine the stiffness and damping of this system.

1.8. A single-degree-of-freedom system is subjected to a sinusoidal force excitation at various frequencies, and the phase angle between its velocity response and force is measured at such two frequencies are as follows:

	Excitation Frequency (rad/s)	$\phi_{v/F}$ (degrees)
1	80	75
2	125	−75

Determine the stiffness, damping and mass of this system if this system consumes 10 mW at an excitation frequency equal to resonance frequency. Assume that the amplitude of the excitation force remains constant for all excitation frequencies.

1.9. A single-degree-of-freedom spring–mass–damper system is subjected to a harmonic force excitation of 0.005 N (rms) constant amplitude at various frequencies, as per the following table. Determine the mass, stiffness and damping coefficient, damping factor, undamped natural frequency, and damped natural frequency of this system.

No.	Frequency (rad/s)	Average Power Consumed (microwatts, μW)
1	100	20 (max)
2	89.75	5

1.10. A single-degree-of-freedom spring–mass–damper system is subjected to a harmonic force excitation of constant rms amplitude of 0.008 N at various frequencies as per the following table:

No.	Frequency (rad/s)	Average Power Consumed (microwatts, μW)	Phase Angle between Velocity Response and Force $(-\phi_m)$
1	100	20	0°
2	95	?	45°

a. Determine the damping factor and the excitation frequency at which the phase angle between velocity response and force is −45°.

b. Determine the mass, stiffness and damping coefficient, undamped natural frequency, and damped natural frequency.

1.11. A single-degree-of-freedom system is subjected to a force excitation of 0.1 N (rms), and the phase angle between its velocity response and force is measured at two frequencies as shown in the following table. The system consumes 10 mW when the force excitation is in phase with velocity and 1 microwatt when the force excitation is in phase with the displacement response.

	Excitation Frequency (rad/s)	$\phi_{v/F}$ (degrees)
1	80	45
2	125	−45

   a. Determine the stiffness, damping, and mass of this system.

   b. Determine the frequency at which one microwatt power is drawn by the system. Assume that the amplitude of the excitation force remains constant for all excitation frequencies.

1.12. Derive Equation 1.84 using Equation 1.96.

1.13. A single-degree-of-freedom system has a natural frequency ω_n rad/s, and when it is subjected to sinusoidal excitation at various frequencies the half-power resonance width is $\omega_n/4$ rad/s. What is the value of damping in terms of mass and stiffness?

1.14. Consider Problem 1.9. At resonance:

   a. What is the work done per cycle against the damping force?

   b. What is the total mechanical energy, E_o, of the oscillator?

   c. What is the logarithmic decrement during free vibration of this oscillator?

1.15. For an SDOF system of natural frequency ω_n rad/s, the half-power frequencies are 0.95 ω_n rad/s and 1.05 ω_n rad/s. Determine the damping factor of this system.

1.16. A single-degree-of-freedom system consists of a spring, mass, and dashpot connected in the usual arrangement. The spring stiffness is 1000 N/m and the static deflection due to the mass is 0.981 mm. Two experiments were conducted on this system. When the mass is given an initial displacement and allowed to execute free vibration, its peak amplitude reduces by 20% in each cycle and the phase angle between velocity and force at an excitation frequency is 25°.

   a. What is the value of mass?

   b. What is the value of damping factor?

   c. What is the magnitude and phase of impedance at this excitation frequency?

   d. What is the rms value of velocity at this excitation frequency, if the rms value of force is kept at 5 mN at all excitation frequencies?

   e. What is the power input to the system at this excitation frequency?

   f. What is the maximum possible value of power input to this system from the given excitation force?

1.17. A single-degree-of-freedom spring–mass–damper system has the following data: spring stiffness 20, kN/m; mass, 0.05 kg; damping coefficient, 20 N-s/m.

   a. Draw the impedance diagram on the complex plane and locate the half-power, stiffness-controlled, and mass-controlled regions.

b. If the mass is subjected to excitation forces of 5 sin (550t + π/6) **N** and 10 cos(700t + π/3) **N**, determine the total power input to the system.

c. For the excitation forces of part b, what is the time-domain velocity response?

1.18. A single-degree-of-freedom spring–mass–damper system has the following data: spring stiffness 10 kN/m; mass 0.05 kg; damping coefficient 10 N-s/m.

a. Determine the undamped natural frequency (rad/s and Hz), damped natural frequency (rad/s and Hz), and damping factor.

b. An electromagnetic exciter is used to apply a force excitation that results in the following acceleration that is measured by mounting an accelerometer on the mass: 2 sin (50t + 200°) + 5 cos (1000t – 150°) m/s².

c. Determine the time-domain equation of the force excitation that caused this acceleration response.

d. What is the total power consumed by the exciter to give the above acceleration response?

e. Plot the phasors of velocity and force for each frequency of part b.

1.19. A single-degree-of-freedom spring–mass–damper system has the following data: spring stiffness, 10 kN/m; mass, 0.05 kg; damping coefficient, 10 N-s/m.

a. Determine the undamped natural frequency (rad/s and Hz), damped natural frequency (rad/s and Hz), and damping factor.

b. An electromagnetic exciter is used to apply a force excitation that results in the following velocity that is measured directly using a transducer: 5 sin (50t + 200°) mm/s. Determine the time-domain equation for the force excitation that caused this velocity response.

c. Using the same amplitude of force of part b but with a phase of 100° with respect to velocity, determine the time-domain acceleration response if the excitation frequency is changed to 150 Hz.

d. Determine the power drawn by the system for the excitation in parts b and c.

1.20. A single-degree-of-freedom spring–mass–damper system has the following data: spring stiffness 10 kN/m; mass 0.05 kg; damping coefficient 10 N-s/m.

a. Determine the undamped natural frequency (rad/s and Hz), damped natural frequency (rad/s and Hz), and damping factor.

b. An electromagnetic exciter is used to apply a force excitation that results in the following displacement response that is measured using a transducer: $y(t)$ = 8 sin (50t – 3°) + 0.005 cos (1000t – 165°) mm. Determine the time-domain equation for the force excitation that caused this displacement response.

c. What is the power consumed by the exciter at each of the excitation frequencies to give the displacement response?

d. Plot the phasors of velocity and force for each frequency of part b.

2

Multidegree-of-Freedom (MDOF) Systems and Longitudinal Vibration in Bars

Before we discuss longitudinal vibration in bars, it would be useful to look at the vibration of multidegree-of-freedom (MDOF) discrete systems first. The governing equations for their motion can be easily formed using Newton's second law and can be analyzed by casting them into the eigenvalue form. In addition, the concept of mode shapes can be easily explained for an MDOF discrete system; the mode shapes presented in the form of phasors can be easily visualized. The mode shapes of a discrete system can be used to obtain a system of uncoupled equations that can be solved to obtain their response. This is known as the modal approach and a similar approach can also be used to obtain the response of continuous systems. Once the physical concepts of natural frequencies and mode shapes for a MDOF discrete system are clearly understood, the same can be easily applied to continuous systems. However, the governing equations for bars are expressed in the form of partial differential equations that are solved in a different way than an MDOF that is represented by a system of ordinary differential equations.

After studying the discrete MDOF systems, longitudinal vibration in bars are studied by using both the wave and modal approaches. The wave approach is studied first. Whenever a disturbance propagates in a medium, if the particles of the medium respond in the direction of travel, then we say it is longitudinal wave propagation. Based on the stress–strain relationship due to longitudinal wave propagation in a bar, the basic wave equation is derived in terms of the longitudinal wave speed. By applying boundary conditions to the equation, the resonant frequencies are determined. The concepts of wave number and driving point impedance, among others, are defined and will be useful in the later chapters. The modal approach is also used by solving the wave equation by using separation of variables to obtain natural frequencies and mode shapes and to further obtain the response by obtaining the uncoupled equations using the orthogonality condition. Thus, by presenting both the modal and wave approach to a longitudinally vibrating bar, the fundamentals of vibration analysis of continuous systems are presented.

Disturbances in fluids (airborne or waterborne sound) always propagate as longitudinal waves, which will be studied in Chapter 3. Longitudinal waves are not very common in solids, although some musical instruments use this principle to generate sound. Since longitudinal wave types are the simplest of all other wave types, it would be useful to study longitudinal wave propagation in bars first, before discussing flexural vibration in beams and plates. In addition, the concept of multiple resonant frequencies and their corresponding mode shapes can be easily explained in the case of longitudinal waves in bars. Due to the similarities between sound waves and longitudinal waves in bars, they complement in each other's understanding.

2.1 Multidegree-of-Freedom Discrete System

Most practical systems have many degrees of freedom that are generally known as multidegree-of-freedom systems, MDOF systems for short. So it is important to understand the dynamics of these systems and to obtain their response to any arbitrary excitation. In this section, we shall look at MDOF systems as a system of masses, springs, and dampers that have linear oscillations. However, the same analysis can also be applied to systems that have torsional oscillations.

The main difference between an MDOF system and a single-degree-of-freedom (SDOF) system is that, in addition to natural frequencies corresponding to each degree of freedom, the degrees of freedom move in a characteristic manner corresponding to each natural frequency, known as mode shapes. Mode shapes play a very important role in the analysis of MDOF systems. By using these mode shapes, one can transform an MDOF system into a number of SDOF systems and then combine the response of each of them to obtain the total MDOF response.

First, an undamped MDOF system without force excitation is studied to introduce the basic concept of multiple natural frequencies and their corresponding mode shapes. A detailed physical understanding of the concept of mode shapes is presented by using complex exponential and mode shapes. These mode shapes that are expressed in the form of a modal matrix can be used to decouple the equations corresponding to the MDOF with force excitation, under some specific conditions known as proportional damping. In proportional damping the damping matrix is proportional to the mass and stiffness matrices. This holds good for the vast majority of the cases and, therefore, the response of these MDOFs can be obtained for any arbitrary excitation. Although we have limited the discussion here to natural frequencies and mode shapes of MDOF discrete systems, the same discussion can be easily carried forward to determine the force response. The discussion presented here can also form the basis for experimental modal analysis.

2.1.1 Undamped MDOF

Consider a two-degree of freedom system shown in Figure 2.1, formed by springs k_1, k_2, k_3, and masses m_1 and m_2. The system has two translational degrees of freedom y_1 and y_2, measured in the vertical direction from their static equilibrium positions.

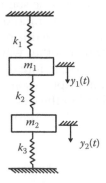

FIGURE 2.1
Two-degrees-of-freedom system.

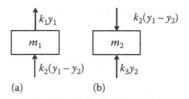

FIGURE 2.2
Free body diagram of masses shown in Figure 2.1 due to initial deflection in the downward direction (assume $y_1 > y_2$).

The first step in analyzing this dynamic system is to determine its natural frequencies and mode shapes during free vibration. For this purpose, we can give a small displacement to both degrees of freedom and deduce its equation of motion by applying Newton's second law of motion.

The free-body diagrams of both masses under the action of spring forces due to initial displacement are shown in Figure 2.2 and the downward direction is taken as positive.

From Figure 2.2a, applying Newton's second law to the first mass

$$m_1\ddot{y}_1 = -k_1 y_1 - k_2(y_1 - y_2)$$ (2.1)

From Figure 2.2b, applying Newton's second law to the second mass

$$m_2\ddot{y}_2 = k_2(y_1 - y_2) - k_3 y_2$$ (2.2)

Rearranging Equations 2.1 and 2.2

$$m_1\ddot{y}_1 + k_1 y_1 + k_2(y_1 - y_2) = 0$$ (2.3)

$$m_2\ddot{y}_2 - k_2(y_1 - y_2) + k_3 y_2 = 0$$ (2.4)

Regrouping terms corresponding to y_1 and y_2

$$m_1\ddot{y}_1 + (k_1 + k_2)y_1 - k_2 y_2 = 0$$ (2.5)

$$m_2\ddot{y}_2 - k_2 y_1 + (k_2 + k_3)y_2 = 0$$ (2.6)

Equations 2.5 and 2.6 can be put in the matrix form as

$$\begin{bmatrix} m_1 & 0 \\ 0 & m_2 \end{bmatrix} \begin{Bmatrix} \ddot{y}_1 \\ \ddot{y}_2 \end{Bmatrix} + \begin{bmatrix} k_1 + k_2 & -k_2 \\ -k_2 & k_2 + k_3 \end{bmatrix} \begin{Bmatrix} y_1 \\ y_2 \end{Bmatrix} = 0$$ (2.7)

The matrix differential equation can be extended to an n number of degrees-of-freedom system as

$$\mathbf{M}\ddot{\mathbf{y}} + \mathbf{K}\mathbf{y} = 0 \tag{2.8}$$

where $\mathbf{M}$ is an $n \times n$ mass matrix, $\mathbf{K}$ is an $n \times n$ stiffness matrix, and $\ddot{\mathbf{y}}$ and $\mathbf{y}$ are $n \times 1$ vectors.

The mass matrix is a diagonal matrix that has value of masses corresponding to each degree of freedom along the diagonal and the off-diagonal elements are zero.

The stiffness matrix can be formed based on the following rules:

1. Sum of all the stiffnesses connected to a degree of freedom form the diagonal terms
2. Negative of the stiffness connecting one degree of freedom to the other degrees of freedom form the off-diagonal elements

Even for dynamic systems whose springs and masses are connected in a different configuration than shown in Figure 2.1, the same aforementioned rules for forming mass and stiffness matrices are applicable. In addition, the same rules of forming the stiffness matrix can be applied to form the damping matrix of an MDOF wherein the damping elements are connected between various degrees of freedom.

2.1.2 Natural Frequencies

When the MDOF vibrates at a natural frequency during free vibration, all the degrees of freedom vibrate with the same frequency. Therefore, the following solution to the motion of a system represented by Equation 2.8 can be assumed

$$\mathbf{y} = \mathbf{Y}\cos\omega t \tag{2.9}$$

where $\mathbf{Y}$ is the vector of peak amplitude of various degrees of freedom, and ω could take values corresponding to any of its natural frequencies.

The second derivative of Equation 2.9 with respect to time is given by

$$\ddot{\mathbf{y}} = -\omega^2 \mathbf{Y}\cos\omega t \tag{2.10}$$

From Equations 2.8 through 2.10

$$[\mathbf{K} - \omega^2\mathbf{M}]\mathbf{Y} = 0 \tag{2.11}$$

Since the displacements cannot be equal to zero during vibration, the only possibility is that

$$\left|\mathbf{K} - \omega^2\mathbf{M}\right| = 0 \tag{2.12}$$

Equation 2.12 represents the determinant of the matrix formed by $\mathbf{K} - \omega^2\mathbf{M}$.

Applying Equation 2.12 to the two-degrees-of-freedom system represented by Equation 2.7

$$\begin{vmatrix} k_1 + k_2 - m_1\omega^2 & -k_2 \\ -k_2 & k_2 + k_3 - m_2\omega^2 \end{vmatrix} = 0 \tag{2.13}$$

Expanding the determinant of Equation 2.13 and collecting terms corresponding to ω^4 and ω^2

$$m_1 m_2\omega^4 - [m_1(k_2 + k_3) + m_2(k_1 + k_2)]\omega^2 + k_1 k_2 + k_2 k_3 + k_1 k_3 = 0 \tag{2.14}$$

Equation 2.14 is a quadratic in ω^2, which can be solved to obtain the resonant frequencies ω_1 and ω_2 that correspond to the positive roots of Equation 2.14.

For an n-degree-of-freedom system, a similar procedure can be adopted to compute natural frequencies ω_i, $i = 1, 2, ..., n$. In addition, Equation 2.11 can be cast as a standard eigenvalue problem; natural frequencies are eigenvalues and mode shapes eigenvectors. Since standard computer programs are available for solving eigenvalue problems, they can be used in the computation of natural frequencies and mode shapes of large MDOF systems.

2.1.3 Mode Shapes

Mode shapes represent relative displacement of each degree of freedom with respect to the other, corresponding to each natural frequency. At a natural frequency, in order that all the degrees of freedom must move with the same frequency, they have to adjust their relative amplitudes. These mode shapes are unique at each natural frequency and similar to natural frequencies; they are characteristics of the system. Therefore, the response of a system to either initial conditions or force excitation or both will depend on them.

Writing Equation 2.11 at each of the natural frequencies

$$\left[\mathbf{K} - \omega_i^2\mathbf{M}\right]\mathbf{Y}_i = 0 \tag{2.15}$$

where, the vector $\mathbf{Y}_i$ contains the peak amplitude of various degrees of freedom at the ith natural frequency.

Equation 2.15 represents a system of linear homogeneous equations. Therefore, it cannot give absolute values of peak amplitudes, but only relative values, which represent mode shapes for the corresponding natural frequency.

Example 2.1

Consider the system shown in Figure 2.1. Let $k_1 = k$, $k_2 = 2k$, $k_3 = 3k$, $m_1 = m$, and $m_2 = 2m$.

    a. Determine the natural frequencies and mode shapes.
    b. Plot the mode shapes.

ANSWERS

a. The mass matrix is given by

$$\mathbf{M} = \begin{bmatrix} m & 0 \\ 0 & 2m \end{bmatrix} = m \begin{bmatrix} 1 & 0 \\ 0 & 2 \end{bmatrix} \tag{1}$$

The stiffness matrix is given by

$$\mathbf{K} = \begin{bmatrix} 3k & -2k \\ -2k & 5k \end{bmatrix} = k \begin{bmatrix} 3 & -2 \\ -2 & 5 \end{bmatrix} \tag{2}$$

From Equation 2.12

$$\left| k \begin{bmatrix} 3 & -2 \\ -2 & 5 \end{bmatrix} - \omega^2 m \begin{bmatrix} 1 & 0 \\ 0 & 2 \end{bmatrix} \right| = 0 \tag{3}$$

Expanding the determinant of Equation 3

$$\omega^4 - \frac{11}{2}\frac{k}{m}\omega^2 + \frac{11}{2}\left(\frac{k}{m}\right)^2 = 0 \tag{4}$$

Solving Equation 4, the undamped natural frequencies are given by

$$\omega_{1,2} = \sqrt{\frac{\dfrac{11}{2}\dfrac{k}{m} \pm \sqrt{\left(\dfrac{11}{2}\dfrac{k}{m}\right)^2 - 4\left[\dfrac{11}{2}\left(\dfrac{k}{m}\right)^2\right]}}{2}} \tag{5}$$

Simplifying Equation 5, the first undamped natural frequency is given by

$$\omega_1 = 1.15\sqrt{\frac{k}{m}} \text{ rad/s} \tag{6}$$

and the second undamped natural frequency is given by

$$\omega_2 = 2.05\sqrt{\frac{k}{m}} \text{ rad/s} \tag{7}$$

Substituting the first resonance frequency in Equation 2.11

$$\left[k \begin{bmatrix} 3 & -2 \\ -2 & 5 \end{bmatrix} - \omega_1^2 m \begin{bmatrix} 1 & 0 \\ 0 & 2 \end{bmatrix} \right] \begin{Bmatrix} Y_{11} \\ Y_{21} \end{Bmatrix} = 0 \tag{8}$$

where Y_{11} and Y_{21} are the peak amplitudes of free vibration of the degrees of freedom at the first resonant frequency.

From Equations 6 and 8

$$Y_{21} = 0.84 Y_{11} \text{ (First mode)} \tag{9}$$

Similarly, corresponding to the second mode

$$\left[k \begin{bmatrix} 3 & -2 \\ -2 & 5 \end{bmatrix} - \omega_2^2 m \begin{bmatrix} 1 & 0 \\ 0 & 2 \end{bmatrix} \right] \begin{Bmatrix} Y_{12} \\ Y_{22} \end{Bmatrix} = 0 \tag{10}$$

where Y_{12} and Y_{22} are the peak amplitudes of free vibration of the degrees of freedom at the second resonant frequency.

From Equations 7 and 10

$$Y_{22} = -0.6 Y_{12} \text{ (Second mode)} \tag{11}$$

b. Since mode shapes are relative amplitudes, the amplitude of one of the degrees can be assumed to be 1 and other amplitudes can be computed in terms of it. Let us assume the first degree of freedom as the reference and assume its peak amplitude as equal to 1 for both the resonant frequencies. The peak relative amplitudes corresponding to free vibration at each mode are assembled into a vector known as a modal vector.

From Equation 9, the modal vector for the first resonant frequency becomes

$$\eta_1 = \begin{Bmatrix} Y_{11} \\ Y_{21} \end{Bmatrix} = \begin{Bmatrix} 1 \\ 0.84 \end{Bmatrix} \tag{12}$$

This modal vector indicates that in the first mode, the peak amplitude of the second degree of freedom is 0.84 times the peak amplitude of the first degree of freedom in the same direction (on the same side as the first degree of freedom above or below the static equilibrium position). The phasor representation of both degrees of freedom, rotating in phase at the first natural frequency, is shown in Figure 2.3.

From Equation 11, the modal vector for the second resonant frequency becomes

$$\eta_2 = \begin{Bmatrix} Y_{12} \\ Y_{22} \end{Bmatrix} = \begin{Bmatrix} 1 \\ -0.6 \end{Bmatrix} \tag{13}$$

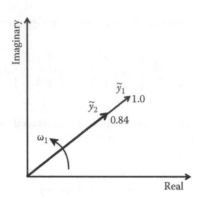

FIGURE 2.3
Mode shape for the first natural frequency on the phasor diagram.

This modal vector indicates that in the first mode, the peak amplitude of the second degree of freedom is 0.6 times the peak amplitude of the first degree of freedom in the opposite direction (on the opposite side as the first degree of freedom above or below the static equilibrium position). The phasor representation of both degrees of freedom, rotating out of phase at the second natural frequency, is shown in Figure 2.4.

The preceding modal vectors corresponding to the first and second natural frequencies can be combined into a matrix known as a modal matrix, given by

$$\boldsymbol{\Psi} = \begin{bmatrix} \boldsymbol{\eta}_1 & \boldsymbol{\eta}_2 \end{bmatrix}$$

$$= \begin{bmatrix} 1 & 1 \\ 0.84 & -0.6 \end{bmatrix} \tag{14}$$

The mode shapes of the first and second natural frequency shown in Figures 2.3 and 2.4 can be combined in a slightly different form as shown in Figure 2.5. The peak deflection of various points on the springs and masses of the

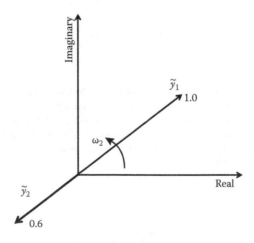

FIGURE 2.4
Mode shape for the second natural frequency on the phasor diagram.

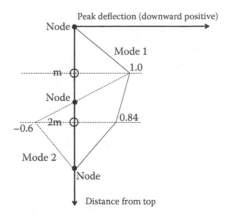

FIGURE 2.5
Peak deflections of various points of the system shown in Example 2.1 during resonant vibration (first and second mode).

two-degrees-of-freedom system are shown in this figure for both modes of vibration. Since the direction of motion and distribution of points on the system are along the same line, the peak deflections are shown along the horizontal direction. In addition to masses, various points on the spring also undergo deflection as shown. In mode 1, both masses move together but with a different peak amplitude. The peak amplitude of spring deflections can also be linearly interpolated but are not generally required for any practical purpose; it is only the peak amplitude of the masses corresponding to each degree of freedom that is important. In mode 2, both masses move 180° out of phase with respect to each other; that is, if mass 1 moves one unit downward, mass 2 moves 0.6 unit upward.

Mode shapes can also pinpoint locations on the system that do not undergo any movement, which are known as node points. For example, in the second mode, a certain point on the middle spring does not move at all. It is known as a node point. But this node point on a spring does not have much practical value since we cannot locate a source of excitation on a spring. Node points on masses, however, have practical significance while locating machinery, since a node point does not transfer any vibration. Although the node points of this problem are not useful, if there are a large number of masses, certain masses can coincide with a node for a mode of vibration. Excitation forces located on such masses do not play any role in exciting that particular mode. As will be seen later, node points can be physically located on continuous systems and used for locating force excitations.

Now that we have established the procedure for determining the undamped natural frequencies and their corresponding mode shapes of an MDOF, they can be used to determine its force response.

In addition, viscous damping elements can be introduced between various degrees of freedom that have damping forces proportional to their relative velocities. The damping matrix can be formed based on similar rules that were used for forming the stiffness matrix.

The general matrix differential equation accounting for mass, damping, and stiffness matrices and force excitation can be written as follows:

$$\mathbf{M\ddot{y}} + \mathbf{R_m\dot{y}} + \mathbf{Ky} = \mathbf{F}(t) \tag{2.16}$$

The mode shapes obtained, similar to those in Example 2.1, can be used for decoupling the above matrix differential equation with force excitation, with the assumption that the damping matrix is proportional to the mass and stiffness matrices. The resulting uncoupled equations can be solved using the techniques outlined in Chapter 1; for harmonic excitation of various degrees of freedom, the complex impedance method can be conveniently used.

2.1.4 Mass Normalized Modes

After obtaining the natural frequencies and mode shapes of the undamped system, we can use them to compute the response of a MDOF that is subject to initial conditions and/ or force excitation. However, the differential equations similar to Equation 2.7 are coupled, that is, the coordinates of various degrees of freedom are present in all the equations. Therefore, the first step is to obtain a set of uncoupled equations by using a transformation using the modal matrix. If the modal matrix is normalized with respect to the mass matrix, the resulting uncoupled equations will be in the standard form. Such normalization is explained as follows.

First the mass matrix [M] is postmultiplied by Ψ and premultiplied by Ψ^T to obtain the following equation:

$$\mathbf{M}_r = \begin{bmatrix} \ddots & & \\ & m_r & \\ & & \ddots \end{bmatrix} = \Psi^T \mathbf{M} \Psi, r = 1, 2, \ldots n \tag{2.17}$$

Now the modal matrix Ψ is normalized with respect to the mass matrix by the following equation:

$$\Phi = \mathbf{M}_r^{-0.5} \Psi \tag{2.18}$$

where Φ is known as the normalized modal matrix with respect to the mass matrix.

Using the transformation

$$\mathbf{y} = \Phi \mathbf{q} \tag{2.19}$$

where $\mathbf{q}$ represents the vector of generalized displacements. Changing the coordinates of Equation 2.16 using Equation 2.19 and premultiplying Equation 2.1 by Φ^T, the following equation is obtained:

$$[I]\ddot{\mathbf{q}} + \left\{ \alpha[I] + \beta \begin{bmatrix} \ddots & & \\ & \omega_r^2 & \\ & & \ddots \end{bmatrix} \right\} \dot{\mathbf{q}} + \begin{bmatrix} \ddots & & \\ & \omega_r^2 & \\ & & \ddots \end{bmatrix} \mathbf{q} = \mathbf{Q} \tag{2.20}$$

where α and β correspond to the proportional damping relation given by

$$[\mathbf{R}_m] = \alpha[\mathbf{M}] + \beta[\mathbf{K}] \tag{2.21}$$

and $\mathbf{Q}$ is the vector containing generalized forces given by

$$\mathbf{Q} = [\mathbf{\Phi}]^T \mathbf{F}(t) \tag{2.22}$$

The uncoupled matrix differential Equation 2.20, which has all the off-diagonal elements of zero, can also be written in the following form:

$$\ddot{q}_r + 2\zeta_r \omega_r \dot{q}_r + \omega_r^2 q_r = Q_r(t), r = 1, 2, \ldots n \tag{2.23}$$

$$2\zeta_r \omega_r = \alpha + \beta \omega_r^2, r = 1, 2, \ldots n \tag{2.24}$$

Equation 2.23 is similar to the equation of a single-degree-of-freedom system and therefore its solution for each of the natural frequencies can be easily obtained.

The relationship between the initial conditions of the original system and the initial conditions that are to be used in Equation 2.23 are as follows:

$$\mathbf{q}(0) = [\phi]^{-1} \mathbf{y}(0); \dot{\mathbf{q}}(0) = [\phi]^{-1} \dot{\mathbf{y}}(0) \tag{2.25}$$

The solutions to each of the generalized coordinates corresponding to each natural frequency are obtained using the initial conditions of equation of Equation 2.25 or the force excitation as the case may be. Actual displacements corresponding to each degree of freedom are then obtained from Equation 2.19 by converting them to natural coordinates.

The main objective of presenting the MDOF discrete system was to clearly show the physics of mode shapes, which play in important role in determining the response at low frequencies. Similar concepts can be extended to continuous systems while studying wave propagation in them.

Example 2.2

A two-degrees-of-freedom system, constructed by a combination of spring–mass–damper systems, is represented by the following matrix differential equation:

$$\begin{bmatrix} 1 & 0 \\ 0 & 1 \end{bmatrix} \begin{Bmatrix} \ddot{y}_1 \\ \ddot{y}_2 \end{Bmatrix} + \begin{bmatrix} 30 & -10 \\ -10 & 30 \end{bmatrix} \begin{Bmatrix} \dot{y}_1 \\ \dot{y}_2 \end{Bmatrix} + \begin{bmatrix} 20 \times 10^3 & -10 \times 10^3 \\ -10 \times 10^3 & 20 \times 10^3 \end{bmatrix} \begin{Bmatrix} y_1 \\ y_2 \end{Bmatrix} = 0$$

Determine the velocity of both degrees of freedom at $t = 0.2$ sec, after only the first degree of freedom is given the following set of initial conditions: displacement, $y_1(0) = 0$; and velocity, $v_1(0) = 10$ m/s.

The mass matrix is given by

$$\mathbf{M} = \begin{bmatrix} 1 & 0 \\ 0 & 1 \end{bmatrix} \tag{1}$$

The damping matrix is given by

$$\mathbf{R}_m = \begin{bmatrix} 30 & -10 \\ -10 & 30 \end{bmatrix} \tag{2}$$

The stiffness matrix is given by

$$\mathbf{K} = \begin{bmatrix} 20 & -10 \\ -10 & 20 \end{bmatrix} 10^3 \tag{3}$$

From Equation 2.12, the undamped natural frequencies of the preceding system are obtained by

$$|\mathbf{K} - \omega^2 \mathbf{M}| = 0$$

$$\left| \begin{bmatrix} 20 & -10 \\ -10 & 20 \end{bmatrix} 10^3 - \omega^2 \begin{bmatrix} 1 & 0 \\ 0 & 1 \end{bmatrix} \right| = 0 \tag{4}$$

Expanding Equation 4

$$\omega^4 - 40 \times 10^3 \omega^2 + 300 \times 10^6 = 0 \tag{5}$$

Solving quadratic Equation 5, the two undamped resonant frequencies of the system are given by

$$\omega_{1,2} = \sqrt{\left(\frac{40 \pm \sqrt{40^2 - 4 \times 300}}{2} \right)} 10^3$$

$$= 100 \, \text{rad/s}$$

$$= 173.2 \, \text{rad/s} \tag{6}$$

From Equation 2.11, the eigenvector for the first undamped resonant frequency is given by

$$\left[\begin{bmatrix} 20 & -10 \\ -10 & 20 \end{bmatrix} 10^3 - 100^2 \begin{bmatrix} 1 & 0 \\ 0 & 1 \end{bmatrix} \right] \left\{ \begin{matrix} Y_{11} \\ Y_{21} \end{matrix} \right\} = 0 \tag{7}$$

If Y_{11} is assumed to be equal to 1, Y_{21} is given by

$$Y_{21} = \frac{20 \times 10^3 - 100^2}{10 \times 10^3} = 1 \tag{8}$$

Similarly, from Equation 2.11, the eigenvector for the second undamped resonant frequency is given by

$$\left[\begin{bmatrix} 20 & -10 \\ -10 & 20 \end{bmatrix} 10^3 - 173.2^2 \begin{bmatrix} 1 & 0 \\ 0 & 1 \end{bmatrix}\right] \begin{Bmatrix} Y_{21} \\ Y_{22} \end{Bmatrix} = 0 \tag{9}$$

If Y_{21} is assumed to be equal to 1, Y_{22} is given by

$$Y_{22} = \frac{20 \times 10^3 - 173.2^2}{10 \times 10^3} = -1 \tag{10}$$

From Equations 8 and 10, the modal matrix is given by

$$\Psi = \begin{bmatrix} 1 & 1 \\ 1 & -1 \end{bmatrix} \tag{11}$$

From Equation 2.17, the generalized mass matrix is given by

$$\mathbf{M}_r = \Psi^T \mathbf{M} \Psi$$

$$= \begin{bmatrix} 1 & 1 \\ 1 & -1 \end{bmatrix} \begin{bmatrix} 1 & 0 \\ 0 & 1 \end{bmatrix} \begin{bmatrix} 1 & 1 \\ 1 & -1 \end{bmatrix} = \begin{bmatrix} 2 & 0 \\ 0 & 2 \end{bmatrix} \tag{12}$$

From Equation 2.18, the mass normalized modal matrix is given by

$$\Phi = \mathbf{M}_r^{-0.5} \Psi = \begin{bmatrix} 2 & 0 \\ 0 & 2 \end{bmatrix}^{-0.5} \begin{bmatrix} 1 & 1 \\ 1 & -1 \end{bmatrix} = \begin{bmatrix} 0.707 & 0.707 \\ 0.707 & -0.707 \end{bmatrix} \tag{13}$$

From Equation 2.21, the damping matrix related to mass and stiffness matrices as

$$\begin{bmatrix} 30 & -10 \\ -10 & 30 \end{bmatrix} = \alpha \begin{bmatrix} 1 & 0 \\ 0 & 1 \end{bmatrix} + \beta \begin{bmatrix} 20 & -10 \\ -10 & 20 \end{bmatrix} 10^3 \tag{14}$$

Solving Equation 14 gives $\alpha = 10$ and $\beta = 0.001$.
From Equation 2.24, the damping factor of the first uncoupled system is given by

$$\zeta_1 = \frac{\alpha + \beta \omega_1^2}{2\omega_1} = \frac{10 + 0.001 \times 100^2}{2 \times 100} = 0.1 \tag{15}$$

And similarly, the damping factor of the second uncoupled system is given by

$$\zeta_2 = \frac{\alpha + \beta \omega_2^2}{2\omega_2} = \frac{10 + 0.001 \times 173.2^2}{2 \times 173.2} = 0.1155 \tag{16}$$

From Equations 15 and 16, the uncoupled equations of motion are given by

$$\ddot{q}_1 + 20\dot{q}_1 + 10^4 q_1 = 0$$
$$\ddot{q}_2 + 40\dot{q}_2 + 3 \times 10^4 q_2 = 0 \tag{17}$$

The damped natural frequencies are given by

$$\omega_{d1} = \omega_1 \sqrt{1 - \zeta_1^2} = 100\sqrt{1 - 0.1^2} = 99.5 \, \text{rad/s}$$
$$\omega_{d2} = \omega_2 \sqrt{1 - \zeta_2^2} = 173.2\sqrt{1 - 0.115^2} = 172.05 \, \text{rad/s} \tag{18}$$

The initial conditions are given by

$$y_1(0) = 0; \quad y_2(0) = 0$$
$$\dot{y}_1(0) = 10 \, \text{m/s}; \quad \dot{y}_2(0) = 0 \tag{19}$$

Since the initial displacements are zero, the initial displacement of the generalized coordinates is also zero.

From Equation 2.25, the initial velocities of the generalized coordinates are given by

$$\dot{\mathbf{q}}(0) = \Phi^{-1}\dot{\mathbf{y}}(0)$$

$$= \begin{bmatrix} 0.707 & 0.707 \\ 0.707 & -0.707 \end{bmatrix}^{-1} \begin{Bmatrix} 10 \\ 0 \end{Bmatrix} = \begin{Bmatrix} 7.07 \\ 7.07 \end{Bmatrix} \text{m/s} \tag{20}$$

Using the impulse response of Equation 1.51, the displacement of the first generalized coordinate is given by

$$q_1(t = 0.2) = \frac{7.07 \times e^{-0.1 \times 100 \times 0.2}}{99.5} = 0.0083 \text{ m} \tag{21}$$

Similarly, using the impulse response of Equation 1.51, the displacement of the second generalized coordinate is given by

$$q_2(t) = \frac{7.07 \times e^{-0.1115 \times 173.2 \times 0.2}}{172.05} = 1.11 \times 10^{-4} \text{ m} \tag{22}$$

From Equations 21 and 22 the displacement response at $t = 0.2$ is given by

$$\mathbf{q}(t = 0.2) = \begin{Bmatrix} 0.0083 \\ 1.11 \times 10^{-4} \end{Bmatrix} \text{m} \tag{23}$$

From Equation 2.25 and Equation 23, the response of the system is given by

$$
\mathbf{y}(t = 0.2) = \begin{bmatrix} 0.707 & 0.707 \\ 0.707 & -0.707 \end{bmatrix} \begin{Bmatrix} 0.0083 \\ 1.11 \times 10^{-4} \end{Bmatrix}
$$

$$
= \begin{Bmatrix} 6 \\ 5.8 \end{Bmatrix} \text{mm}
$$

(24)

Using the impulse response function of Equation 1.51, the velocity of the first generalized response is given by

$$
\dot{q}_1(0.2) = \frac{7.07 \times e^{-0.1 \times 100 \times 0.2}}{99.5} \left(\frac{-0.1 \times 100 \sin 99.5 \times 0.2}{+99.5 \cos 99.5 \times 0.2} \right)
$$

$$
= 0.3926 \text{ m/s}
$$

(25)

Similarly, using the impulse response function of Equation 1.51, the velocity of the second generalized response is given by

$$
\dot{q}_2(t = 0.2) = \frac{7.07 e^{-0.1115 \times 173.2 \times 0.2}}{172.05} \left(\frac{-0.1115 \times 173.2 \sin 172.05 \times 0.2}{+172.05 \cos 172.05 \times 0.2} \right)
$$

$$
= -0.1303 \text{ m/s}
$$

(26)

From Equations 25 and 26, the generalized velocities at $t = 0.2$ can be obtained as

$$
\dot{\mathbf{q}}(t = 0.2) = \begin{Bmatrix} 0.3926 \\ -0.1303 \end{Bmatrix} \text{m/s}
$$

(27)

From Equations 2.25 and 27 the velocity response is given by

$$
\dot{\mathbf{y}}(t = 0.2) = \begin{bmatrix} 0.707 & 0.707 \\ 0.707 & -0.707 \end{bmatrix} \begin{Bmatrix} 0.3926 \\ -0.1303 \end{Bmatrix}
$$

$$
= \begin{Bmatrix} 0.1856 \\ 0.3695 \end{Bmatrix} \text{m/s}
$$

(28)

Example 2.3

If there are many signals representing a time–domain signal formed by a combination of several signals of the same frequency but different magnitude and phase, determine an equivalent single signal.

Let the signals be defined as follows:

$$
\tilde{y}_1 = y_1 e^{j\phi_1} e^{j\omega t}, \ \tilde{y}_2 = y_2 e^{j\phi_2} e^{j\omega t} \dots \tilde{y}_n = y_n e^{j\phi_n} e^{j\omega t}
$$

(1)

Let the equivalent signal be defined as

$$\tilde{y} = y e^{j\phi} e^{j\omega t} \tag{2}$$

Since the time-domain signal is a combination of several components of Equation 1, a sum of these should be equal to Equation 2. Therefore, from Equations 1 and 2

$$y e^{j\phi} e^{j\omega t} = y_1 e^{j\phi_1} e^{j\omega t} + y_2 e^{j\phi_2} e^{j\omega t} \cdots + y_n e^{j\phi_n} e^{j\omega t} \tag{3}$$

Expanding Equation 3 and equating real and imaginary parts on both sides, the magnitude is given by

$$y = \sqrt{(y_1 \cos\phi_1 + y_2 \cos\phi_2 + \cdots y_n \cos\phi_n)^2 + (y_1 \sin\phi_1 + y_2 \sin\phi_2 + \cdots y_n \sin\phi_n)^2} \tag{4}$$

And the phase

$$\phi = \tan^{-1}\left(\frac{y_1 \sin\phi_1 + y_2 \sin\phi_2 + \cdots y_n \sin\phi_n}{y_1 \cos\phi_1 + y_2 \cos\phi_2 + \cdots y_n \cos\phi_n} \right) \tag{5}$$

Equations 4 and 5 can be used in Equation 2 to get the equivalent signal.

Example 2.4

Consider a two-degrees-of-freedom spring–mass–damper system that has the following mass normalized modal matrix $\mathbf{\Phi} = \begin{bmatrix} 0.6 & 0.8 \\ 0.6 & -0.4 \end{bmatrix}$ and the undamped natural frequencies are 165 and 240 rad/s. The damping matrix is proportional to the mass and stiffness matrix: $[\mathbf{R}_m] = \alpha[\mathbf{M}] + \beta[\mathbf{K}]$, where $\alpha = 0.02$ and $\beta = 0.001$.

If a force excitation $2\sin(180t + 50°)$ N is applied to the first degree of freedom, determine the velocity response of both degrees of freedom and the total power input to the system, and write down the mass, stiffness, and damping matrices.

The mass normalized modal matrix is given by

$$\mathbf{\Phi} = \begin{bmatrix} 0.6 & 0.8 \\ 0.6 & -0.4 \end{bmatrix} \tag{1}$$

The undamped natural frequencies are given as $\omega_1 = 165$ rad/s and $\omega_2 = 240$ rad/s. The proportional damping constants as per Equation 2.21 are $\alpha = 0.02$ and $\beta = 0.001$.

From Equation 2.24, the damping factor of the first resonance of the uncoupled equations is given by

$$\zeta_1 = \frac{\alpha + \beta\omega_1^2}{2\omega_1} = \frac{0.02 + 0.001(165)^2}{2 \times 165} = 0.0826 \tag{2}$$

Similarly, the damping factor of the second resonance of the uncoupled equations is given by

$$\zeta_2 = \frac{\alpha + \beta \omega_2^2}{2\omega_2} = \frac{0.02 + 0.001(240)^2}{2 \times 240} = 0.12 \tag{3}$$

From Equation 2.22, the generalized force is given by

$$\mathbf{Q} = \Phi^T \mathbf{f} = \begin{bmatrix} 0.6 & 0.6 \\ 0.8 & -0.4 \end{bmatrix} \begin{Bmatrix} f_1(t) \\ 0 \end{Bmatrix} \tag{4}$$

where

$$f_1(t) = 2\sin(180t + 50°) \tag{5}$$

From Equations 4 and 5, the generalized force corresponding to each degree of freedom is given by

$$\begin{aligned} Q_1(t) &= 0.6 f_1(t) \\ Q_2(t) &= 0.8 f_1(t) \end{aligned} \tag{6}$$

From Equation 2.23, the uncoupled complex differential equations of the system are given by

$$\begin{aligned} \ddot{\tilde{q}}_1 + 2\zeta_1 \omega_1 \dot{\tilde{q}}_1 + \omega_1^2 \tilde{q}_1 &= 0.6 \tilde{f}_1(t) \\ \ddot{\tilde{q}}_2 + 2\zeta_2 \omega_2 \dot{\tilde{q}}_2 + \omega_2^2 \tilde{q}_2 &= 0.8 \tilde{f}_1(t) \end{aligned} \tag{7}$$

where

$$\tilde{f}_1(t) = 2e^{j180t} e^{50\pi/180} \tag{8}$$

The complex velocities of the principle coordinates are given by

$$\dot{\tilde{q}}_1 = \frac{\tilde{Q}_1}{\tilde{Z}_1}$$

$$\dot{\tilde{q}}_2 = \frac{\tilde{Q}_2}{\tilde{Z}_2} \tag{9}$$

The complex impedances of Equation 9 are given by

$$\bar{Z}_1 = 2\zeta_1\omega_1 + j\left(\Omega_1 - \frac{\omega_1^2}{\Omega_1}\right)$$

$$= 2 \times 0.0826 \times 165 + j\left(180 - \frac{165^2}{180}\right) \tag{10}$$

$$= 27.25 + j28.75\, rayl$$

$$\bar{Z}_2 = 2\zeta_2\omega_2 + j\left(\Omega_1 - \frac{\omega_2^2}{\Omega_1}\right)$$

$$= 2 \times 0.12 \times 240 + j\left(180 - \frac{240^2}{180}\right) \tag{11}$$

$$= 57.62 - j140 \text{ rayl}$$

From Equations 9 through 11, the generalized velocities are given by

$$\dot{\tilde{q}}_1 = \frac{\tilde{Q}_1}{\bar{Z}_1} = \frac{2 \times 0.6 \times e^{j180t}e^{50\pi/180}}{27.25 + j28.75} = 0.03\angle 3.46° \text{ m/s}$$

$$\dot{\tilde{q}}_2 = \frac{\tilde{Q}_2}{\bar{Z}_2} = \frac{2 \times 0.8 \times e^{j180t}e^{50\pi/180}}{57.62 - j140} = 0.01\angle 117.62° \text{ m/s} \tag{12}$$

From differentiating Equation 2.19,

$$\left\{ \begin{array}{c} \dot{\tilde{y}}_1 \\ \dot{\tilde{y}}_2 \end{array} \right\} = \left[\begin{array}{cc} 0.6 & 0.8 \\ 0.6 & -0.4 \end{array} \right] \left\{ \begin{array}{c} 0.03\angle 3.46° \\ 0.01\angle 117.62° \end{array} \right\}$$

$$= \left\{ \begin{array}{c} 0.018\angle 3.46° + 0.008\angle 117.62° \\ 0.018\angle 3.46° - 0.004\angle 117.62° \end{array} \right\} e^{j180t} \text{ m/s} \tag{13}$$

From Equation 13, the time-domain responses are given by

$$\dot{y}_1 = 0.018\sin(180t + 3.46) + 0.008\sin(180t + 117.62)\,\text{m/s}$$

$$\dot{y}_2 = 0.018\sin(180t + 3.46) - 0.004\sin(180t + 117.62)\,\text{m/s} \tag{14}$$

By using the results of Example 2.3, the velocities can be written in a phasor form as

$$\dot{\tilde{y}}_1 = 0.0166e^{\frac{j60\pi}{180}} e^{j180t} \text{ m/s}$$

$$\dot{\tilde{y}}_2 = 0.0203e^{-\frac{j83\pi}{180}} e^{j180t} \text{ m/s} \tag{15}$$

From Equation 1.96 (Chapter 1), and Equations 5 and 15, the average power drawn by the system is given by

$$
\begin{aligned}
< W > &= \frac{1}{2} \mathrm{Re}\left(\tilde{F}_1 \dot{\tilde{y}}_1^* \right) \\
&= \frac{1}{2} \mathrm{Re}\left[e^{j180t} e^{\frac{j50\pi}{180}} \left(0.0166 e^{-\frac{j60\pi}{180}} e^{-j180t} \text{ m/s} \right) \right] \\
&= 16 \text{ mW}
\end{aligned}
$$

(16)

From Equation 2.20, the mass normalized modal matrix satisfies the following equation:

$$
\mathbf{\Phi}^\mathsf{T} \mathbf{M} \mathbf{\Phi} = \mathbf{I}
$$

(17)

Premultiplying by $[\mathbf{\Phi}^\mathsf{T}]^{-1}$ and postmultiplying by $[\mathbf{\Phi}]^{-1}$ Equation 17 becomes

$$
\mathbf{M} = \mathbf{\Phi} = [\mathbf{\Phi}^\mathsf{T}]^{-1} \mathbf{I} [\mathbf{\Phi}]^{-1}
$$

$$
= \begin{bmatrix} 0.6 & 0.6 \\ 0.8 & -0.4 \end{bmatrix}^{-1} \begin{bmatrix} 1 & 0 \\ 0 & 1 \end{bmatrix} \begin{bmatrix} 0.6 & 0.8 \\ 0.6 & -0.4 \end{bmatrix}^{-1} = \begin{bmatrix} 1 & 0 \\ 0 & 1.92 \end{bmatrix}
$$

(18)

From Equation 2.20, the stiffness matrix satisfies the following equation:

$$
\mathbf{\Phi}^\mathsf{T} \mathbf{K} \mathbf{\Phi} = \begin{bmatrix} \omega_1^2 & 0 & 0 \\ 0 & \ddots & 0 \\ 0 & 0 & \omega_n^2 \end{bmatrix}
$$

(19)

Premultiplying by $[\mathbf{\Phi}^\mathsf{T}]^{-1}$ and post multiplying by $[\mathbf{\Phi}]^{-1}$ Equation 19 becomes

$$
\mathbf{K} = [\mathbf{\Phi}^\mathsf{T}]^{-1} \mathbf{I} [\mathbf{\Phi}]^{-1}
$$

$$
= \begin{bmatrix} 0.6 & 0.6 \\ 0.8 & -0.4 \end{bmatrix}^{-1} \begin{bmatrix} 165^2 & 0 \\ 0 & 240^2 \end{bmatrix} \begin{bmatrix} 0.6 & 0.8 \\ 0.6 & -0.4 \end{bmatrix}^{-1} = \begin{bmatrix} 48.4 & -23.2 \\ -23.2 & 73.6 \end{bmatrix} \text{kN/m}
$$

(20)

From Equation 2.21, the damping matrix of the system is given by

$$
\mathbf{R}_\mathrm{m} = \alpha \mathbf{M} + \beta \mathbf{K}
$$

$$
= 0.02 \begin{bmatrix} 1 & 0 \\ 0 & 1.92 \end{bmatrix} + 0.001 \times 10^3 \begin{bmatrix} 48.403 & -23.19 \\ -23.19 & 73.611 \end{bmatrix}
$$

$$
= \begin{bmatrix} 48.42 & -23.2 \\ -23.2 & 73.65 \end{bmatrix} \text{N-s/m}
$$

(21)

Example 2.5

A two-degrees-of-freedom system shown in Figure 2.6 is subjected to force excitation as shown; m = 1 kg and k = 10000 N/m. The values of the viscous damping coefficients are chosen such that the damping matrix $[\mathbf{R_m}] = 0.02[\mathbf{M}] + 0.0008[\mathbf{K}]$. Determine the natural frequencies and mode shapes, the velocity response of each degree of freedom, and the total power input to the system.

Based on the rules presented in Section 2.1.1, the mass matrix of the system is given by

$$\mathbf{M} = m \begin{bmatrix} 1 & 0 \\ 0 & 2 \end{bmatrix} \tag{1}$$

and the stiffness matrix of the system is given by

$$\mathbf{K} = k \begin{bmatrix} 5 & -2 \\ -2 & 8 \end{bmatrix} \tag{2}$$

From Equation 2.12, the undamped natural frequencies of the preceding system is obtained by

$$\left| \mathbf{K} - \omega^2 \mathbf{M} \right| = 0$$

$$\left| k \begin{bmatrix} 5 & -2 \\ -2 & 8 \end{bmatrix} - \omega^2 m \begin{bmatrix} 1 & 0 \\ 0 & 2 \end{bmatrix} \right| = 0 \tag{3}$$

Expanding the determinant of Equation 3, the following characteristic equation is obtained

$$m^2 \omega^4 - 9\omega^2 km + 18k^2 = 0 \tag{4}$$

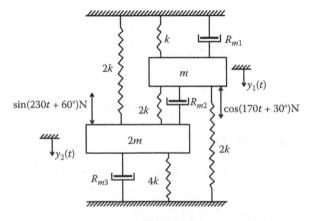

FIGURE 2.6
Two-degrees-of-freedom system (Example 2.5).

Solving Equation 4 for positive real values of ω, the two undamped natural frequencies of the system are obtained as

$$\omega_{1,2} = \sqrt{\frac{3k}{m}}, \sqrt{\frac{6k}{m}} \text{ rad/s}$$

$$= 173.21, 245 \text{ rad/s}$$

(5)

From Equation 2.11, the eigenvector equation for the system corresponding to the first undamped frequency can be expressed as

$$\begin{bmatrix} 5k - \omega_1^2 m & -2k \\ -2k & 8k - 2\omega_1^2 m \end{bmatrix} \begin{Bmatrix} Y_{11} \\ Y_{21} \end{Bmatrix}$$

(6)

For $\omega_1 = \sqrt{\dfrac{3k}{m}}$, assuming $Y_{11} = 1$, $Y_{21} = 1$ will satisfy Equation 6.

Similarly, the eigenvector equation for the system for the second undamped frequency can be expressed as

$$\begin{bmatrix} 5k - \omega_2^2 m & -2k \\ -2k & 8k - 2\omega_2^2 m \end{bmatrix} \begin{Bmatrix} Y_{21} \\ Y_{22} \end{Bmatrix}$$

(7)

For $\omega_1 = \sqrt{\dfrac{6k}{m}}$, assuming $Y_{21} = 1$, $Y_{22} = -0.5$ will satisfy Equation 6.

Combining the eigenvectors obtained into the eigenvector matrix or modal matrix

$$\Psi = \begin{bmatrix} 1 & 1 \\ 1 & -0.5 \end{bmatrix}$$

(8)

From Equations 2.17, and Equations 1 and 8, the generalized mass matrix is given by

$$\mathbf{M_r} = \Psi^T \mathbf{M} \Psi$$

$$= m \begin{bmatrix} 1 & 1 \\ 1 & -0.5 \end{bmatrix} \begin{bmatrix} 1 & 0 \\ 0 & 2 \end{bmatrix} \begin{bmatrix} 1 & 1 \\ 1 & -0.5 \end{bmatrix} = m \begin{bmatrix} 3 & 0 \\ 0 & 1.5 \end{bmatrix}$$

(9)

From Equation 2.18 and Equation 9, the mass normalized eigen matrix is given by

$$\Phi = \Psi \mathbf{M_r}^{-0.5} = \frac{1}{\sqrt{m}} \begin{bmatrix} 1 & 1 \\ 1 & -0.5 \end{bmatrix} \begin{bmatrix} \dfrac{1}{\sqrt{3}} & 0 \\ 0 & \dfrac{1}{\sqrt{1.5}} \end{bmatrix} = \frac{1}{\sqrt{m}} \begin{bmatrix} 0.58 & 0.82 \\ 0.58 & -0.41 \end{bmatrix}$$

(10)

Since $m = 1$, the mass normalized eigen matrix of equation 10 becomes

$$\Phi = \begin{bmatrix} 0.58 & 0.82 \\ 0.58 & -0.41 \end{bmatrix} \tag{11}$$

From Equation 2.24, the damping factor corresponding to the first resonant frequency of the uncoupled system is given by

$$\zeta_1 = \frac{\alpha + \beta\omega_1^2}{2\omega_1} = \frac{0.02 + 0.0008(173.21)^2}{2 \times 173.21} = 0.0693 \tag{12}$$

Similarly, the damping factor corresponding to the second resonant frequency of the uncoupled system is given by

$$\zeta_2 = \frac{\alpha + \beta\omega_2^2}{2\omega_2} = \frac{0.02 + 0.0008(245)^2}{2 \times 245} = 0.0980 \tag{13}$$

From Equation 2.22, the generalized force is given by

$$Q = \Phi^T f = \begin{bmatrix} 0.58 & 0.58 \\ 0.82 & -0.41 \end{bmatrix} \begin{Bmatrix} \cos(170t + 30°) \\ \sin(230t + 60°) \end{Bmatrix} \tag{14}$$

From Equation 14, the generalized force corresponding to each degree of freedom is given by

$$Q_1(t) = 0.58\cos(170t + 30°) + 0.58\sin(230t + 60°)\,N$$
$$Q_2(t) = 0.82\cos(170t + 30°) - 0.41\sin(230t + 60°)\,N \tag{15}$$

From Equation 2.23, the uncoupled complex differential equations of the system are given by

$$\ddot{\tilde{q}}_1 + 2\zeta_1\omega_1\dot{\tilde{q}}_1 + \omega_1^2\tilde{q}_1 = \tilde{Q}_1(t)$$
$$\ddot{\tilde{q}}_2 + 2\zeta_2\omega_2\dot{\tilde{q}}_2 + \omega_2^2\tilde{q}_2 = \tilde{Q}_2(t) \tag{16}$$

where

$$\tilde{Q}_1(t) = \tilde{Q}_{11}(t) + \tilde{Q}_{12}(t) = 0.58e^{j170t}e^{j\frac{30\pi}{180}} + 0.58e^{j230t}e^{j\frac{60\pi}{180}}$$
$$\tilde{Q}_2(t) = \tilde{Q}_{21}(t) + \tilde{Q}_{22}(t) = 0.82e^{j170t}e^{j\frac{30\pi}{180}} - 0.41e^{j230t}e^{j\frac{60\pi}{180}} \tag{17}$$

It should be noted that there are two excitation frequencies: $\Omega_1 = 170$ rad/s and $\Omega_2 = 230$ rad/s.

The complex velocities of the principle coordinates are given by

$$\dot{\tilde{q}}_1 = \frac{\tilde{Q}_{11}}{\overline{Z}_{11}} + \frac{\tilde{Q}_{12}}{\overline{Z}_{12}}$$

$$\dot{\tilde{q}}_2 = \frac{\tilde{Q}_{21}}{\overline{Z}_{21}} + \frac{\tilde{Q}_{22}}{\overline{Z}_{22}}$$

(18)

The various impedances of Equation 18 are given by

$$\overline{Z}_{11} = 2\zeta_1\omega_1 + j\left(\Omega_1 - \frac{\omega_1^2}{\Omega_1}\right)$$

$$= 2 \times 0.0693 \times 173.21 + j\left(170 - \frac{173.21^2}{170}\right)$$

$$= 24.02 - j6.47 = 24.88\angle - 15.08° \, \text{rayl}$$

(19)

$$\overline{Z}_{12} = 2\zeta_1\omega_1 + j\left(\Omega_2 - \frac{\omega_1^2}{\Omega_2}\right)$$

$$= 2 \times 0.0693 \times 173.21 + j\left(230 - \frac{173.21^2}{230}\right)$$

$$= 24.02 + j99.57 = 102.42\angle 76.44° \, \text{rayl}$$

(20)

$$\overline{Z}_{21} = 2\zeta_2\omega_2 + j\left(\Omega_1 - \frac{\omega_1^2}{\Omega_1}\right)$$

$$= 2 \times 0.0980 \times 245 + j\left(170 - \frac{245^2}{170}\right)$$

$$= 48.02 - j182.94 = 189.14\angle - 75.29° \, \text{rayl}$$

(21)

$$\overline{Z}_{22} = 2\zeta_2\omega_2 + j\left(\Omega_2 - \frac{\omega_2^2}{\Omega_2}\right)$$

$$= 2 \times 0.0980 \times 245 + j\left(230 - \frac{245^2}{230}\right)$$

$$= 48.02 - j30.87 = 57.09\angle - 32.73° \, \text{rayl}$$

(22)

The complex velocities of the two masses are given by

$$\dot{\tilde{\mathbf{y}}} = \boldsymbol{\Phi}\dot{\tilde{\mathbf{q}}}$$

(23)

From Equations 11, 18, and 23

$$
\left\{ \begin{array}{c} \dot{\tilde{y}}_1 \\ \dot{\tilde{y}}_2 \end{array} \right\} = \left[\begin{array}{cc} 0.58 & 0.82 \\ 0.58 & -0.41 \end{array} \right] \left\{ \begin{array}{c} \dfrac{\tilde{Q}_{11}}{\tilde{Z}_{11}} + \dfrac{\tilde{Q}_{12}}{\tilde{Z}_{12}} \\[4mm] \dfrac{\tilde{Q}_{21}}{\tilde{Z}_{21}} + \dfrac{\tilde{Q}_{22}}{\tilde{Z}_{22}} \end{array} \right\} \tag{24}
$$

From Equations 17, 19 through 22, and 24

$$
\begin{aligned}
\dot{\tilde{y}}_1 &= 0.0135\angle 45.08°e^{j170t} + 0.00356\angle 105.29°e^{j170t} \\
&\quad + 0.00328\angle -16.44°e^{j230t} - 0.00588\angle 92.73°e^{j230t} \\
\dot{\tilde{y}}_2 &= 0.0135\angle 45.08°e^{j170t} + 0.0032845\angle -16.44°e^{j170t} \\
&\quad + 0.00178\angle 105.29°e^{j230t} + 0.002944\angle 92.73°e^{j230t}
\end{aligned} \tag{25}
$$

By using the actual components of the force excitations and combining same frequency components (see Example 2.3), the time-domain velocities of the two masses are given by

$$
\begin{aligned}
\dot{y}_1 &= 0.0154\cos(170t + 34°) + 0.0025\sin(230t - 105°)\,\text{m/s} \\
\dot{y}_2 &= 0.0154\cos(170t + 34°) + 0.0013\sin(230t + 75°)\,\text{m/s}
\end{aligned} \tag{26}
$$

The given complex harmonic forces can be defined as

$$
\begin{aligned}
\tilde{F}_1 &= e^{j170t}e^{\frac{j34\pi}{180}} \\
\tilde{F}_2 &= e^{j230t}e^{\frac{j60\pi}{180}}
\end{aligned} \tag{27}
$$

From Equation 26, the complex velocities are given by

$$
\begin{aligned}
\dot{\tilde{y}}_1 &= 0.0154e^{j170t}e^{\frac{j34\pi}{180}} + 0.0025e^{j230t}e^{\frac{-j105\pi}{180}} \\
\dot{\tilde{y}}_2 &= 0.0154e^{j170t}e^{\frac{j34\pi}{180}} + 0.0013e^{j230t}e^{\frac{j75\pi}{180}}
\end{aligned} \tag{28}
$$

From Equation 1.51, and Equations 27 and 28, the average power consumed by the system is given by

$$
\begin{aligned}
<W> &= \frac{1}{2}\operatorname{Re}\left(\tilde{F}_1\dot{\tilde{y}}_1 + \tilde{F}_2\dot{\tilde{y}}_2\right) \\
&= \frac{1}{2}\operatorname{Re}\left[e^{j170t}e^{\frac{j30\pi}{180}}\left(0.0154e^{-j170t}e^{-\frac{j34\pi}{180}}\right)\right] \\
&\quad + \frac{1}{2}\operatorname{Re}\left[e^{j230t}e^{\frac{j60\pi}{180}}\left(0.0013e^{-j230t}e^{-\frac{j75\pi}{180}}\right)\right] \\
&= 8.30 \text{ mW}
\end{aligned}
\tag{29}
$$

Now that the basic concepts of natural frequencies and mode shapes are clearly understood, longitudinal waves in bars is taken up next. After deriving the basic wave equation, the modal approach is studied first to explain mode shapes due to longitudinal vibration. Then, the concept of wave propagation is explained by using the complex differential equation. Wave number, longitudinal speed, and input impedance are explained with respect to these waves that are of prime importance in the later chapters. Finally, the orthogonality condition for longitudinal vibration is derived that can be used to uncouple the equations of motion in terms of generalized coordinate, generalized mass, and generalized force. These concepts are useful in later chapters.

2.2 Longitudinal Waves in Bars

2.2.1 Wave Equation

A bar can be defined as a long rod of uniform cross section. Consider a long uniform bar in which longitudinal wave propagation is in progress as shown in Figure 2.7. Consider a section at location x and another section at $x + \Delta x$. Due to wave propagation, the section at x moves a distance u_x, and the section at $x + \Delta x$ moves a distance $u_{x+\Delta x}$. Therefore, Δx is the original distance between the sections when there was no wave propagation. The new length due to wave propagation is given by

$$
\Delta x' = (u_{x+\Delta x} + x + \Delta x) - (u_x + x)
\tag{2.26}
$$

The strain induced due to wave propagation is given by

$$
\varepsilon = \frac{\Delta x' - \Delta x}{\Delta x}
\tag{2.27}
$$

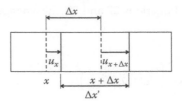

FIGURE 2.7
Longitudinal wave in a bar.

From Equations 2.26 and 2.27, the strain is given by

$$\varepsilon = \frac{(u_{x+\Delta x} + x + \Delta x) - (u_x + x) - \Delta x}{\Delta x} \tag{2.28}$$

From Equation 2.28, the limiting value of the strain as $\Delta x \to 0$ is given by

$$\varepsilon = \lim_{\Delta x \to 0} \frac{u_{x+\Delta x} - u_x}{\Delta x} = \frac{\partial u}{\partial x} \tag{2.29}$$

The function $u(x)$, therefore, represents the displacement field of the bar due to wave propagation.

If E is the Young's modulus of elasticity of the bar material, the stress at any location of the bar due to longitudinal wave propagation is given by

$$\sigma = E\varepsilon = E\frac{\partial u}{\partial x} \tag{2.30}$$

Therefore, the stress and strain are in the same direction.

The net force on the element between x and $x + \Delta x$ is given by

$$F = (\sigma_{x+\Delta x} - \sigma_x)A \tag{2.31}$$

From Equations 2.30 and 2.31, the net force becomes

$$F = EA\left(\frac{\partial u}{\partial x}\bigg|_{x+\Delta x} - \frac{\partial u}{\partial x}\bigg|_x\right) \tag{2.32}$$

Based on the definition of the second derivative of the displacement function

$$\frac{\partial^2 u}{\partial x^2} = \lim_{\Delta x \to 0} \frac{\left(\frac{\partial u}{\partial x}\bigg|_{x+\Delta x} - \frac{\partial u}{\partial x}\bigg|_x\right)}{\Delta x} \tag{2.33}$$

Equation 2.32 becomes

$$F = EA \frac{\partial^2 u}{\partial x^2} \Delta x \tag{2.34}$$

If ρ is the mass density of the bar material, the mass of the aforementioned element is given by

$$dm = \rho A \Delta x \tag{2.35}$$

And the acceleration of the element is given by

$$a = \frac{\partial^2 u}{\partial t^2} \tag{2.36}$$

Combining Equations 2.34 through 2.36 through Newton's law

$$\frac{\partial^2 u}{\partial t^2} = \frac{E}{\rho} \frac{\partial^2 u}{\partial x^2} \tag{2.37}$$

The longitudinal wave speed is defined as

$$c_L = \sqrt{\frac{E}{\rho}} \tag{2.38}$$

From Equations 2.37 and 2.38

$$\frac{\partial^2 u}{\partial x^2} = \frac{1}{c_L^2} \frac{\partial^2 u}{\partial t^2} \tag{2.39}$$

Equation 2.39 represents the partial differential equation for longitudinal wave propagation in a rod, which is similar to the airborne sound wave equation that will be derived in Chapter 3.

2.2.2 Solution to the Longitudinal Wave Equation

Equation 2.39 can be converted into a complex partial differential equation as

$$\frac{\partial^2 \tilde{u}}{\partial x^2} = \frac{1}{c_L^2} \frac{\partial^2 \tilde{u}}{\partial t^2} \tag{2.40}$$

Since the displacement function $\tilde{u}$ is a function of x and t, it can be written as a product of two complex functions: one a function of x and the other a function of t. The space-dependent function will be related to the mode shapes of the bar at various resonance frequencies for a specific set of boundary conditions. This method of solving partial differential equations is known as separation of variables.

Writing $\tilde{u}$ as a product of functions of x and t

$$\tilde{u}(x,t) = \tilde{\phi}(x)\tilde{q}(t) \tag{2.41}$$

Differentiating Equation 2.41 twice with respect to x

$$\frac{\partial^2 \tilde{u}}{\partial x^2} = \tilde{\phi}''(x)\tilde{q}(t) \tag{2.42}$$

Differentiating Equation 2.41 twice with respect to time

$$\frac{\partial^2 \tilde{u}}{\partial t^2} = \tilde{\phi}(x)\ddot{\tilde{q}}(t) \tag{2.43}$$

From Equations 2.40, 2.42, and 2.43

$$c_L^2 \tilde{\phi}''(x)\tilde{q}(t) = \tilde{\phi}(x)\ddot{\tilde{q}}(t) \tag{2.44}$$

Rearranging Equation 2.44

$$\frac{\tilde{\phi}''(x)}{\tilde{\phi}(x)} = \frac{\ddot{\tilde{q}}(t)}{c_L^2 \tilde{q}(t)} \tag{2.45}$$

Since Equation 2.45 contains ratios of two functions, which are dependent on x and t, they should be equal to a constant that can be expressed as

$$\frac{\tilde{\phi}''(x)}{\tilde{\phi}(x)} = \frac{\ddot{\tilde{q}}(t)}{c_L^2 \tilde{q}(t)} = -\beta^2 \tag{2.46}$$

where β is an arbitrary constant. From Equation 2.46, two independent equations that are, respectively, dependent on x and t can be obtained.

The space-dependent equation is given by

$$\tilde{\phi}''(x) + \beta^2 \tilde{\phi}(x) = 0 \tag{2.47}$$

Assuming a solution of the form

$$\tilde{\phi}(x) = \overline{\phi} e^{jkx} \tag{2.48}$$

where k is defined as the wave number (spatial frequency) expressed as radians per meter (rad/m), Equation 2.47 becomes

$$(-k^2 + \beta^2)\bar{\phi}e^{jkx} = 0 \tag{2.49}$$

The nontrivial solution for Equation 2.49 is

$$k = \pm\beta \tag{2.50}$$

The time-dependent Equation 2.46 is now given by

$$\ddot{\tilde{q}}(t) + \beta^2 c_L^2 \tilde{q}(t) = 0 \tag{2.51}$$

Assuming a solution of the form

$$\tilde{q}(t) = \bar{q}e^{j\omega t} \tag{2.52}$$

where ω is the angular frequency in radians per second (rad/s), Equation 2.51 becomes

$$\left(-\omega^2 + \beta^2 c_L^2\right)\bar{q}e^{j\omega t} = 0 \tag{2.53}$$

The nontrivial solution of Equation 2.53 is

$$\omega = \pm\beta c_L \tag{2.54}$$

From Equations 2.50 and 2.54 and considering that the wave number and frequency can only be positive

$$k = \frac{\omega}{c_L} = \frac{2\pi f}{c_L} = \frac{2\pi}{\lambda} \tag{2.55}$$

where λ is the wavelength. Equation 2.55 is a very important one that relates wave number, wavelength, angular frequency, and longitudinal wave speed. Wave number is also known as spatial frequency, as it represents the spatial variation of vibration amplitude, which will be clear after obtaining complete solutions for longitudinal vibration.

In terms of wave number k, Equation 2.47 becomes

$$\tilde{\phi}''(x) + k^2\tilde{\phi}(x) = 0 \tag{2.56}$$

The general solution of Equation 2.56 is given by

$$\tilde{\phi}(x) = \bar{\phi}_1 e^{jkx} + \bar{\phi}_2 e^{-jkx} \tag{2.57}$$

From Equations 2.57 and 2.52, Equation 2.41 becomes

$$\tilde{u}(x,t) = \left(\bar{\phi}_1 e^{jkx} + \bar{\phi}_2 e^{-jkx}\right)\bar{q}e^{j\omega t} \tag{2.58}$$

Combining the constants and rearranging, Equation 2.58 becomes

$$\tilde{u}(x,t) = \bar{A}e^{j(\omega t + kx)} + \bar{B}e^{j(\omega t - kx)} \tag{2.59}$$

From Equation 2.55, Equation 2.59 becomes

$$\tilde{u}(x,t) = \bar{A}e^{j\omega\left(t + \frac{x}{c_L}\right)} + \bar{B}e^{j\omega\left(t - \frac{x}{c_L}\right)} \tag{2.60}$$

Before taking up further discussion of Equation 2.60, let us consider a function of the form $f\left(t - \dfrac{x}{c}\right)$, where x and t are independent parameters and c is a constant that has dimensions of velocity. A typical shape of the function $f\left(t - \dfrac{x}{c}\right)$ at $t = t_o$ is shown in Figure 2.8 and is indicated by a solid line. In a short instant of time Δt, it moves to another position such that corresponding points are displaced by x_o, which is indicated by the dashed line. If the dashed line and the solid line are to have an identical shape, then $\Delta t = x_o/c$. This situation represents a disturbance moving in the positive direction. That is, $f\left(t - \dfrac{x}{c}\right)$ represents a wave moving to the right. Similarly, $f\left(t + \dfrac{x}{c}\right)$ represents a wave moving to the left.

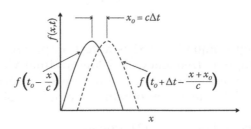

FIGURE 2.8
Shifting of a displacement function.

Now, returning to Equation 2.60, since it is similar to functions $f\left(t - \dfrac{x}{c}\right)$ and $f\left(t + \dfrac{x}{c}\right)$, it represents a combination of a left moving and a right moving waves. The proportion of each of them will depend on the boundary conditions.

2.3 Fixed-Fixed Bars

2.3.1 Natural Frequencies

Figure 2.9 shows a bar fixed at both ends that will not allow any displacement in any direction at these fixed points. The boundary conditions are given by

$$\tilde{u}(0,t) = \tilde{u}(\ell,t) = 0 \tag{2.61}$$

Applying the boundary conditions of Equation 2.61 to Equation 2.60

$$\tilde{u}(0,t) = \bar{A}e^{j\omega t} + \bar{B}e^{j\omega t} = 0 \tag{2.62}$$

$$\tilde{u}(\ell,t) = \bar{A}e^{j\omega\left(t + \frac{\ell}{c_L}\right)} + \bar{B}e^{j\omega\left(t - \frac{\ell}{c_L}\right)} = 0 \tag{2.63}$$

From Equation 2.62

$$\bar{A} = -\bar{B} \tag{2.64}$$

From Equation 2.64, Equation 2.63 becomes

$$\bar{A}e^{j\omega t}\left(e^{\frac{j\omega\ell}{c_L}} - e^{\frac{j\omega\ell}{c_L}}\right) = 0 \tag{2.65}$$

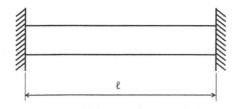

FIGURE 2.9
Fixed-fixed bar.

By using the following trigonometric property

$$\sin\frac{\omega\ell}{c_L} = \frac{e^{\frac{j\omega\ell}{c_L}} - e^{\frac{j\omega\ell}{c_L}}}{2j} \qquad (2.66)$$

Equation 2.65 can be simplified as

$$\bar{A}e^{j\omega t}(2j)\sin\frac{\omega\ell}{c_L} = 0 \qquad (2.67)$$

Equation 2.67 is satisfied for the following values of $\dfrac{\omega\ell}{c_L}$:

$$\frac{\omega\ell}{c_L} = n\pi \qquad (2.68)$$

where n takes the integer values of 1, 2, 3.

From Equation 2.68 it is clear that the bar with fixed-fixed boundary conditions can support only certain values of angular frequencies ω. These are known as natural frequencies or resonant frequencies of the bar and they are given by

$$\omega_n = \frac{n\pi c_L}{\ell}, n = 1,2,3,\ldots \qquad (2.69)$$

For typical values of longitudinal wave speed, most of the resonant frequencies of a bar will be beyond the audible frequency range. Hence, longitudinal wave propagation does not produce significant airborne sound.

From Equations 2.55 and 2.69, the wave number corresponding to various natural frequencies is given by

$$k_n = \frac{n\pi}{\ell} \qquad (2.70)$$

2.3.2 Mode Shapes

Using Equations 2.62, 2.64, 2.68, and 2.58, the displacement function for a fixed-fixed for a wave number is given by

$$\tilde{u}_n(x,t) = \bar{A}_n e^{j\omega_n t}(2j)\sin k_n x \qquad (2.71)$$

Since there are theoretically infinite wave numbers that can be supported by the bar in a longitudinal bar, the total displacement will be a sum of the displacement for each wave number. Therefore, the total displacement due to the action of all the natural frequencies is given by

$$\tilde{u}(x, t) = \sum_{n=1}^{\infty} \bar{A}_n e^{j\omega_n t}(2j) \sin k_n x \tag{2.72}$$

The actual displacement is given by either the real or imaginary part of Equation 2.72. Taking the real part of Equation 2.72 gives

$$u(x, t) = \text{Re}\left\{ \sum_{n=1}^{\infty} \bar{A}_n e^{j\omega_n t}(2j) \sin k_n x \right\}$$

$$= \sum_{n=1}^{\infty} (C_n \cos \omega_n t + D_n \sin \omega_n t) \sin k_n x \tag{2.73}$$

C_n and D_n are constants that will have to be evaluated based on initial conditions.

The spatial component of Equation 2.73 defines mode shapes given by

$$\phi_n(x) = \sin k_n x \tag{2.74}$$

Equation 2.74 is plotted in Figure 2.10 for the first four resonant frequencies that represents the relative displacement of various sections of the bar vibrating at the resonance frequencies. The number of nodes, other than those introduced by the boundary conditions, is one less than the mode number.

Since the movement of various points of the longitudinal bar is in the direction of its length, it is much easier to visualize Equation 2.74 in the following figures.

Figure 2.11 represents the first mode of vibration of the longitudinal bar that has fixed-fixed boundary conditions. The peak amplitude of vibration for the first mode (Figure 2.10a) of selected points on the bar is shown. The length of the arrow indicates the magnitude and the direction indicates whether all of them reach peak amplitude along the same direction or not. Since all the arrows point towards right, all particles of the bar move in the same direction, but with different peak amplitudes and the amplitudes are zero at both the ends due to boundary conditions. During the other half of the cycle, the same points move in the opposite direction with the same relative peak amplitude.

The peak amplitudes of a fixed-fixed bar during the second mode of vibration are shown in Figure 2.12. It has a node at the center in addition to those imposed due to boundary conditions on either end. The peak amplitude is a maximum at one-quarter length and three-quarters length; however, particles on the left side of the bar are always 180 degrees out of phase with those on the right-hand side. The peak amplitudes increase from one node and then decrease toward the other node.

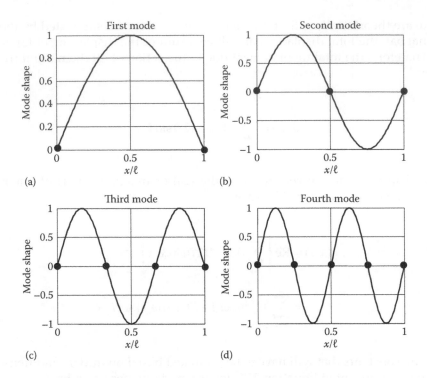

FIGURE 2.10
Longitudinal mode shapes of a fixed-fixed bar.

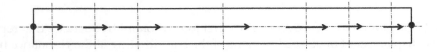

FIGURE 2.11
Peak amplitudes of a fixed-fixed beam in longitudinal bar at the first resonant frequency.

FIGURE 2.12
Peak amplitudes of a fixed-fixed beam in longitudinal bar at the second resonant frequency.

Example 2.6

A steel bar of length 1 m and diameter 50 mm is clamped at both the ends. The given parameters are as follows: length of the bar, ℓ = 1 m; diameter, d = 50 mm; Young's modulus, E = 200 GPa, ρ = 7800 kg/m^3 for steel.

   a. What is the longitudinal wave speed?
   b. What are the first four resonance frequencies in hertz (Hz)?
   c. If a displacement of 1 mm is given at the midportion of the bar in the longitudinal direction as the initial condition, determine the equation for response.

ANSWERS

a. The longitudinal wave speed is given by

$$c_L = \sqrt{\frac{E}{\rho}} = \sqrt{\frac{200 \times 10^9}{7800}} = 5064 \, \text{m/s} \tag{1}$$

b. The boundary conditions are as follows: displacements at the fixed ends are zero.

$$u(0,t) = 0$$
$$u(\ell,t) = 0 \tag{2}$$

The general equation for the longitudinal vibration of a bar in the complex form is given by

$$\tilde{u}(x,t) = \bar{A}e^{j(\omega t + kx)} + \bar{B}e^{j(\omega t - kx)} \tag{3}$$

Applying the boundary condition of Equation 2 at $x = 0$ to Equation 3

$$\bar{A} = -\bar{B} \tag{4}$$

Applying the second boundary condition of Equation 2 at $x = l$ to Equation 3

$$\bar{A}(e^{jk\ell} - e^{-jk\ell}) = 0 \tag{5}$$

Equation 5 leads to the following condition for determining the natural frequencies of the bar with clamped end conditions:

$$\sin k_n \ell = 0, \quad k_n \ell = n\pi, n = 1,2,3,... \tag{6}$$

By using the relationship between the wave number, frequency, and longitudinal speed, the natural frequency in hertz (Hz) is given by

$$f_n = \frac{nc_L}{2\ell}, n = 1,2,3... \tag{7}$$

where n is the mode number.

Using Equation 7, the natural frequencies of the first four modes are given by

$f_1 = 2532 \, \text{Hz}$
$f_2 = 5064 \, \text{Hz}$
$f_3 = 7596 \, \text{Hz}$
$f_4 = 10128 \, \text{Hz}$

c. The initial condition is given by $u(\ell/2,0) = 1$ mm.

The equation for free vibration of a longitudinal bar in terms of its modes of vibration is given by

$$\tilde{u}(x,t) = \sum_{n=1}^{\infty} \tilde{A}_n e^{j(\omega_n t + k_n x)} + \tilde{B}_n e^{j(\omega_n t - k_n x)} \tag{8}$$

From Equation 4, constants of Equation 9 are related as

$$\bar{A}_n = -\bar{B}_n \tag{9}$$

From Equations 8 and 9, and using the relation between the cosine and complex exponential, the general equation for displacement can be expressed in terms of a single constant, C_n, as

$$\tilde{u}(x,t) = \sum_{n=1}^{\infty} \tilde{C}_n e^{j\omega_n t} \sin k_n x \tag{10}$$

If the middle bar is given an initial displacement of δ, assuming that this displacement linearly reduces to zero on both ends, the following initial conditions result:

$$u(x,0) = \frac{2\delta x}{\ell}, 0 \le x \le \ell/2$$

$$u(x,0) = 2\delta\left(\frac{x}{\ell} - 1\right), \ell/2 \le x \le \ell \tag{11}$$

It is important to note that the initial condition has to be applied for every location of the bar as expressed in Equation 10. By applying these initial conditions at $t = 0$ to Equation 10, the value of C_n, which can be obtained similar to the method of obtaining Fourier coefficients, is given by

$$\tilde{C}_n = \frac{2}{\ell} \int_0^{\ell/2} \frac{2\delta x}{\ell} \sin k_n x\, dx + \frac{2}{\ell} \int_{\ell/2}^{\ell} 2\delta\left(\frac{x}{\ell} - 1\right) \sin k_n x\, dx \tag{12}$$

Integrating by parts, Equation 12 can be simplified as

$$\tilde{C}_n = C_n = \frac{4\delta}{\ell^2}\left\{ \begin{array}{l} -\dfrac{\ell}{2k_n}\cos\dfrac{k_n\ell}{2} + \dfrac{\sin\dfrac{k_n\ell}{2}}{k_n^2} - \dfrac{1}{k_n}\left[\ell\cos k_n\ell - \dfrac{\ell}{2}\cos\dfrac{k_n\ell}{2}\right] \\[3mm] + \dfrac{1}{k_n^2}\left[\sin k_n\ell - \sin\dfrac{k_n\ell}{2}\right] \end{array} \right\}$$

$$- \frac{4\delta}{\ell}\left\{ -\frac{1}{k_n}\left[\cos k_n\ell - \cos\frac{k_n\ell}{2}\right] \right\} \tag{13}$$

Example 2.7

A bar of length ℓ is clamped at one end and presses against a spring at the other as shown in Figure 2.13. If $\ell = 300$ mm, the diameter is 50 mm, $E = 200$ GPa, and $\rho = 7800$ kg/m^3, determine the spring stiffness such that it doubles the first combined resonance frequency with respect to the first resonance frequency of the bar without it (spring). The boundary conditions at the left end remain the same in both cases.

WITHOUT SPRING

The boundary condition at the fixed end is that the displacement is zero, given by

$$\tilde{u}(0,t) = 0 \tag{1}$$

The boundary condition at the free end is given by

$$\left.\frac{\partial \tilde{u}}{\partial x}\right|_{x=\ell} = 0 \tag{2}$$

The general equation for longitudinal vibration of a bar is given by

$$\tilde{u}(x,t) = \bar{A}e^{j(\omega t + kx)} + \bar{B}e^{j(\omega t - kx)} \tag{3}$$

The first derivative of displacement with respect to location along the bar is given by

$$\frac{\partial \tilde{u}(x,t)}{\partial x} = jk\left(\bar{A}e^{j(\omega t + kx)} - \bar{B}e^{j(\omega t - kx)}\right) \tag{4}$$

From the first boundary condition

$$\bar{A} = -\bar{B} \tag{5}$$

From the second boundary condition

$$jk\left(\bar{A}e^{jk\ell} - \bar{B}e^{-jk\ell}\right)e^{j\omega t} = 0 \tag{6}$$

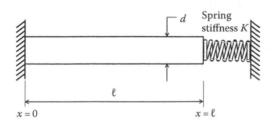

FIGURE 2.13
Clamped-free bar butted against a spring (Example 2.7).

From Equations 5 and 6

$$e^{jk\ell} + e^{-jk\ell} = 0 \Rightarrow \cos k\ell = 0 \tag{7}$$

Equation 7 is satisfied by various values of the wave number k. Thus we have the following wave numbers that satisfy the given boundary conditions:

$$k_n\ell = (2n-1)\frac{\pi}{2}, n = 1,2,3\ldots \tag{8}$$

From Equation 2.55 and Equation 8, the first resonance frequency of the given bar without spring is given by

$$\omega_1 = \frac{\pi c_L}{2\ell} \tag{9}$$

WITH SPRING

The boundary condition at the fixed end is that the displacement is zero, given by

$$\tilde{u}(0,t) = 0 \tag{10}$$

The boundary condition at the free end is given by

$$EA\frac{\partial \tilde{u}}{\partial x}\bigg|_{x=\ell} = Ku(\ell,t) \tag{11}$$

From Equations 3 and 10

$$\bar{A} = -\bar{B} \tag{12}$$

From Equations 3, 4, 11, and 12

$$jkEA\bar{A}\left(e^{jk\ell} + e^{-jk\ell}\right)e^{j\omega t} = K\bar{A}e^{j\omega t}\left(e^{jk\ell} - e^{-jk\ell}\right) \tag{13}$$

Simplifying Equation 13

$$\frac{kEA}{K} = \tan k\ell \tag{14}$$

The first resonance frequency of the bar with spring that is twice that of the bar without spring is given by

$$\omega_1' = 2\omega_1$$

$$= \frac{\pi c_L}{\ell} = k_1' c_L \tag{15}$$

where k_1' is the wavenumber of the bar with spring at the free end. From Equation 15, $k_1' = \frac{\pi}{\ell}$. Substituting this wavenumber in Equation 14, the stiffness value of the spring that can double the resonance frequency of a cantilever bar is infinite, that is, equivalent to a clamped boundary condition.

2.4 Free-Free Bar

Figure 2.14 shows a bar with free-free end conditions that are obtained by hanging the bar by means of two strings. This arrangement almost ensures that the bar is free to vibrate in the longitudinal direction at both ends.

2.4.1 Natural Frequencies

Due to free-free end conditions, the strain is zero at both ends and it can be expressed as

$$\frac{\partial \tilde{u}}{\partial x} = 0 \text{ at } x = 0 \text{ and } x = \ell \tag{2.75}$$

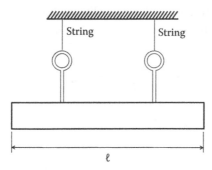

FIGURE 2.14
Bar with free-free end conditions.

From Equation 2.59, the strain equation is given by

$$\frac{\partial \tilde{u}}{\partial x} = e^{j\omega t}\left(\frac{j\omega}{c_L}\right)\left(\bar{A}e^{\frac{j\omega x}{c_L}} - \bar{B}e^{-\frac{j\omega x}{c_L}}\right)$$ (2.76)

Applying the first boundary condition of Equation 2.75, the constants are related as

$$\bar{A} = \bar{B}$$ (2.77)

Using Equation 2.77 and applying the second boundary condition of Equation 2.75 to Equation 2.76

$$\left.\frac{\partial \tilde{u}}{\partial x}\right|_{x=\ell} = \bar{A}e^{j\omega t}\left(\frac{j\omega}{c_L}\right)\left(e^{\frac{j\omega\ell}{c_L}} - e^{-\frac{j\omega\ell}{c_L}}\right) = 0$$ (2.78)

Using Equation 2.66, Equation 2.78 becomes

$$\left.\frac{\partial \tilde{u}}{\partial x}\right|_{x=\ell} = 2j\bar{A}e^{j\omega t}\left(\frac{j\omega}{c_L}\right)\sin\left(\frac{\omega\ell}{c_L}\right) = 0$$ (2.79)

which gives

$$\sin\left(\frac{\omega\ell}{c_L}\right) = 0 \text{ or } \frac{\omega\ell}{c_L} = n\pi$$ (2.80)

Therefore, a bar with free-free boundary conditions has the same natural frequencies as that of a fixed-fixed bar, given by

$$\omega_n = \frac{n\pi c_L}{\ell}, n = 1, 2, 3, \ldots$$ (2.81)

2.4.2 Mode Shapes

Using Equation 2.80, the complex displacement function in the *n*th for a bar with free-free boundary conditions is given by

$$\tilde{u}_n(x,t) = \bar{A}_n e^{j\omega_n t}\left(e^{\frac{j\omega_n x}{c_L}} + e^{-\frac{j\omega_n x}{c_L}}\right)$$ (2.82)

By using the following trigonometric relation

$$\cos \frac{\omega_n x}{c_L} = \left(\frac{e^{\frac{j\omega_n x}{c_L}} + e^{-\frac{j\omega_n x}{c_L}}}{2} \right)$$

(2.83)

Equation 2.83 becomes

$$\tilde{u}_n(x,t) = 2\bar{A}_n e^{j\omega_n t} \cos \frac{\omega_n x}{c_L}$$

(2.84)

The actual displacement response in the nth mode of the free-free bar in longitudinal vibration is given by either the real or imaginary part of Equation 2.84. Let us take the real part as

$$\tilde{u}_n(x,t) = (A_n \cos \omega_n t + B_n \sin \omega_n t) \cos k_n x$$

(2.85)

where $\cos k_n x$ is the mode shape of vibration for the nth mode given by

$$\phi_n(x) = \cos k_n x$$

(2.86)

Equation 2.86 represents the relative displacement of various points of the bar during various resonant frequencies and is presented in Figure 2.15. The number of nodes is equal to the mode number.

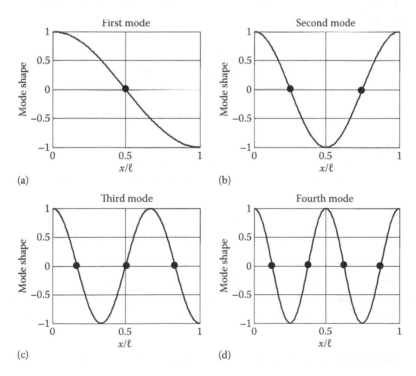

FIGURE 2.15
Mode shapes for longitudinal vibration of a free-free bar.

Since the motion is along the length of the bar, it would be easier to understand the physics of mode shapes from the following figures: Figures 2.16 and 2.17.

Figure 2.16 shows the peak amplitudes of relative motion during the first resonant frequency. There is a single node at the center; the relative peak amplitudes are in opposite direction on either side of this node. The peak amplitudes are a maximum at the ends and keep reducing toward the center node.

Figure 2.17 shows peak amplitudes of relative displacement of a free-free beam during second resonant frequency. There are two nodes and the peak amplitude of displacement is a maximum at the free ends. The peak amplitude from left to right keeps on decreasing till the first node, and then it is out of phase and keeps on increasing, and then again decreasing till the next node. After the node it reverses direction and the peak amplitude keeps increasing and reaches a maximum at the right free end.

The total displacement due to all the modes of vibration of the free-free bar is given by

$$u(x,t) = \sum_{n=1}^{\infty} (A_n \cos \omega_n t + B_n \sin \omega_n t) \cos k_n x \tag{2.87}$$

Example 2.8

An aluminum rod is of 20 mm diameter and length 0.4 m. What will be the fundamental frequency of free vibrations and displacement amplitude of the free ends, if the displacement amplitude of the rod at its center is 1 mm, and given length of the beam ℓ, 0.4 m; Young's modulus of elasticity, $E = 80$ GPa; and density, $\rho = 2800$ kg/m^3.

Longitudinal wave speed, c_L, is given by

$$c_L = \sqrt{\frac{E}{\rho}} = \sqrt{\frac{80 \times 10^9}{2800}} = 5345\,\text{m/s} \tag{1}$$

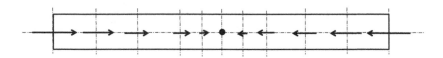

FIGURE 2.16
Peak amplitudes of a free-free longitudinal bar at the first resonant frequency.

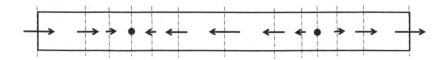

FIGURE 2.17
Peak amplitudes of a free-free in longitudinal bar at the second resonant frequency.

From Equation 2.81, the following condition for determining the natural frequencies of the bar with free-free end conditions can be written as

$$\sin k_n \ell = 0, k_n \ell = n\pi, n = 1,2,3,\ldots \tag{2}$$

By using the relationship between the wave number, frequency, and longitudinal speed, the natural frequency in hertz (Hz) is given by

$$f_n = \frac{nc_L}{2\ell}, n = 1,2,3,\ldots \tag{3}$$

Using Equation 3, the fundamental frequency is $f_1 = 5345$ Hz.

The equation for free vibration of a longitudinal bar in terms of its modes of vibration is given by

$$\tilde{u}(x,t) = \sum_{n=1}^{\infty} \tilde{A}_n e^{j(\omega_n t + k_n x)} + \tilde{B}_n e^{j(\omega_n t - k_n x)} \tag{4}$$

From Equation 2.7, constants of the Equation 4 are related as

$$\tilde{A}_n = \tilde{B}_n \tag{5}$$

From Equations 4 and 5, and using the relation between the cosine and complex exponential, the general equation for displacement can be expressed in terms of a single constant, C_n, as

$$\tilde{u}(x,t) = \sum_{n=1}^{\infty} \tilde{C}_n e^{j\omega_n t} \cos k_n x \tag{6}$$

If the middle bar is given an initial displacement of δ, assuming that this displacement linearly reduces to zero on both ends, the following initial conditions result:

$$u(x,0) = \frac{2\delta x}{\ell}, 0 \leq x \leq \ell/2$$
$$u(x,0) = 2\delta\left(\frac{x}{\ell} - 1\right), \ell/2 \leq x \leq \ell \tag{7}$$

It is important to note that the initial condition has to be applied for every location of the bar as expressed in Equation 7. By applying these initial conditions at $t = 0$ to Equation 11, the value of C_n, which can be obtained similar to the method of obtaining Fourier coefficients, is given by

$$\tilde{C}_n = \frac{2}{\ell} \int_0^{\ell/2} \frac{2\delta x}{\ell} \cos k_n x \, dx + \frac{2}{\ell} \int_{\ell/2}^{\ell} 2\delta\left(\frac{x}{\ell} - 1\right) \cos k_n x \, dx \tag{8}$$

Integrating by parts, Equation 8 can be simplified as

$$\tilde{C}_n = C_n = \frac{4\delta}{\ell^2}\left\{ \frac{\ell}{2k_n}\sin\frac{k_n\ell}{2} + \frac{\cos\frac{k_n\ell}{2}-1}{k_n^2} \right\}$$

$$+\frac{4\delta}{\ell}\left\{ \frac{1}{2k_n}\sin\frac{k_n\ell}{2} + \frac{\sin k_n\ell - \sin\frac{k_n\ell}{2}}{\ell k_n^2} \right\}$$

(9)

The constant in Equation 9 can be computed for various modes of vibration (at least 1 through 5). When substituted in Equation 6, its real part gives the response at various locations of the bar.

Example 2.9

A long thin bar of length L is driven by a longitudinal force $F\cos\omega t$ at $x = 0$ and is free at $x = \ell$ (Figure 2.18).

a. Derive the equation that gives the amplitude of the standing waves set up in the bar.
b. What is the input mechanical impedance?
c. What is the input mechanical impedance of a similar bar of infinite length?
d. If the material of the bar is steel, the length is 1 m, the diameter is 25 m, and the amplitude of the driving force is 10 N, plot the amplitude of the driven end of part a from 200 to 2000 Hz.

ANSWERS

a. Let E be the Young's modulus of the bar material and A the area of cross section.
   The boundary condition at the left at which the reaction of the bar is balanced by the external force and is given by

$$-EA\frac{\partial \tilde{u}}{\partial x}\bigg|_{x=0} = Fe^{j\omega t}$$

(1)

And since the strain at $x = L$ is zero, the boundary condition at the right is given by

$$\frac{\partial \tilde{u}}{\partial x}\bigg|_{x=\ell} = 0$$

(2)

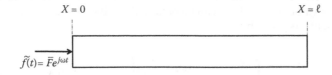

FIGURE 2.18
A longitudinal bar free at both ends driven by a complex harmonic force at $x = 0$.

The boundary conditions in Equations 1 and 2 can be applied to a general equation of longitudinal bar vibration to determine the constants.

From Equation 2.59, the general response of a longitudinal bar is given by

$$\tilde{u}(x,t) = \bar{A}e^{j(\omega t + kx)} + \bar{B}e^{j(\omega t - kx)} \tag{3}$$

The spatial derivative Equation 3 that relates to strain is given by

$$\frac{\partial \tilde{u}}{\partial x} = jk\left(\bar{A}e^{j(\omega t + kx)} - \bar{B}e^{j(\omega t - kx)}\right) \tag{4}$$

From Equations 1 and 4

$$-EA(jk)e^{j\omega t}(\bar{A} - \bar{B}) = Fe^{j\omega t} \tag{5}$$

Equation 5 simplifies to

$$\bar{A} - \bar{B} = -\frac{F}{jEAk} \tag{6}$$

From Equations 2 and 4

$$\bar{A}e^{jk\ell} - \bar{B}e^{-jk\ell} = 0 \tag{7}$$

Equations 6 and 7 form a system of linear equations given by

$$\begin{bmatrix} 1 & -1 \\ e^{jk\ell} & e^{-jk\ell} \end{bmatrix} \begin{Bmatrix} \bar{A} \\ \bar{B} \end{Bmatrix} = \begin{Bmatrix} -\dfrac{F}{jEAk} \\ 0 \end{Bmatrix} \tag{8}$$

Equation 8 can be solved to obtain the following values of constants:

$$\bar{A} = -\frac{Fe^{-jk\ell}}{2EAk \sin k\ell} \tag{9}$$

$$\bar{B} = -\frac{Fe^{jk\ell}}{2EAk \sin k\ell} \tag{10}$$

From Equations 3, 9, and 10, the displacement response due to a longitudinal bar excited by a harmonic force at one end is given by

$$\tilde{u}(x,t) = \frac{Fe^{j\omega t}}{jEAk \sin k\ell}\left(e^{-jk(\ell-x)} + e^{jk(\ell-x)}\right) \tag{11}$$

By using the relationship between exponential and cosine functions, Equation 11 becomes

$$\tilde{u} = -\frac{Fe^{j\omega t}\cos k(\ell - x)}{EAk\sin k\ell} \tag{12}$$

From Equation 12, the amplitude of longitudinal vibrations along the bar is given by

$$U(x) = \frac{F\cos k(\ell - x)}{EAk\sin k\ell} \tag{13}$$

b. The equation for input impedance is given by

$$Z_i = \frac{F(x=0)}{\tilde{v}(x=0)} \tag{14}$$

From Equation 12, the velocity of longitudinal vibrations is given by

$$\tilde{v} = \frac{\partial \tilde{u}}{\partial t} = \frac{-j\omega Fe^{j\omega t}\cos k(\ell - x)}{EAk\sin k\ell} \tag{15}$$

From Equation 15, the velocity at $x = 0$ is given by

$$\tilde{v}(0,t) = \frac{-j\omega Fe^{j\omega t}\cos k\ell}{EAk\sin k\ell} \tag{16}$$

From Equations 14 and 16, the input impedance is given by

$$Z_i = \frac{EAk\tan k\ell}{-j\omega} \tag{17}$$

In terms of longitudinal wave speed, Equation 17 becomes

$$Z_i = j\rho c_L A \tan k\ell \tag{18}$$

For an infinite rod $k\ell = \infty$. Therefore, rewrite Equation 18 as

$$Z_i = \rho c_L A \left(\frac{e^{jk\ell} - e^{-jk\ell}}{e^{jk\ell} + e^{-jk\ell}} \right) \tag{19}$$

Since the negative exponential becomes zero and the positive exponentials cancel each other, the input impedance of an infinite rod is given by

$$Z_i = \rho c_L A \tag{20}$$

ρc_L is called the characteristic impedance of the rod in longitudinal vibration.

c. $\ell = 1$ m, $d = 25$ mm, $F = 10$ N, $\rho = 7800$ kg/m³, $E = 200$ GPa

The longitudinal wave speed is given by

$$c_L = \sqrt{\frac{E}{\rho}} = \sqrt{\frac{200 \times 10^9}{7800}} = 5064 \, \text{m/s} \tag{21}$$

The wave number is given by

$$k = \frac{\omega}{c_L} = \frac{2\pi f}{c_L} \tag{22}$$

The area of cross section is given by

$$A = \frac{\pi \times 0.025^2}{4} = 4.908 \times 10^{-4} \, \text{m}^2 \tag{23}$$

From Equation 13, the amplitude of longitudinal vibration of the driven end is given by

$$U(k) = \frac{10 \cos k}{200 \times 10^9 \times 4.908 \times 10^{-4} k \sin k} = \frac{1}{10 k \tan k} \mu\text{m} \tag{24}$$

Equation 24 is plotted in Figure 2.19.

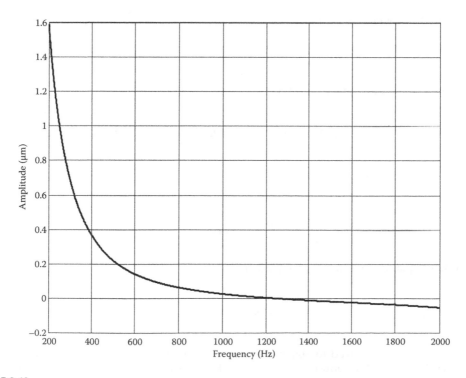

FIGURE 2.19

Response of a free-free longitudinal rod of length 1 m and diameter 25 mm to a harmonic excitation of 10 N amplitude from 200 to 2000 Hz.

Example 2.10

Determine the natural frequencies of a longitudinal bar with a mass at one end and fixed at the other end.

The boundary condition at the fixed end is that the displacement is zero, given by

$$\tilde{u}(0,t) = 0 \tag{1}$$

The boundary condition at the free end is given by

$$-EA\frac{\partial \tilde{u}}{\partial x}\bigg|_{x=\ell} = M\frac{\partial^2 \tilde{u}}{\partial t^2}\bigg|_{x=\ell} \tag{2}$$

The general equation for longitudinal vibration of a bar is given by

$$\tilde{u}(x,t) = \bar{A}e^{j(\omega t + kx)} + \bar{B}e^{j(\omega t - kx)} \tag{3}$$

The first derivative of displacement with respect to location along the bar is given by

$$\frac{\partial \tilde{u}(x,t)}{\partial x} = jk\left(\bar{A}e^{j(\omega t + kx)} - \bar{B}e^{j(\omega t - kx)}\right) \tag{4}$$

From the first boundary condition

$$\bar{A} = -\bar{B} \tag{5}$$

From the second boundary condition

$$-jkEA\left(\bar{A}e^{jk\ell} - \bar{B}e^{-jk\ell}\right)e^{j\omega t} = -M\omega^2\left(\bar{A}e^{jk\ell} + \bar{B}e^{-jk\ell}\right)e^{j\omega t} \tag{6}$$

From Equations 5 and 6

$$\tan k\ell = \frac{EAk}{M\omega^2} \tag{7}$$

Using Equations 2.38 and 2.55, Equation 7 can be simplified as

$$k\ell \tan k\ell = \frac{m}{M} \tag{8}$$

Equation 8 can be solved to obtain the natural frequencies. When $M = 0$, the system shown in Figure 2.20 reduces to a fixed-free beam, and when $M = \infty$, it reduces to a fixed-fixed beam.

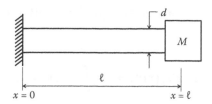

FIGURE 2.20
Bar clamped at one end and mass loaded at the other end (Example 2.10).

2.5 Orthogonality Condition

If the response of the bar in longitudinal vibration has to be solved for force excitation using modal approach, the partial differential equation has to be converted into a system of uncoupled equations by using the property of mode shapes. The orthogonality condition derived in this section provides the basis for obtaining uncoupled equations. This process of obtaining uncoupled equations is very similar to the uncoupled equations of discrete MDOF systems using the modal matrix.

From Equation 2.56, the following equation can be written for the nth mode:

$$\phi_n''(x) = -k_n^2 \phi_n(x) \tag{2.88}$$

Multiplying both sides of Equation 2.88 by ϕ_m (mode shape with index m) and integrating

$$\int_0^\ell \phi_n''(x)\phi_m(x)\,dx = -k_n^2 \int_0^\ell \phi_n(x)\phi_m(x)\,dx \tag{2.89}$$

By integrating the left-hand side of Equation 2.89 by parts

$$\int_0^\ell \phi_n''(x)\phi_m(x)\,dx = \phi_n'(x)\phi_m(x)\Big|_0^\ell - \phi_n(x)\phi_m'(x)\Big|_0^\ell + \int_0^\ell \phi_n(x)\phi_m''(x)\,dx \tag{2.90}$$

Because of the self-adjoint property of boundary conditions, the first and second terms of the right-hand side of Equation 2.90 vanish, resulting in

$$\int_0^\ell \phi_n''(x)\phi_m(x)\,dx = \int_0^\ell \phi_n(x)\phi_m''(x)\,dx \tag{2.91}$$

By interchanging the indices m and n, an equation similar to Equation 2.89 can be written as

$$\int_0^\ell \phi_m''(x)\phi_n(x)\,dx = -k_m^2 \int_0^\ell \phi_m(x)\phi_n(x)\,dx \tag{2.92}$$

Subtracting Equations 2.89 and 2.92, and using Equation 2.91

$$\left(k_m^2 - k_n^2\right)\int_0^\ell A\phi_m(x)\phi_n(x)\,dx = 0 \tag{2.93}$$

The area, A, is intentionally included as part of the integral of Equation 2.93 that will prove to be useful later.

From Equation 2.55, Equation 2.93 can be written as

$$\frac{\left(\omega_m^2 - \omega_n^2\right)}{E}\int_0^\ell \rho A\phi_m(x)\phi_n(x)\,dx = 0 \tag{2.94}$$

For $m = n$, Equation 2.94 is satisfied irrespective of the value of the integral. For $m \neq n$, the integral should be zero. This leads to the definition of the following orthogonality condition:

$$\int_0^\ell \rho A\phi_m(x)\phi_n(x)\,dx = 0 \quad m \neq n$$
$$\tag{2.95}$$
$$\int_0^\ell \rho A\phi_n^2(x)\,dx = m_n \quad for\ m = n$$

where m_n is known as the generalized mass corresponding to the nth mode.

2.6 Force Excitation

Let the longitudinal bar in Figure 2.7 be subjected to force excitation $f(x,t)$, which has the units of force per unit length. Adding this external excitation to the Equations 2.34 through 2.36, the force balance equation becomes

$$EA\frac{\partial^2 u}{\partial x^2}\Delta x + f(x,t)\Delta x = \rho A\frac{\partial^2 u}{\partial t^2}\Delta x \tag{2.96}$$

Equation 2.96 simplifies to

$$EA\frac{\partial^2 u}{\partial x^2} + f(x,t) = \rho A\frac{\partial^2 u}{\partial t^2} \tag{2.97}$$

The displacement response for force response can be assumed as

$$u(x,t) = \sum_{n=1}^{\infty} \phi_n(x)q_n(t) \tag{2.98}$$

where $\phi_n(x)$ represents mode shapes and $q_n(t)$ are the generalized displacements. They also can be called time-dependent values that define the linear combination of mode shapes. We shall eventually see that they are coordinates corresponding to a system of uncoupled equations that can be solved to obtain the solution due to force excitation using Equation 2.98.

From Equation 2.98, Equation 2.97 becomes

$$EA\sum_{n=1}^{\infty} \phi_n''(x)q_n(t) + f(x,t) = \rho A\sum_{n=1}^{\infty} \phi_n(x)\ddot{q}_n(t) \tag{2.99}$$

Using Equation 2.88, Equation 2.99 becomes

$$EA\sum_{n=1}^{\infty} -k_n^2\phi_n(x)q_n(t) + f(x,t) = \rho A\sum_{n=1}^{\infty} \phi_n(x)\ddot{q}_n(t) \tag{2.100}$$

By using the relationship between the longitudinal wave speed, natural frequency, and wave number, Equation 2.100 becomes

$$\sum_{n=1}^{\infty} \rho A\left[\ddot{q}_n(t) + \omega_n^2 q_n(t)\right]\phi_n(x) = f(x,t) \tag{2.101}$$

Multiplying both sides of Equation 2.101 by $\phi_m(x)$ and integrating along the beam length

$$\sum_{n=1}^{\infty} \left[\ddot{q}_n(t) + \omega_n^2 q_n(t)\right]\int_0^\ell \rho A\phi_n(x)\phi_m(x)\,dx = \int_0^\ell f(x,t)\phi_m(x)\,dx \tag{2.102}$$

Using the orthogonality condition of Equation 2.95, Equation 2.102 is valid only for $n = m$. Therefore, Equation 2.102 becomes

$$\ddot{q}_n(t) + \omega_n^2 q_n(t) = \frac{\displaystyle\int_0^\ell f(x,t)\phi_n(x)\,dx}{\displaystyle\int_0^\ell \rho A\phi_n^2(x)\,dx} = Q_n(t) \tag{2.103}$$

where $Q_n(t)$ is the generalized force given by

$$Q_n(t) = \frac{\int_0^\ell f(x,t)\phi_n(x)\,dx}{m_n} \tag{2.104}$$

By solving Equation 2.103, the generalized coordinate can be obtained, which is substituted in Equation 2.98 for obtaining the displacement response of a longitudinal bar for any external force excitation.

Example 2.11

An aluminium rod ($E = 80$ GPa and density 2800 kg/m³) of 20 mm diameter and length 0.4 m is fixed at one end and free at the other end.

a. Determine its natural frequencies and mode shapes.
b. Determine its generalized mass.
c. If it is subjected to a simple harmonic force of amplitude 5 N (rms) and frequency 100 Hz at its free end, determine its generalized force, and what will be the rms velocity response of the free end by using the normal mode method?

From Equation 2.56, the general equation for longitudinal vibration is given by

$$\phi''(x) + k^2\phi(x) = 0 \tag{1}$$

The general solution of Equation 1 is given by

$$\phi(x) = A\cos kx + B\sin kx \tag{2}$$

The derivative of Equation 2 with respect to location is given by

$$\phi'(x) = k(-A\sin kx + B\cos kx) \tag{3}$$

The boundary conditions are displacement $u(0,t) = 0$ at the fixed end and strain $\dfrac{\partial u(\ell,t)}{\partial x} = 0$. Using these boundary conditions in Equations 2 and 3

$$A = 0;\quad \cos k\ell = 0; \tag{4}$$

$\cos k\ell = 0$ is satisfied for several values of k as follows:

$$k_n = \frac{(2n-1)\pi}{2\ell},\, n = 1,2,3\ldots \tag{5}$$

From Equation 2.55 and Equation 5, the resonant frequencies are given by

$$\omega_n = \frac{(2n-1)\pi c_L}{2\ell} \tag{6}$$

For each of the resonant frequencies of Equation 6, the corresponding mode shapes are given by

$$\phi_n(x) = B_n \sin k_n x \tag{7}$$

B_n is only a scaling factor that can be taken as 1.

From Equation 2.95 and Equation 7, the generalized mass is given by

$$
\begin{aligned}
m_n &= \int_0^\ell \rho A \sin^2(k_n x)\,dx \\
&= \frac{\rho A}{2}\left(x - \frac{\sin 2k_n x}{2k_n}\right)\Bigg|_0^\ell \tag{8} \\
&= \frac{\rho A \ell}{2}
\end{aligned}
$$

The excitation force at the free end can be represented as

$$f(x,t) = 5\delta\langle x - \ell \rangle \cos \omega t \tag{9}$$

From Equation 2.104 and Equation 9, the generalized force is given by

$$
\begin{aligned}
Q_n(t) &= \frac{1}{m_n}\int_0^\ell f(x,t)\phi_n(x)\,dx \\
&= \frac{10}{\rho AL}\int_0^L \delta\langle x - \ell\rangle \cos \omega t \sin k_n x\,dx \tag{10} \\
&= \frac{10\cos \omega t \sin k_n \ell}{\rho A\ell} = \frac{10\cos \omega t(-1)^{n+1}}{\rho A\ell} \quad n = 1,2,3\ldots
\end{aligned}
$$

From Equation 2.103 and Equation 10, the uncoupled equations for vibration of the given longitudinal bar are given by

$$\ddot{q}_n(t) + \omega_n^2 q_n(t) = \frac{10\cos \omega t(-1)^{n+1}}{\rho A\ell} \quad n = 1,2,3\ldots \tag{11}$$

The solution to Equation 11 is given by

$$q_n(t) = \frac{10\cos\omega t(-1)^{n+1}}{\rho A \ell \left(\omega_n^2 - \omega^2\right)} \tag{12}$$

From Equation 2.98 and Equation 12, the general equation for displacement of the given bar is given by

$$u(x,t) = \frac{10\cos\omega t}{\rho A \ell} \sum_{n=1}^{\infty} \frac{(-1)^{n+1}\sin k_n x}{\left(\omega_n^2 - \omega^2\right)} \quad n = 1,2,3... \tag{13}$$

Differentiating Equation 13 with respect to time, the general equation for velocity of the given bar is given by

$$v(x,t) = u(x,t) = -\frac{10\omega\sin\omega t}{\rho A \ell} \sum_{n=1}^{\infty} \frac{(-1)^{n+1}\sin k_n x}{\left(\omega_n^2 - \omega^2\right)} \quad n = 1,2,3... \tag{14}$$

From Equation 14, the velocity at $x = \ell$ is given by

$$v(\ell,t) = -\frac{10\omega\sin\omega t}{\rho A \ell} \sum_{n=1}^{\infty} \frac{(-1)^{2(n+1)}}{\left(\omega_n^2 - \omega^2\right)} \quad n = 1,2,3... \tag{15}$$

Computing terms up to the third mode for the resonant frequencies of $\omega_1 = 20991$ rad/s, $\omega_2 = 62997$ rad/s, and $\omega_3 = 104995$ rad/s, the rms value of velocity is 48 µm/s.

2.7 Conclusions

The discussion presented in this chapter should enable a clear physical interpretation of the basic concept of mode shapes and traveling waves. These will form the basis for studying a more difficult wave type—the flexural wave—in beams and plates, which will eventually help understand the vibro-acoustic interaction in plates and shells. Both the mode and the wave approaches were utilized to derive their best advantages; the physical understanding obtained from these approaches would be useful for solving more difficult problems in vibro-acoustics. In addition, it should be noted that one should be able to comfortably move from the wave approach to the modal approach and vice versa, and be able to use the complex form of equations to derive their best advantages.

PROBLEMS

2.1. The spring–mass model shown in the following figure is used for modeling packages. If $m = 1$ kg and $k = 2$ kN/m, determine the undamped natural frequencies of the system and the corresponding mode shapes.

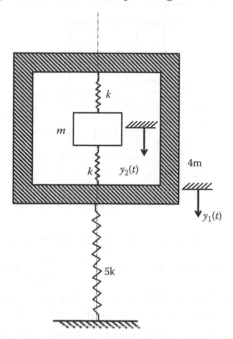

2.2. Consider the spring-mass system shown in the following figure. If $m = 1$ kg and $k = 5$ kN/m, determine the two natural frequencies and the corresponding mode shapes.

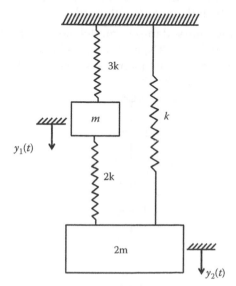

2.3. Consider the two-degrees-of-freedom system shown in the following figure. Assume $k_1 = k_2 = k_3 = 20$ kN/m and $m_1 = m_2 = 1$ kg. Determine the value of damping coefficients so the damping matrix is proportional to the stiffness and mass matrices as follows: $[\mathbf{R_m}] = 0.2[\mathbf{M}] + 0.006[\mathbf{K}]$. Also compute the undamped natural frequencies and damping factors of the uncoupled system.

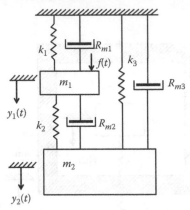

2.4. Consider a two-degrees-of-freedom spring–mass–damper system that has the mass normalized modal matrix of $\phi = \begin{bmatrix} 0.6 & 0.8 \\ 0.6 & -0.4 \end{bmatrix}$ and the undamped natural frequencies are 170 and 250 rad/s. The damping matrix is proportional to the mass and stiffness matrix as follows: $[\mathbf{R_m}] = 0.01[\mathbf{M}] + 0.001[\mathbf{K}]$.

   a. If a force excitation $2\sin(190t + 30°)$ N is applied to the first degree of freedom, determine the velocity response of both degrees of freedom and the total power input to the system.

   b. Write down the mass, stiffness, and damping matrices, and one of the configurations in which they can be attached.

2.5. Consider the spring–mass–damper model shown in the following figure. Assume $k_1 = 40$ kN/m, $k_2 = 10$ kN/m, and $k_3 = 40$ kN/m; $m_1 = 4$ kg and $m_2 = 4$ kg; and $[\mathbf{R_m}] = [\mathbf{M}] + 0.001[\mathbf{K}]$. If the first mass is given an impulse input, what is the response of each mass at $t = 0.1$ s?

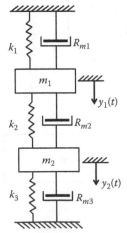

2.6. Consider Example 2.5. Write down the voltage and current analog of this system, and determine the velocity response and power input to the system.

2.7. Consider the following spring–mass–damper model. Assume $k_1 = 40$ kN/m, $k_2 = 10$ kN/m, and $k_3 = 40$ kN/m; $m_1 = 4$ kg and $m_2 = 4$ kg; and $[\mathbf{R_m}] = [\mathbf{M}] + 0.001[\mathbf{K}]$. If an excitation force $f(t) = 10\sin 100t$ N is applied to the first degree of freedom as shown, determine the velocity response of both masses by using any of the electric analogs (voltage or current).

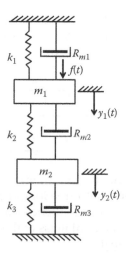

2.8. A rigid bar of length $\ell = 2$ m and mass $m = 6$ kg is attached to spring–dampers at each end as shown the following figure. If $R_{m1} = R_{m2} = 500$ N-s/m and $k_1 = k_2 = 50$ kN/m, determine the maximum displacement response of the system to a unit step input at its center of gravity. Check whether the system has proportional damping. If not, assume proportional damping and compute the displacement response and then compare it with the actual response using any of the numerical methods.

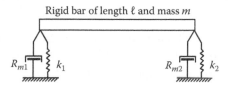

Rigid bar of length ℓ and mass m

2.9. Consider Example 2.2. If the second mass is subjected to a sinusoidal excitation of $\sin 150t$, determine the peak velocity amplitude of both the masses. What is the power input to the system? Use the impedance method to solve this problem.

2.10. Consider a steel bar ($E = 200$ GPa and $\rho = 7800$ kg/m³) of length $\ell = 1$ m and diameter 25 mm that terminates with a mass at one end and a spring at the other end (see following figure). Assume the mass (M) to be 25% of the shaft mass and the spring stiffness to be 50% of the bar stiffness in the axial direction. Determine the first five resonant frequencies of longitudinal vibration.

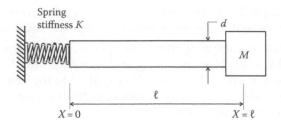

2.11. Consider a clamped-clamped bar of diameter d in longitudinal vibration as shown in the following figure. Determine an equivalent string that has the same natural frequencies as this system.

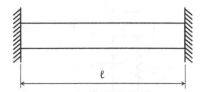

2.12. A steel bar of 2 m length and 40 mm diameter, as shown in the following figure, is assumed to be excited with free longitudinal vibrations. Determine the general equation for its resonant frequencies and compute its first natural frequency if each mass is half the total mass of the shaft. Assume $E = 200$ GPa and density as 7800 kg/m³ for steel.

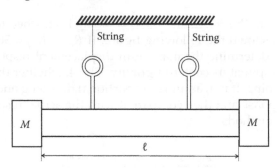

2.13. An aluminium rod ($E = 80$ GPa and density $= 2800$ kg/m³) of 20 mm diameter and length 0.4 m is fixed at one end and free at the other end.

   a. Determine its natural frequencies and mode shapes.

   b. Determine its generalized mass.

   c. If it is subjected to a simple harmonic force of amplitude 5 N (rms) and frequency 100 Hz at its free end, determine its generalized force, and what will be the rms velocity response of the free end by using the normal mode method?

2.14. Consider the steel bar ($E = 200$ GPa and $\rho = 7800$ kg/m³) shown in the following figure of 2 m length and diameter 25 mm that is clamped at the center and the two ends are free. Assume that the clamping distance is negligible and determine the equation for its resonant frequencies in longitudinal vibration.

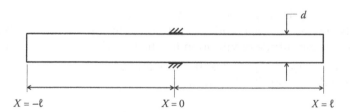

2.15. Consider a spring–mass system of mass m and spring stiffness k.

    a. Prove that one-third of the spring mass should be added to the mass m to get an accurate value of natural frequency by using the energy method.

    b. Prove part a by assuming standing waves in the spring–mass system.

2.16. A spring of stiffness k and mass per unit length ρ_ℓ is allowed to perform longitudinal vibrations along its length. Prove that the speed of wave propagation is given by $v = \sqrt{\dfrac{k}{\rho_\ell}}$.

2.17. Refer to Problem 2.11. Is this equation always true if the mass at the end of the spring is removed and we are left only with the mass of the spring?

2.18. A bar of diameter d and length ℓ is clamped at one end and left free at the other end. A harmonic force of excitation frequency ω is applied at the free end in the longitudinal direction as shown in the following figure. Determine the input impedance at the free end and the input impedance if the length ℓ becomes extremely long.

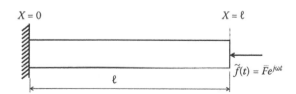

2.19. Consider the steel bar ($E = 200$ GPa and $\rho = 7800$ kg/m³) shown in the following figure of length ℓ 1 m and diameter 25 mm that terminates with a spring–mass–damper system at the right end and driven by a harmonic excitation at the left end. Assume the mass M to be 25% of the shaft mass and the spring stiffness to be 50% of the bar stiffness in the axial direction, and the damping coefficient is such that the damping factor of the independent SDOF is 0.15. Determine the equation for power input at various frequencies.

2.20. Consider a steel bar ($E = 200$ GPa and $\rho = 7800$ kg/m³) shown in the following figure of length ℓ 1 m and diameter 25 mm.

a. If its free tip is displaced by a distance of 2 mm in the axial direction by applying a static force and let go to perform longitudinal vibration, determine the amplitude of vibration for the first five modes of vibration.

b. Determine the amplitudes if the static force of 6000 N is applied and removed, very similar to part a.

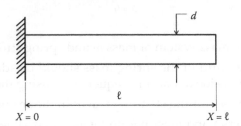

3

Airborne Sound

Sound can be defined as a disturbance that moves in a solid, liquid, or gaseous medium. But in this chapter we shall be concerned about sound that propagates in a compressible medium such as air, which is known as airborne sound. However, it is common in everyday usage to call airborne sound as just sound and we shall do the same throughout. Since airborne sound contributes significantly to sound-induced hearing loss and comes in the way of audio communication, it has received maximum attention. Airborne sound can be produced either by vibration of a solid surface or due to flow. But in this chapter, we shall only be concerned about the basic phenomenon of airborne sound, how it propagates and how it is quantified. In later chapters, we will connect how a vibrating surface results in airborne sound.

When a disturbance passes through air, it changes the local pressure and density along its path as well as the particle velocity. Furthermore, the disturbance travels at a definite speed that is characteristic of the medium, which is also the speed of sound. We will first study the changes that occur in a medium due to propagation of disturbance and then relate them through the wave equation. For this we shall assume that the propagation is along one direction only. However, the same theory can be extended to the case of airborne sound propagation in all the three dimensions, which is discussed in Chapter 7.

After deriving the one-dimensional wave equation, its solution will be obtained for known boundary conditions for plane and spherical waves. These are the two most important cases that explain the airborne wave propagation mechanism and their important characteristics that are of practical value. The resulting changes in dynamic pressure and particle velocity and their phase relationship due to airborne sound are helpful in this characterization. This airborne sound propagation will eventually be linked to characteristics of its source in later chapters.

Quantification of airborne sound is an important step in identifying frequency and magnitude of the dynamic pressure. A microphone is the transducer that is used for measuring dynamic pressure. Details of a microphone and sound intensity probe, which is built using two closely spaced microphones, are discussed in this chapter along with quantifying parameters, such as sound pressure level and sound intensity level. The frequency analysis of airborne sound using octave and 1/3 octave band analysis is covered that will help identify the source of noisy machines. The spectrum of airborne sound can also be used to obtain a single overall sound pressure level based on our hearing sensitivity that will be useful in an objective comparison of noisy machines. Time-averaged sound pressure levels that are important for computing exposure of industrial workers to airborne sound and for evaluating environment noise pollution are also discussed in this chapter.

3.1 Piston Propagated Disturbance

A piston does not normally generate an acoustic disturbance nor does it need a constrained tube to travel. But in order to explain the physical changes that occur in the medium through which a disturbance travels, it is useful to imagine an arrangement in which a piston at one end of the tube, moving at a certain velocity, introduces a disturbance that travels down the tube. Such an arrangement is shown in Figure 3.1 in which there is a piston at one end of a long tube and the piston is allowed to move in the forward direction.

Figure 3.1a shows a stationary piston behind a fluid contained within the tube. In Figure 3.1b the piston moves with a velocity u. Although physically it would take a certain amount of time for a stationary piston to reach a velocity u, we can assume that it has happened in a negligibly short period of time. As the piston moves with a velocity u, the fluid in close vicinity of the piston gets locally compressed and attains the velocity u. As the piston continues to move with the velocity u, an increased thickness of the fluid layer gets locally compressed, which attains the velocity u as shown in Figure 3.1c. Now if the piston stops, the locally compressed layer of thickness, L, moves down the tube with a velocity c_o, which can be assumed as the speed of sound, although we are yet to prove it. The moving disturbance increases the local velocity of particles of the medium to u as it propagates down the tube and u is known as particle velocity. This is shown in Figure 3.1d.

The disturbance propagation of a compressed fluid in the positive direction is shown in Figure 3.1. Although it helps us to understand the physics of propagation, it does not give a complete picture of the mechanism of airborne sound transmission, because airborne sound contains both compressed and rarefied sections of the fluid. However, the same principle explained earlier can be used to demonstrate how a rarefied portion of the fluid moves in the positive direction; if the piston is moved in the reverse direction, a rarefied block can similarly move in the positive direction. Finally, it would be necessary to generate many such disturbances that generate layers of the medium that are both compressed and rarefied, attaining local velocities in both directions, which can be accomplished by a pulsating piston. Only such a disturbance can create the sensation of hearing. Hence, we shall modify our experiment shown in Figure 3.1 to generate a disturbance, which is closer to an airborne sound.

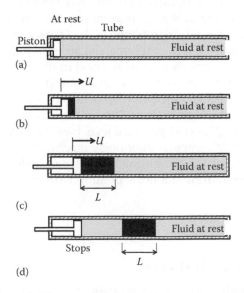

FIGURE 3.1
Disturbance generated by a piston propagating along a tube.

3.2 Pulsating Piston

Figure 3.2 shows an experiment to simulate the generation of airborne sound. In this experiment, it is assumed that the piston oscillates with a time period, T, and it is assumed that at $t = 0$ the piston has attained a velocity u_{max} and that this velocity decreases to zero at $t = T/4$, attains a negative value $-u_{min}$ at $t = T/2$, reverses the direction to reach zero velocity at $t = 3T/4$, and once again reaches its maximum value at $t = T$. As the piston changes its velocity, the disturbance propagates down the tube and it creates regions of compression and rarefaction. At $t = 0$, a layer of locally compressed fluid moves down the tube and when the piston reduces its velocity and stops at $t = T/4$, the compressed layer continues to move down with the speed of sound. When the piston moves in the negative direction, it tries to pull the layer of fluid toward the left and creates a locally rarefied thickness of fluid. When the piston stops at $t = 3T/4$, the locally rarefied and compressed sections continue to move down the tube. When the piston reverses direction toward increasing values of velocity, it results in a locally compressed layer. The variation of piston velocity and the pressure variation at various locations of the tube is as shown in Figure 3.2. If the piston continues to operate, it can continuously create similar disturbance patterns throughout the tube.

Since the aforementioned disturbance propagates at the speed of sound, c_o, in one cycle of piston movement, it creates regions of compression and rarefaction along a length of the tube, $c_o T$, as shown in Figure 3.2. If the piston was to operate further, similar patterns will be repeated along the tube. If the tube end has a facility to absorb all the acoustic disturbances, which is known as anechoic termination, the spatial distribution of the disturbance will form continuously as long as the piston is pulsating.

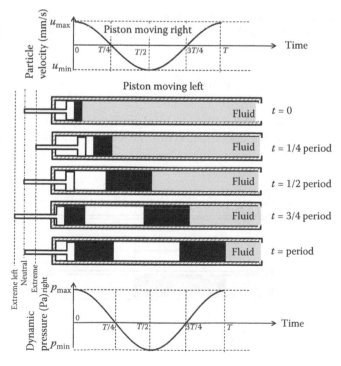

FIGURE 3.2
Pulsating piston of period T creating regions of compression and rarefaction.

3.3 Pressure Density Relationships due to Acoustic Disturbance

After visualizing the spatial distribution of regions of compression and rarefaction in a pulsating tube, we shall now look at the relationship between pressure and density variations due to the passage of a disturbance. The acoustic disturbance is assumed to pass for a very short period of time and will not allow any heat transfer. Therefore, the passage of an acoustic wave can be assumed as an isentropic process, which follows the following law that relates pressure and density:

$$\frac{p}{\rho^{\gamma}} = \text{constant} \tag{3.1}$$

where γ is the ratio of specific heats.

Consider the piston and tube arrangement shown in Figure 3.3 wherein a piston of constant velocity, u, moves for t seconds. By the time the piston moves a distance of ut, during the same period of time the disturbance would have moved by $c_o t$. The disturbance increases the local particle velocity to u and also increases the local density and pressure.

Let the total pressure that is the sum of ambient and dynamic pressure at any location and instant be given by

$$p(x,t) = p_o + p_d(x,t) \tag{3.2}$$

The dynamic pressure is assumed to vary both as a function of time and along various locations of the tube length. And the total density at any location along the tube length and instant of time is given by

$$\rho(x,t) = \rho_o + \rho_d(x,t) \tag{3.3}$$

It is clear from Equations 3.2 and 3.3 that the total pressure and density and the dynamic variation are functions of space and time. With that understanding we can drop the use

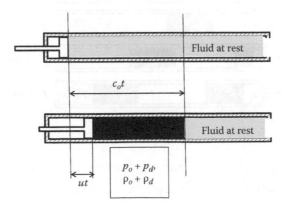

FIGURE 3.3
Particle velocity and speed of sound.

of (x,t) in further derivations for ease of writing. In addition, it should be clear that the dynamic variation of pressure and density is extremely small in comparison to their nominal ambient values; this assumption will allow for neglecting the products of such terms in later derivations that will follow in this section.

The total pressure and density of a fluid through which a disturbance passes can be expressed as the following function:

$$p = f(\rho) \tag{3.4}$$

Equation 3.4 can be represented in Figure 3.4. The relationship between the total pressure and density, due to the passage of a disturbance, is a nonlinear relationship as shown. By assuming that the second-order terms are negligible, the pressure density relationship can be expanded using Taylor's series around the ambient pressure and density as

$$p = f(\rho) = f(\rho_o) + \rho_d \frac{df(\rho_o)}{d\rho} \tag{3.5}$$

The term $\dfrac{df(\rho_o)}{d\rho}$ of Equation 3.5 is the slope of the pressure–density relationship evaluated around ambient values of pressure and density.

Comparing Equations 3.2 and 3.5, the dynamic component of pressure change during an acoustic disturbance is given by

$$p_d = \rho_d \frac{df(\rho_o)}{d\rho} \tag{3.6}$$

Therefore, the dynamic pressure and density are related by the slope of the pressure–density variation curve.

From Equations 3.1 and 3.4, the slope of the pressure–density variation around the nominal pressure and density is given by

$$\frac{df(\rho_o)}{d\rho} = \frac{\gamma p_o}{\rho_o} \tag{3.7}$$

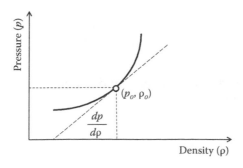

FIGURE 3.4
Relationship between pressure and density.

From Figure 3.3, the mass of the uncompressed fluid is given by

$$dm = \rho_o c_o t S \tag{3.8}$$

Since the entire compressed fluid is moving with a velocity u, the rate of change of momentum is equal to the net force applied by the piston of area S that increases the dynamic pressure. It is given by

$$p_d S = \frac{d}{dt}(\rho_o c_o t S u) \tag{3.9}$$

From Equation 3.9, the dynamic pressure is given by

$$p_d = \rho_o c_o u \tag{3.10}$$

Applying conservation of mass to the fluid before and after compression, the following equation results:

$$\rho_o c_o t S = (\rho_o + \rho_d)(c_o t - ut)S \tag{3.11}$$

Since dynamic variation of density is extremely small, by neglecting insignificant terms, Equation 3.11 for particle velocity becomes

$$u = \frac{\rho_d c_o}{\rho_o} \tag{3.12}$$

From Equations 3.10 and 3.12, the dynamic pressure is given by

$$p_d = c_o^2 \rho_d \tag{3.13}$$

And from Equations 3.6 and 3.13, the speed of sound is given by

$$c_o^2 = \frac{df(\rho_o)}{d\rho} \tag{3.14}$$

From Equations 3.7 and 3.14, the velocity sound, ambient pressure, density, and ratio of specific heats are related by

$$c_o^2 = \frac{\gamma p_o}{\rho_o} \tag{3.15}$$

In terms of the absolute ambient temperature, T_o, and gas constant, R, of the fluid medium, Equation 3.15 for the speed of sound can be expressed as

$$c_o = \sqrt{\gamma R T_o}$$

(3.16)

Using Equation 3.16, the speed of sound in a gaseous medium of known properties of ratio of specific heats and gas constant can be determined at any temperature.

3.4 One-Dimensional Wave Equation

An elemental layer of fluid, initially having a thickness of dx and surface area S is shown in Figure 3.5. All distances are measured from a fixed frame of reference. As an acoustic disturbance passes, after a small increment of time, t, the left face moves from x to a new position $x + \xi$, where ξ is the instantaneous particle displacement. The right face moves from $x + dx$ to a new position $x + dx + \xi + \dfrac{\partial \xi}{\partial x} dx$. The fluid moves from one position to another as a result of forces. It is assumed that frictional forces are negligible, so only forces acting on the fluid element are pressure forces.

3.4.1 Conservation of Mass

Due to the passage of an acoustic wave through the aforementioned control volume, there will be a net mass flow into it due to the change in particle velocity at the left and right cross sections. The inflow mass rate is given by

$$\dot{m}_{in} = \rho u S - \left[\rho u + \frac{\partial (\rho u)}{\partial x} dx \right] S = -\frac{\partial (\rho u)}{\partial x} dx S$$

(3.17)

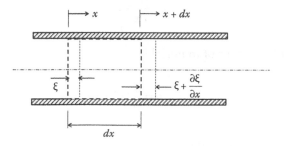

FIGURE 3.5
A control volume in a tube through which an acoustic wave passes.

Density is a sum of the dynamic density and nominal density, and in addition, the dynamic density, ρ_d, and particle velocity, u, are extremely small. Due to these reasons, Equation 3.17 becomes

$$\dot{m}_{in} = -\rho_o \frac{\partial u}{\partial x} dxS \tag{3.18}$$

The dynamic density within the control volume changes due to the passage of acoustic wave, which results in an increase of mass. Its rate of change is given by

$$\dot{m}\Delta_V = \frac{\partial \rho}{\partial t} dxS = \frac{\partial \rho_d}{\partial t} dxS \tag{3.19}$$

Since the total mass in flow rate must be equal to the rate of increase of mass within the control volume, Equations 3.18 and 3.19 are equal. Equating them

$$\frac{\partial \rho_d}{\partial t} + \rho_o \frac{\partial u}{\partial x} = 0 \tag{3.20}$$

Equation 3.20 is known as the linearized equation of mass conservation.

3.4.2 Conservation of Momentum

The conservation of momentum principle can be applied to the control volume of Figure 3.5 as follows. The net force applied in the positive x direction to the change in pressure is given by

$$F_{net} = p(x,t)S - \left(p(x,t) + \frac{\partial p(x,t)}{\partial x} dx \right)S = \frac{-\partial p(x,t)}{\partial x} dxS \tag{3.21}$$

From Equation 3.2, noting that the ambient pressure remains constant, Equation 3.21 results in

$$F_{net} = \frac{-\partial p_d}{\partial x} dxS \tag{3.22}$$

The particle velocity of the layer of material is

$$u = \frac{\partial \xi}{\partial t} \tag{3.23}$$

Acceleration of the layer of material is

$$a = \frac{\partial u}{\partial t} = \frac{\partial^2 \xi}{\partial t^2} \tag{3.24}$$

The mass of the layer is given by

$$dm = \rho_o\, S dx \tag{3.25}$$

Since the net force must be equal to the differential mass times acceleration, from Equations 3.22, 3.24, and 3.25 the following equation results:

$$\frac{\partial p_d}{\partial x} + \rho_o \frac{\partial u}{\partial t} = 0 \tag{3.26}$$

3.4.3 Wave Equation

Partial differentiation of Equation 3.20 with respect to time results in

$$\frac{\partial^2 p_d}{\partial t^2} + \rho_o \frac{\partial^2 u}{\partial x \partial t} = 0 \tag{3.27}$$

And the partial differentiation of Equation 3.26 with respect to x gives

$$\frac{\partial^2 p_d}{\partial x^2} + \rho_o \frac{\partial^2 u}{\partial x \partial t} = 0 \tag{3.28}$$

Subtracting Equations 3.27 and 3.28

$$\frac{\partial^2 p_d}{\partial t^2} - \frac{\partial^2 p_d}{\partial x^2} = 0 \tag{3.29}$$

From Equation 3.13 that relates dynamic density, dynamic pressure, and velocity of sound, Equation 3.29 becomes

$$\frac{\partial^2 p_d}{\partial x^2} = \frac{1}{c_o^2} \frac{\partial^2 p_d}{\partial t^2} \tag{3.30}$$

Since dynamic pressure and particle displacement are related to each other, the following equation can be obtained as

$$\frac{\partial^2 \xi}{\partial x^2} = \frac{1}{c_o^2} \frac{\partial^2 \xi}{\partial t^2} \tag{3.31}$$

3.4.4 Integral Equations

From some of the equations derived earlier, the following equations can be obtained that relate dynamic pressure, particle displacement, and particle velocity, which are as useful in many applications.

From Equations 3.23, 3.26, and 3.31

$$\frac{\partial p_d}{\partial x} = -\rho_o c_o^2 \frac{\partial^2 \xi}{\partial x^2} \tag{3.32}$$

Integrating Equation 3.32 with respect to x

$$p_d = -\rho_o c_o^2 \frac{\partial \xi}{\partial x} \tag{3.33}$$

From Equations 3.23 and 3.26

$$\frac{\partial p_d}{\partial x} + \rho_o \frac{\partial^2 \xi}{\partial t^2} = 0 \tag{3.34}$$

Integrating Equation 3.34 with respect to time

$$\frac{\partial \xi}{\partial t} = u(x,t) = \frac{-1}{\rho_o} \int \frac{\partial p_d}{\partial x} dt \tag{3.35}$$

Integrating Equation 3.33 with respect to x

$$\xi = \frac{-1}{\rho_o c_o^2} \int p_d(x,t) dx \tag{3.36}$$

It is important to note that the integration constants are zero in all the preceding equations since the variables are dynamic quantities.

3.5 Wave Equation Solution for Plane Waves

Consider the case of a single frequency sound wave being generated on one side of a vibrating rigid wall, as shown in Figure 3.6. This is one of the simplest sources that can demonstrate the production of plane waves and all the parameters are expressed in terms of complex quantities. Since we assume that the resultant wave propagation is governed by the wave equation (Equation 3.31) for particle displacement, which is a linear differential equation, the principle of separation of variables can be used. Therefore, particle displacement in the complex form can be written as a product of two space and time functions, given by

$$\tilde{\xi}(x,t) = \overline{\psi}(x) e^{j\omega t} \tag{3.37}$$

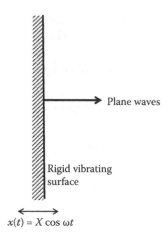

FIGURE 3.6
Plane wave generation by a rigid vibrating wall.

where $\bar{\psi}(x)$ is the spatial variation of the particle displacement due to wave propagation at the speed of sound. From Equation 3.37, the second time-derivative of particle displacement is given by

$$\frac{\partial^2 \tilde{\xi}}{\partial t^2} = -\omega^2 \bar{\psi}(x) e^{j\omega t} \tag{3.38}$$

From Equation 3.37, the second spatial-derivative of the particle displacement is given by

$$\frac{\partial^2 \tilde{\xi}}{\partial x^2} = \bar{\psi}''(x) e^{j\omega t} \tag{3.39}$$

From Equations 3.37, 3.38, and 3.39 in Equation 3.31

$$\bar{\psi}''(x) e^{j\omega t} = \frac{1}{c_o^2} \cdot (-\omega^2) \bar{\psi}(x) e^{j\omega t} \tag{3.40}$$

which gives the one-dimensional Helmholtz's equation

$$\bar{\psi}''(x) + k_o^2 \bar{\psi}(x) = 0 \tag{3.41}$$

where k_o is the wave number of the airborne sound at the frequency ω, given by

$$k_o = \frac{\omega}{c_o} \tag{3.42}$$

As discussed earlier in Chapter 2, the wave number is also known as spatial frequency and is related to the wavelength as $k_o = 2\pi/\lambda_o$. Therefore, wave number is an important parameter to express wave motion.

Equation 3.41 in terms of wave number is given by

$$\bar{\psi}''(x) + k_o^2\,\bar{\psi}(x) = 0 \tag{3.43}$$

The general solution of Equation 3.43 is given by

$$\bar{\psi}(x) = \bar{A}e^{-jk_ox} + \bar{B}\,e^{jk_ox} \tag{3.44}$$

From Equations 3.37 and 3.44

$$\tilde{\xi}(x,t) = \bar{A}\,e^{j(\omega t - k_ox)} + \bar{B}e^{j(\omega t + k_ox)} \tag{3.45}$$

We have earlier seen an equation similar to Equation 3.45 for longitudinal wave propagation in bars. The first term of Equation 3.45 represents a right-moving wave and the second term represents a left-moving wave at a frequency ω. Since there is no reflection in this case, there is no left-moving wave and hence $\bar{B} = 0$.

At the surface of the rigid vibrating surface, the particle displacement must be the same as the vibrating surface as shown in Figure 3.6.

$$\tilde{\xi}(o,t) = \bar{X}e^{j\omega t} = \bar{A}e^{j(\omega t - k_ox)}\Big|_{x=0} \tag{3.46}$$

$$\tilde{\xi}(x,t) = \bar{X}e^{j(\omega t - k_ox)} \tag{3.47}$$

The particle displacement is either the real or imaginary part of Equation 3.47 and by considering the real part

$$\xi = X\cos(\omega t - k_ox - \phi) \tag{3.48}$$

The complex particle velocity is given by

$$\tilde{u}(x,t) = \frac{\partial\tilde{\xi}}{dt} = j\omega\bar{X}e^{j(\omega t - k_ox)} \tag{3.49}$$

Since $j = 1\angle\dfrac{\pi}{2}$, Equation 3.49 becomes

$$\tilde{u}(x,t) = \omega\bar{X}e^{j(\omega t - k_ox + \pi/2)} \tag{3.50}$$

The particle velocity is either the real or imaginary part of Equation 3.50 and the real part is given by

$$u(x,t) = \omega X \cos(\omega t - k_o x + \pi/2 - \phi) \tag{3.51}$$

From Equation 3.33, the complex dynamic pressure is given by

$$\tilde{p}_d(x,t) = -\rho_o c_o^2 \frac{\partial \tilde{\xi}}{\partial x} = -\rho_o c_o^2 \overline{X}(-k_o j)e^{j(\omega t - k_o x)} \tag{3.52}$$

Equation 3.52 simplifies to

$$\tilde{p}_d(x,t) = j\rho_o c_o^2 \overline{X} k_o e^{j(\omega t - k_o x)} \tag{3.53}$$

Since $k_o = \dfrac{\omega}{c_o}$, Equation 3.53 becomes

$$\tilde{p}_d(x,t) = \rho_o c_o \omega \overline{X} e^{j(\omega t - k_o x + \pi/2)} \tag{3.54}$$

From Equations 3.50 and 3.54

$$\frac{\tilde{u}(x,t)}{\tilde{p}(x,t)} = \frac{1}{\rho_o c_o} \tag{3.55}$$

Equation 3.55 can also be written as

$$\frac{\tilde{p}(x,t)}{\tilde{u}(x,t)} = \rho_o c_o = Z_o \tag{3.56}$$

$\rho_o c_o$ is known as the characteristic impedance. Since $\angle \tilde{p}/\tilde{u} = 0$, pressure and velocity are always in phase for a plane wave.

Example 3.1

The particle displacement of a right traveling wave is described by the following expression: $\tilde{\xi}(x,t) = 10e^{j(200\pi t - 1.8265x)}$ μm. Evaluate the particle velocity, the wavelength, and the period of the wave (note that x is in meters).

By comparing the given equation for particle displacement with Equation 3.45, the frequency of the acoustic wave can be deduced as 100 Hz.

The equation for particle displacement is given by

$$\tilde{\xi}(x,t) = 10e^{j(200\pi t - 1.8265x)} \, \mu m \tag{1}$$

Differentiating Equation 1, we get the equation for particle velocity as

$$\tilde{u} = 10 \times 200\pi e^{j(200\pi t - 1.8265x)} \, \mu m/s \tag{2}$$

$$= 6.3 e^{j(200\pi t - 1.8265x)} \, mm/s$$

From the given particle displacement equation, the wave number $k_o = 1.8265$ rad/m. Since there is no data regarding temperature and density of the medium and that we know the wave number and frequency, the speed of sound can be computed from Equation 3.42 as

$$c_o = \frac{\omega}{k_o} = \frac{200\pi}{1.8265} = 344 \, m/s \tag{3}$$

The wavelength can be either determined using the speed of sound and frequency or the wave number alone. Using the wave number, from Equation 2.55 the wavelength becomes

$$\lambda_o = \frac{2\pi}{k_o} = \frac{2\pi}{1.8265} = 3.44 \, m \tag{4}$$

The time period of oscillation of the given acoustic wave is given by

$$\tau = \frac{1}{f} = \frac{1}{100} = 0.01 \, s \tag{5}$$

Example 3.2

A plane sound wave in air of 100 Hz has a peak acoustic pressure amplitude of 2 Pa. Determine the

  a. Root mean square (rms) pressure
  b. Particle velocity amplitude
  c. Particle displacement amplitude
  d. Wavelength

The following is given:

  Room temperature, $t = 22°C$
  Frequency, $f = 100$ Hz
  Peak value of dynamic pressure, $p_d = 2$ Pa
  Gas constant of air, $R = 286.7$ J/kg-K
  Ambient pressure, $P = 100$ kPa
  Absolute room temperature, $T = 273 + 22 = 295$ K
  Ratio of specific heats for air, $\gamma = 1.4$

From Equation 3.16, the velocity of sound in air is given by

$$c_o = \sqrt{\gamma R T} = \sqrt{1.4 \times 286.7 \times 295} = 344.10 \, m/s \tag{1}$$

Using the perfect gas equation, the nominal density of air is given by

$$\rho_o = \frac{P}{RT} = \frac{100 \times 10^3}{286.7 \times 295} = 1.18 \, \text{kg/m}^3 \tag{2}$$

ANSWERS

a. The rms value of dynamic pressure is given by

$$p_{rms} = \frac{p_d}{\sqrt{2}} = \frac{2}{\sqrt{2}} = \sqrt{2} \, \text{Pa} \tag{3}$$

b. Since the ratio of dynamic pressure and particle velocity are equal to the characteristic impedance for planes, from Equation 3.56 the particle velocity peak amplitude is given by

$$u = \frac{p_d}{\rho_o c_o} = \frac{2}{1.18 \times 344.10} = 4.93 \, \text{mm/s} \tag{4}$$

c. The particle displacement amplitude is given by

$$x = \frac{u}{2\pi f} = \frac{4.93 \times 10^{-3}}{2 \times \pi \times 100} = 7.83 \, \mu\text{m} \tag{5}$$

d. The wavelength is given by

$$\lambda_o = \frac{c_o}{f} = \frac{344.10}{100} = 3.44 \, \text{m} \tag{6}$$

From the previous sections we have understood that airborne sound propagation is produced due to the dynamic variation of local pressure and particle velocity. Among these two parameters, the dynamic pressure can be easily measured compared to particle velocity. Therefore, quantification of sound is mainly carried out by measuring the dynamic pressure. This chapter deals with various aspects of expressing this dynamic pressure. However, in recent times particle velocity can also be measured.

3.6 Sound Pressure Level

The dynamic pressure varies from 20 μPa (rms) at the threshold of hearing to 200 Pa (rms) at the threshold of pain. Hence, on a linear scale these extreme values differ by a ratio of 10 million. Since most of the airborne sound we ordinarily deal with is within this range, it is necessary to express the dynamic pressure on a different scale that can express the aforementioned extreme values on the same order of magnitude. It is obtained by expressing it on the logarithmic scale. Due to similar difficulties arising in expressing many other dynamic quantities like voltage and current, a standard form of logarithmic scale is defined in terms of decibels and one bel is defined as the logarithmic ratio of the mean square value of a dynamic quantity to a reference mean square value and one decibel is 10 times a bel. Thus the dynamic pressure is expressed in terms of sound pressure level in decibels (dB), defined by

$$L_p = 10 \log \frac{p_{rms}^2}{p_{ref}^2}, dB \tag{3.57}$$

where p_{rms}^2 is the mean square dynamic sound pressure, defined as follows:

$$p_{rms}^2 = \frac{1}{T} \int_0^T p^2(t)\, dt \tag{3.58}$$

and p_{ref} is the rms reference sound pressure and is taken as 20 μPa (20×10^{-6} N/m²), which corresponds to the rms sound pressure of a pure tone of 1000 Hz at the normal threshold of hearing. All the sound measuring equipment give readings as defined by Equation 3.57. These sound pressure levels can be based on their measurement over a wider frequency range, or in octave bands, or 1/3 octave bands, or narrow bands, depending upon the type of instrument and application.

Figure 3.7 shows the dynamic pressure variation of a pure tone that is imposed on the atmospheric pressure. Note that dynamic pressure is only a small fraction of the atmospheric pressure. Typical sound pressure levels are presented in Table 3.1.

Figure 3.8 shows how a single frequency sound wave, whose dynamic pressure variation was shown in Figure 3.7, can be produced using a tuning fork. A tuning fork vibrates at a single frequency, which is the natural frequency of its arms, and produces regions of compression and rarefaction in its path. These regions are representative of a dynamic pressure that changes above and below the atmospheric pressure at a frequency. In addition to the time variation of dynamic pressure represented in Figure 3.7, a single frequency sound wave also shows variation as a function of distance. The distance between regions of maximum compression or rarefaction remains fixed for a single frequency sound wave and is

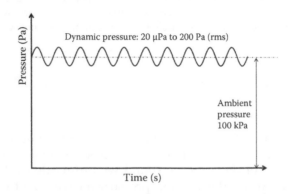

FIGURE 3.7
Dynamic pressure at a frequency superimposed on atmospheric pressure.

TABLE 3.1

Typical Sound Pressure Levels

Examples	Sound Pressure Level (dB)	Sound Pressure (Pa)
30 m from jet aircraft	140	200
Threshold of pain	130	
	120	20
Noisy machine	110	
Ballroom	100	2
	90	
Traffic noise	80	0.2
	70	
Speech	60	0.02
	50	
	40	0.002
Residence (night)	30	
Background recording studio	20	0.0002
	10	
Threshold of hearing	0	0.00002

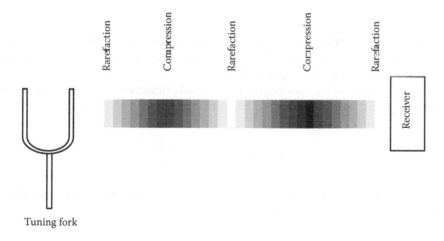

FIGURE 3.8
Single frequency sound generation by a tuning fork.

known as the wavelength. The wavelength and frequency of a single frequency sound wave are related to the speed of sound, c_o. The dynamic variation of pressure is sensed by our hearing mechanism that gives the sensation of hearing. In a general situation, sound waves of various frequencies from various sources are produced. But we generally have to pay attention to the most dominant ones. However, the sound pressure level at a location can still be computed by using Equation 3.57 by using the rms value of dynamic pressure.

3.6.1 Addition of Sound Pressure Levels

When there are two machines that are radiating airborne sound, it is of interest to determine the total sound pressure level, based on individual sound pressure levels of each machine. When there is more than one source and both sources are incoherent, the mean square sound pressures get added; incoherence means that the sources produce sound independent of each other. In order to determine the total sound pressure level, one has to add the mean square pressures of all the sound sources and use Equation 3.57.

For example, consider two incoherent sound sources with overall sound pressure levels of L_{p1} and L_{p2}, respectively. Applying Equation 3.57 to each of these sound sources, we get

$$L_{p1} = 10 \log \frac{p_1^2}{p_{ref}^2} \tag{3.59}$$

and

$$L_{p2} = 10 \log \frac{p_2^2}{p_{ref}^2} \tag{3.60}$$

The total sound pressure level due to both sources is given by

$$L_{pt} = 10 \log \frac{p_1^2 + p_2^2}{p_{ref}^2} \tag{3.61}$$

Equation 3.61 can be further extended to compute the combined level of any number of sources. However, a much simpler method of adding two sources at a time can be obtained, which is more useful to a practicing engineer.

Let ΔL_{12} be the difference in sound pressure levels between L_{p1} and L_{p2}. Assuming the first level to be greater than the second, this difference can be expressed as

$$\Delta L_{12} = L_{p1} - L_{p2} = 10 \log \frac{p_1^2}{p_2^2} \tag{3.62}$$

Let ΔL_x be the number of decibels that should be added to the larger sound pressure level to obtain L_{pt}, which can be expressed as

$$L_{pt} = L_{p1} + \Delta L_x \tag{3.63}$$

Equation 3.63 can be expressed as

$$\Delta L_x = L_{pt} - L_{p1} = 10 \log \frac{p_1^2 + p_2^2}{p_1^2} \tag{3.64}$$

From Equations 3.62 and 3.64, ΔL_x is given by

$$\Delta L_x = 10 \log \left(1 + 10^{-\frac{\Delta L_{12}}{10}} \right) \tag{3.65}$$

TABLE 3.2

Sum of Two Sound Sources

Difference between Two Decibel Levels to be Added ΔL_{12} (dB)	Amount to be Added to Larger Level to Obtain Decibel Sum ΔL_x(dB) (Equation 3.65)
0	3.0
1	2.6
2	2.1
3	1.8
4	1.4
5	1.2
6	1.0
7	0.8
8	0.6
9	0.5
10	0.4

For sound pressure level differences of 0 to 10 dB, Equation 3.65 is tabulated in Table 3.2. Therefore, if there are two sound sources that have the same sound pressure level, the combined effect will be an increase of 3 dB. For example, two sources of 60 dB each will result in a combined level of 63 dB and so on. On the other extreme, if the difference in sound pressure levels is larger than 10 dB, then the smaller source will not make much difference to the combined levels. Table 3.2 can be used to add two sources at a time, which can be extended to add any number of sources. Equation 3.65 can also be plotted on a graph, as shown in Figure 3.9, which can be used to add two sources at a time.

Example 3.3

The local regulations at a certain place prescribe that the sound pressure levels (from all sources) should not exceed 45 dB during night and 50 dB during the day. The background levels, when measured without any machine operating, are 40 dB during night and 45 dB during day. If a certain machine has a sound pressure level of 40 dB, how many such similar machines can operate during the day and how many during the night, without violating the local regulations?

(The overall sound pressure levels are subjected to A-weighting correction discussed in Section 3.10.2 to correct for the hearing sensitivities of humans. But the basic procedure of adding the contribution from various sources will, however, remain the same.)

$$L_{p,allowable,day} = 50 \text{ dB}; L_{p,allowable,night} = 45 \text{ dB}; L_{p,machine} = 40 \text{ dB}$$

$$L_{p,background,day} = 45 \text{ dB}; L_{p,background,night} = 40 \text{ dB}$$

Let n_d be the number of machines that can be allowed during day and n_n during night.

The total sound pressure level will be the logarithmic sum of the sound pressure levels of the background and that of the machine, which should not be greater than the allowable levels.

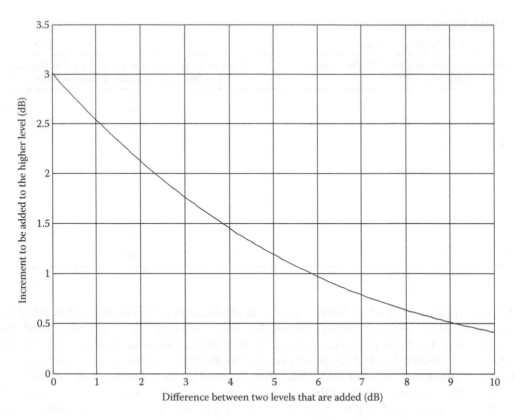

FIGURE 3.9
Adding two sound pressure levels.

The sum of the mean square pressures of the machine and background sound during the day is given by

$$10^{\frac{L_{p,background,day}}{10}} + n_d 10^{\frac{L_{p,machine}}{10}} = 10^{\frac{L_{p,allowable,day}}{10}} \tag{1}$$

From Equation 1, the number of machines that can be operated during day is given by

$$n_d = \frac{10^{\frac{50}{10}} - 10^{\frac{45}{10}}}{10^{\frac{40}{10}}} = 6 \text{ machines} \tag{2}$$

The sum of the mean square pressures of the machine and background sound during night is given by

$$10^{\frac{L_{p,background,night}}{10}} + n_n 10^{\frac{L_{p,machine}}{10}} = 10^{\frac{L_{p,allowable,night}}{10}} \tag{3}$$

From Equation 2, the number of machines that can be operated during night is given by

$$n_n = \frac{10^{\frac{45}{10}} - 10^{\frac{40}{10}}}{10^{\frac{40}{10}}} = 2 \text{ machines} \qquad (4)$$

Therefore, machines that work 24/7 that produce constant sound pressure level should be carefully chosen to meet the local regulations during both day and night. In addition, the background levels should also be accounted for.

3.6.2 Difference between Two Sound Sources

In many situations we want to identify the contribution from two different sound sources; it may so happen that we may not be able to switch off one of the machines. In such a case we can only operate both machines or can switch off the other machine. Based on the sound pressure level measurement of these situations—both working and one of them switched off—we can estimate the sound pressure level of the machine that can be switched off.

Let L_{p1} be the sound pressure level of the machine that cannot be switched off and L_{p2} be the sound pressure level of the machine that can be switched off; and let L_{pt} be the sound pressure level when both machines are working. So we can only measure L_{p1} or L_{pt}.

Applying Equation 3.57 when only first machine is operating, we get

$$L_{p1} = 10 \log \frac{p_1^2}{p_{ref}^2} \qquad (3.66)$$

And when both machines are operating, we get

$$L_{pt} = 10 \log \frac{p_t^2}{p_{ref}^2} \qquad (3.67)$$

Based on these measurements, L_{p2} can be obtained by subtracting the square of the measured dynamic pressures. It can be expressed as

$$L_{p2} = 10 \log \left(\frac{p_t^2 - p_1^2}{p_{ref}^2} \right) \qquad (3.68)$$

Although Equation 3.68 can be used to logarithmically subtract sound levels, similar to the addition of two sound sources, there is a simpler way of subtracting the effect of two sound courses.

Let ΔL be the difference between the sound pressure levels measured when both machines are operating and when only the first machine is operating. It can be expressed as

$$\Delta L = L_{pt} - L_{p1} = 10 \log \frac{p_1^2 + p_2^2}{p_1^2} \tag{3.69}$$

Let ΔL_x be the number of decibels that has to be subtracted from L_{pt} to obtain L_{p2}, which can be expressed as

$$L_{p2} = L_{pt} - \Delta L_x \tag{3.70}$$

Equation 3.70 can be expressed as

$$\Delta L_x = L_{pt} - L_{p2} = 10 \log \frac{p_1^2 + p_2^2}{p_2^2} \tag{3.71}$$

From Equation 3.69, the following ratio of the square of dynamic pressures can be obtained

$$\frac{p_2^2}{p_1^2} = 10^{\frac{\Delta L}{10}} - 1 \tag{3.72}$$

Substituting Equation 3.72 in Equation 3.71

$$\Delta L_x = 10 \log \left(\frac{10^{\frac{\Delta L}{10}}}{10^{\frac{\Delta L}{10}} - 1} \right) \tag{3.73}$$

Equation 3.73 is not valid for $\Delta L = 0$. Physically it means that if there is no change in the overall levels due to switching on both sound sources, then the second sound source is insignificant and its sound level is more than 10 dB less than the first source. Equation 3.73 is tabulated for various values of ΔL in Table 3.3 and the same equation is shown in Figure 3.10.

3.6.3 Equivalent Sound Pressure Level, L_{eq}

When there are a number of sound sources that operate at different times and varying loads, the resulting sound pressure levels vary with time. This is particularly true in the case of urban noise pollution. Therefore, it is necessary to define a new term that would take the time average of sound pressure levels over a period of time. This is known as equivalent sound pressure level, which is defined as

$$L_{eq} = 10 \log \left[\frac{1}{t_2 - t_1} \int_{t_1}^{t_2} \frac{p_{rms}^2}{p_{ref}^2} dt \right] \tag{3.74}$$

TABLE 3.3

Difference of Two Sound Sources

$\Delta L = L_{total} - L_{p1}$	ΔL_x, Amount to be Subtracted from L_{total} to Get L_{p2} (Equation 3.73)
0	Greater than 10
1	6.8
2	4.5
3	3.0
4	2.2
5	1.6
6	1.2
7	1.0
8	0.7
9	0.6
10	0.5

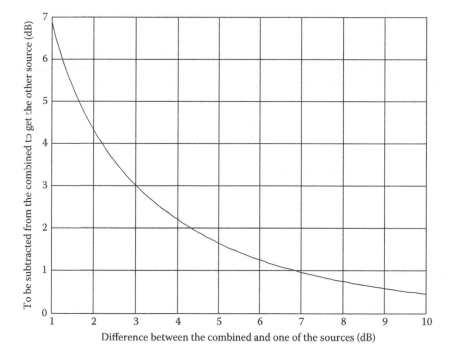

FIGURE 3.10
Difference between two signals.

Equivalent sound pressure level can be measured using sound level meters that have an integrator that can be used to compute the time-averaged sound pressure levels using Equation 3.74. Such meters are useful for continuous measurement of urban noise pollution. However, equivalent sound pressure levels can also be computed by knowing the time duration of sound pressure levels measured at discrete time instants. This approach of computing equivalent sound pressure levels is particularly useful in determining whether a machine of known sound pressure level can be used for a certain period of time without violating the local regulations.

TABLE 3.4

Sound Pressure Data of Example 3.4

No.	SPL, L_{pi} dB	Number of Times, n_i
1	67	6
2	62	17
3	57	51
4	52	52
5	47	147
6	42	96
7	37	11
		380

Example 3.4

The sound pressure level readings in Table 3.4 were taken for a certain period, T, during night around an industry; the readings were taken periodically and the number of times they occurred were also recorded. Determine the $L_{eq,\,T}$ value in decibels.

(The overall sound pressure levels are subjected to A-weighting correction discussed in Section 3.10.2 to correct for the hearing sensitivities of humans. But the basic procedure of computing equivalent sound pressure levels will, however, remain the same.)

The mean square pressure ratio corresponding to a sound pressure level is given by

$$\frac{p^2}{p_{ref}^2} = 10^{\frac{L_{pi}}{10}} \tag{1}$$

The fraction of the time of a sound pressure level L_{pi} is given by

$$x_i = \frac{n_i}{\sum\limits_{i=1}^{N} n_i} \tag{2}$$

where N is the number of sound pressure level readings under consideration; $N = 7$ in this case.

From Equations 1 and 2, fraction of the mean square pressure ratio is given by

$$\alpha_i = \frac{n_i 10^{\frac{L_{pi}}{10}}}{\sum\limits_{i=1}^{N} n_i} \tag{3}$$

The equivalent sound pressure level is given by

$$L_{p_{eq}} = 10 \log \sum\limits_{i=1}^{N} \alpha_i \tag{4}$$

Equations 1 through 3 are used to complete Table 3.5 and then L_{eq} is calculated using Equation 4.

TABLE 3.5

Equivalent Sound Pressure Calculation of Example 3.4

No.	SPL, L_{pi} dBA	Number of Times, n_i	$\dfrac{p^2}{p_{ref}^2} \times 10^6$ (Equation 1)	x_i (Equation 2)	$\alpha_i \times 10^4$ (Equation 3)
1	67	6	5.0119	0.0158	7.9135
2	62	17	1.5849	0.0447	7.0903
3	57	51	0.5012	0.1342	6.7265
4	52	52	0.1585	0.1368	2.1688
5	47	147	0.0501	0.3868	1.9388
6	42	96	0.0158	0.2526	0.4004
7	37	11	0.005	0.0289	0.0145
		380			

$$L_{eq,T} = 10\log \sum_{i=1}^{N} \alpha_i = 54\ \mathrm{dB}$$

3.7 Sound Power Level

The sound pressure level of a machine that is radiating sound is not only dependent on the machine but also on the surroundings in which the machine is placed. The same machine, when placed in different surroundings, might show different sound pressure levels. This makes it difficult for objective evaluation of a machine for its noise radiation in different environments in the customers' locations. On the contrary, sound power is actually the absolute sound radiating capability of a machine. Therefore, in order to get an objective comparison of the sound radiated by various machines, the sound power level is defined. There is no direct method of measuring sound power levels; they are determined based on the measurement of their sound pressure levels in specific environments. Sound from machines is generally specified in terms of the sound power levels in the international standards. The sound power level is defined by the following equation:

$$L_w = 10\log \frac{W}{W_{ref}}, dB \tag{3.75}$$

where W is the sound power emitted by the machine and W_{ref} is the reference power equal to 10^{-12} watt (1 pW). This value of W_{ref} is directly related to the reference value of threshold of dynamic pressure, p_{ref}. (Detailed proof of the pico watt value for W_{ref} will be presented in Chapter 8, Section 8.9.) Since the sound power level is also a logarithmic quantity, when there is more than one sound source, sound power of all the machines can be added and the combined sound power level can be obtained using Equation 3.61. Or by considering the sound power levels of two sources at a time, the combined sound power level of the sources can be graphically obtained by using Figure 3.9.

The sound power levels of a machine can be determined by measuring its sound pressure levels in the following specific environments: (a) anechoic rooms that are treated with acoustic absorption materials so there are no reflections from the walls and roof, and (b) reverberant rooms that produce a uniform sound field in most parts of the room.

We will discuss these aspects further in Chapter 8 on room acoustics. Sound power can also be measured using sound intensity measurement in any environment having steady background sound; this helps in situ measurement of sound power without elaborate, expensive facilities.

3.8 Sound Intensity Level

The sound intensity, at a given point in a sound field, in a specified direction is defined as the average sound power passing through a unit area perpendicular to the surface at that point. For a spherical sound wave propagating in a free field (free of reflections), as shown in Figure 3.11, the sound intensity along the direction of wave propagation is given by

$$I = \frac{W}{4\pi r^2} \ \text{W/m}^2 \tag{3.76}$$

The sound intensity level can be expressed in decibels as follows.

$$L_I = 10 \log \frac{I}{I_{ref}} \tag{3.77}$$

where I is the sound intensity level at a point in a given direction and $I_{ref} = 10^{-12}$ watts/m² (1 pW/m²), which is related to the reference power. A detailed derivation to prove this will be presented in Chapter 8, Section 8.6.

The time-averaged sound intensity is given by

$$\langle I(t) \rangle = \frac{1}{T} \int_0^T (\text{Re} \ \tilde{p}(t))[\text{Re} \ \tilde{u}(t)] \, dt \tag{3.78}$$

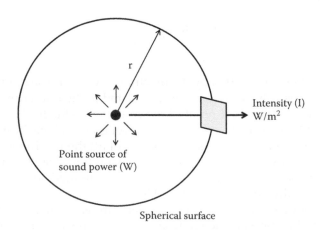

Spherical surface

FIGURE 3.11
Sound intensity and sound power.

For airborne sound waves, pressure is synonymous with force and sound intensity with power input. Therefore, replacing force by pressure, velocity by particle velocity, and power by sound intensity in Equation 1.96 (see Chapter 1), the time-averaged sound intensity is given by

$$\langle I(t) \rangle = \frac{1}{2} \operatorname{Re}\{ \tilde{p} \, \tilde{u}^* \} \tag{3.79}$$

3.8.1 Sound Intensity of Plane Waves

From Equation 3.54, for plane waves produced by a large rigid piston, harmonically vibrating with a peak amplitude X,

$$\tilde{p}(x,t) = \rho_o c_o \omega X e^{j(\omega t - k_o + \pi/2)} \tag{3.80}$$

The complex velocity is given by dividing Equation 3.80 by characteristic impedance

$$\tilde{u}(x,t) = \frac{\tilde{p}(x,t)}{\rho_o c_o} = \omega X e^{j(\omega t - k_o + \pi/2)} \tag{3.81}$$

The complex conjugate of velocity of Equation 3.81 is given by

$$\tilde{u}^*(x,t) = \omega X e^{-j(\omega t - k_o + \pi/2)} \tag{3.82}$$

From Equations 3.79, 3.80, and 3.82, the sound intensity of plane wave is given by

$$\langle I(t) \rangle = \frac{1}{2} \operatorname{Re}\{ \rho_o c \omega X e^{j(\omega t - k_o + \pi/2)} \omega X e^{-j(\omega t - k_o + \pi/2)} \}$$

$$= \frac{\rho_o c_o (\omega X)^2}{2} \tag{3.83}$$

From Equation 3.80, the rms pressure is given by

$$p_{rms}^2 = \frac{(\rho_o c_o \omega X)^2}{2} \tag{3.84}$$

From Equations 3.83 and 3.84, the time-averaged sound intensity is given by

$$\langle I(t) \rangle = \frac{p_{rms}^2}{\rho_o c_o} \tag{3.85}$$

3.9 Spherical Waves

3.9.1 Spherical Wave Equation

Figure 3.12 shows a sound wave propagating as a spherical wave front in which the intensity of sound reduces with distance. This is the most practical form of sound propagation but more difficult to analyze than plane waves. However, at large distances from the sound source, spherical waves can be approximated as plane waves.

The wave equation (Equation 3.30) can be expressed in the following spherical coordinates in which pressure varies with the radial coordinate is given by

$$\frac{1}{r^2}\frac{\partial}{\partial r}\left(r^2\frac{\partial p_d}{\partial r}\right) = \frac{1}{c_o^2}\frac{\partial^2 p_d}{\partial t^2} \tag{3.86}$$

Equation 3.86 can be written in an alternate form and in a complex form as

$$\frac{\partial^2}{\partial r^2}(r\tilde{p}_d) = \frac{1}{c_o^2}\frac{\partial^2(r\tilde{p}_d)}{\partial t^2} \tag{3.87}$$

3.9.2 Solution of Spherical Wave Equation

Let us assume the solution for complex dynamic pressure is

$$\tilde{p}_d(r,t) = \overline{\psi}(r)e^{j\omega t} \tag{3.88}$$

From Equations 3.87 and 3.88

$$\frac{d^2}{dr^2}(r\overline{\psi}(r)) = -\frac{\omega^2}{c_o^2}r\,\overline{\psi}(r) \tag{3.89}$$

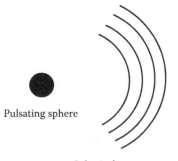

Pulsating sphere

Spherical waves

FIGURE 3.12
Spherical waves from a sound source.

Since $k_o = \dfrac{\omega}{c_o}$ is the wave number, Equation 3.89 becomes

$$\frac{d^2}{dr^2}(r\bar{\psi}(r)) + k_o^2(r\bar{\psi}(r)) = 0 \tag{3.90}$$

The solution of Equation 3.90 becomes

$$r\bar{\psi}(r) = \bar{A}e^{-jk_o r} + \bar{B}e^{jk_o r} \tag{3.91}$$

From Equations 3.59 and 3.62, the dynamic pressure is given by

$$\tilde{p}_d(r,t) = \frac{1}{r}\left(Ae^{-jk_o r} + Be^{jk_o r}\right)e^{j\omega t} \tag{3.92}$$

Equation 3.92 can be written as

$$\tilde{p}_d(r,t) = \frac{1}{r}\left(\bar{A}e^{+j(\omega t - k_o r)} + \bar{B}e^{j(\omega t + k_o r)}\right) \tag{3.93}$$

By extending the similar argument used for plane waves, the first term refers to a radially outward moving wave and the second term refers to a radially inward moving wave.

For the case of a spherical wave moving radially outward from the source without reflection, $\bar{B} = 0$, the complex dynamic pressure becomes

$$\tilde{p}_d(r,t) = \frac{\bar{A}}{r}e^{j(\omega t - k_o r)} \tag{3.94}$$

If the vibrating amplitude of a pulsating sphere is $\bar{X}$ at frequency ω, the complex dynamic pressure of Equation 3.65 is given by

$$\tilde{p}_d(r,t) = \frac{\overline{X}}{r}e^{j(\omega t - k_o r)} \tag{3.95}$$

The dynamic pressure is given by either the real or imaginary part of Equation 3.95 that are, respectively, given by

$$p_d(r,t) = \frac{X}{r}\cos(\omega t - k_o r - \phi) \quad \text{or} \quad \frac{X}{r}\sin(\omega t - k_o r - \phi) \tag{3.96}$$

where ϕ is the phase difference between dynamic pressure and vibration of the rigid piston.

The rms value of dynamic pressure is given by

$$p_{rms} = \frac{X}{\sqrt{2}r} \tag{3.97}$$

3.9.3 Particle Velocity of Spherical Waves

Based on Equation 3.35, the complex particle velocity for a spherical wave is given by

$$\tilde{u}(r,t) = -\frac{1}{\rho_o}\int \frac{\partial \tilde{p}_d(r,t)}{\partial r}dt \tag{3.98}$$

Differentiating Equation 3.96 with respect to r,

$$\frac{\partial \tilde{p}_d}{\partial r} = \frac{-\overline{X}}{r^2}e^{j(\omega t - k_o r)} + \frac{\overline{X}}{r}(-k_o j)e^{j(\omega t - k_o r)} = \frac{-\overline{X}}{r}e^{j(\omega t - k_o r)}\left\{\frac{1}{r} + k_o j\right\} \tag{3.99}$$

From Equations 3.98 and 3.99

$$\tilde{u}(r,t) = \frac{\overline{X}}{r\rho_o}\left\{\frac{1}{r} + k_o j\right\}\int e^{j(\omega t - k_o r)}dt$$

$$= \frac{\overline{X}}{j\omega r\rho_o}\left\{\frac{1}{r} + k_o j\right\}e^{j(\omega t - k_o r)} \tag{3.100}$$

From Equations 3.95 and 3.100

$$\frac{\tilde{p}_d(r,t)}{\tilde{u}(r,t)} = \frac{j\omega\rho_o r}{1 + jk_o r} \tag{3.101}$$

By multiplying and dividing the right-hand side of Equation 3.101 by $(1 - jk_o r)$

$$\frac{\tilde{p}_d(r,t)}{\tilde{u}(r,t)} = \frac{\omega\rho_o r(j + k_o r)}{1 + (k_o r)^2} \tag{3.102}$$

Since $\omega = k_o c_o$, Equation 3.102 becomes

$$\frac{\tilde{p}_d(r,t)}{\tilde{u}(r,t)} = \frac{\rho_o c_o\, k_o r(j + k_o r)}{1 + (k_o r)^2} \tag{3.103}$$

The magnitude of the ratio of complex dynamic pressure to particle velocity is given by $(Z_o = \rho_o c_o)$

$$\left|\frac{\tilde{p}(r,t)}{\tilde{u}(r,t)}\right| = |\overline{Z}_S| = \frac{Z_o \times k_o r}{1 + (k_o r)^2}\cdot\sqrt{1 + (k_o r)^2} = \frac{Z_o k_o r}{\sqrt{1 + (k_o r)^2}} \tag{3.104}$$

From Equation 3.104, the normalized impedance is given by

$$\left|\frac{\overline{Z}_S}{Z_o}\right| = \frac{k_o r}{\sqrt{1 + (k_o r)^2}} \tag{3.105}$$

Equation 3.105 is shown in Figure 3.13, and as shown, the magnitude of the spherical wave approaches the characteristic impedance away from the source; high frequency spherical waves approach the characteristic impedance much closer to the source than low frequency spherical waves.

From Equation 3.102, the phase between dynamic pressure and particle velocity is given by

$$\angle \frac{\tilde{p}_d(r,t)}{\tilde{u}(r,t)} = \phi_o = \tan^{-1}\left(\frac{1}{k_o r}\right) \tag{3.106}$$

Equation 3.106 is shown in Figure 3.14. If $k_o r \ll 1$, that is, very close to the source, $|\overline{Z}_S| \approx Z_o k_o r \angle \frac{\tilde{p}_d(r,t)}{\tilde{u}(r,t)} \approx 90°$, and far away from the source the phase angle is close to zero, that is, pressure and velocity are out of phase. If $k_o r \gg 1$, that is, very far from the source, $|\overline{Z}_S| \approx Z_o \angle \frac{\tilde{p}_d(r,t)}{\tilde{u}(r,t)} = 0°$, and pressure and velocity are in-phase, they behave like plane waves. Therefore, from Figures 3.13 and 3.14, it can be inferred that spherical waves behave like plane waves far away from the source; the exact location, however, depends on the frequency.

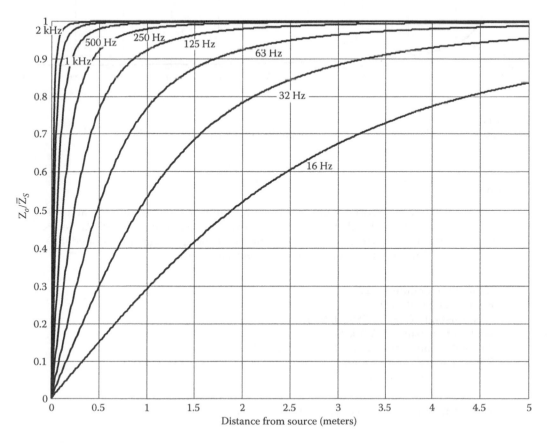

FIGURE 3.13
Normalized impedance versus distance from a point source for spherical waves.

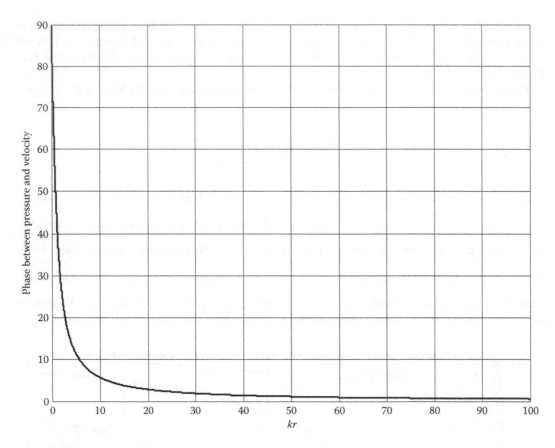

FIGURE 3.14
Phase angle of spherical wave impedance versus of product of wavenumber and radius, *kr*.

Example 3.5

A spherical source of airborne sound radiates uniformly at 22°C and 101 kPa at a frequency of 300 Hz. At a radial distance of 500 mm from the source, if $p_{rms} = 1.97$ Pa, determine the rms acoustic particle velocity, magnitude of the spherical wave impedance, and the phase difference between pressure and velocity, given

>   Room temperature, $t = 22$°C
>   Frequency, $f = 300$ Hz
>   Gas constant of air, $R = 286.7$ J/kg-K
>   Ambient pressure, $P = 101$ kPa
>   Absolute room temperature, $T = 273 + 22 = 295$ K
>   Ratio of specific heats for air, $\gamma = 1.4$
>   Distance from the source r = 500 mm

From Equation 3.16, velocity of sound in air at the given temperature is given by

$$c_o = \sqrt{\gamma R T} = \sqrt{1.4 \times 286.7 \times 295} = 344.10 \text{ m/s} \tag{1}$$

The nominal density of air is given by

$$\rho_o = \frac{P}{RT} = \frac{101 \times 10^3}{286.7 \times 295} = 1.19 \text{ kg/m}^3 \qquad (2)$$

From Equation 3.42, the wave number is given by

$$k_o = \frac{2\pi f}{c_o} = \frac{2\pi \times 300}{344.10} = 5.48 \text{ rad/m} \qquad (3)$$

From Equation 3.105, the magnitude of spherical wave impedance is given by

$$Z_s = \frac{\rho_o c_o k_o r}{\sqrt{1 + (k_o r)^2}} = \frac{1.19 \times 344.10 \times 5.48 \times 0.5}{\sqrt{1 + (5.48 \times 0.5)^2}} = 386 \text{ rayl} \qquad (4)$$

From Equation 3.104, the rms value of particle velocity is given by

$$u_{rms} = \frac{p_{rms}}{Z_s} = \frac{1.97}{386} = 5.1 \text{ mm/s} \qquad (5)$$

From Equation 3.106, the phase angle between dynamic pressure and particle velocity at 300 Hz is given by

$$\phi_o = \tan^{-1}\left(\frac{1}{k_o r}\right) = \tan^{-1}\left(\frac{1}{5.48 \times 0.5}\right) = 20° \qquad (6)$$

3.9.4 Sound Intensity of Spherical Waves

The dynamic pressure equation for spherical waves produced by a pulsating sphere at a frequency ω of peak amplitude X is given by

$$\tilde{p}_d(r,t) = \frac{\overline{X}}{r} e^{j(\omega t - k_o r)} \qquad (3.107)$$

From Equation 3.104 the particle velocity is given by

$$\tilde{u}(r,t) = \frac{\tilde{p}_d(r,t)}{\overline{Z}_S} \qquad (3.108)$$

where the impedance of spherical waves is given by

$$\overline{Z}_S = |Z_S| \angle \phi_o \qquad (3.109)$$

From Equation 3.104 the magnitude of spherical wave impedance is given by

$$|Z_S| = \frac{Z_o k_o r}{\sqrt{1 + (k_o r)^2}} \tag{3.110}$$

And from Equation 3.106 the phase of spherical wave impedance is given by

$$\angle \phi_o = \tan^{-1}\left\{ \frac{1}{k_o r} \right\} \tag{3.111}$$

From Equation 3.108 the complex conjugate of particle velocity is given by

$$\tilde{u}^*(r,t) = \frac{\tilde{p}_d{}^*(r,t)}{\bar{Z}_S^*} = \frac{\bar{X}}{r} \frac{e^{-j(\omega t - k_o r)}}{|Z_S|} e^{j\phi} \tag{3.112}$$

From Equation 3.79 the time-averaged sound intensity of a spherical wave is given by

$$\langle I(t) \rangle = \frac{1}{2} \mathrm{Re}\{\tilde{p}_d\, \tilde{u}^*\}$$

$$= \frac{1}{2}\mathrm{Re}\left\{ \frac{\bar{X}}{r} e^{j(\omega t - k_o r)} \frac{\bar{X}^*}{r} \frac{e^{-j(\omega t - k_o r)}}{|Z_S|} e^{j\phi_o} \right\} \tag{3.113}$$

$$= \frac{1}{2} \frac{X^2}{r^2 |Z_S|} \cos \phi_o$$

From Equation 3.106, the cosine of the spherical wave impedance angle is given by

$$\cos \phi_o = \frac{k_o r}{\sqrt{1 + (k_o r)^2}} \tag{3.114}$$

From Equations 3.113 and 3.114, the average sound intensity of a spherical wave is given by

$$\langle I(t) \rangle = \left(\frac{X}{r\sqrt{2}} \right)^2 \frac{\sqrt{1 + (k_o r)^2}}{Z_o k_o r} \frac{k_o r}{\sqrt{1 + (k_o r)^2}}$$

$$= \left(\frac{X}{r\sqrt{2}} \right)^2 \frac{1}{Z_o} \tag{3.115}$$

From Equation 3.107 the mean square pressure of a spherical wave is given by

$$p_{rms}^2 = \frac{1}{2}\left(\frac{X}{r}\right)^2 \qquad (3.116)$$

From Equations 3.115 and 3.116, the time-averaged intensity of a spherical wave is given by

$$<I(t)>= \frac{p_{rms}^2}{\rho_o c_o} \qquad (3.117)$$

From equations 3.85 and 3.117, we find that the time-averaged intensity is the same for both cases. However, the mean square pressure for spherical waves keeps changing with distance, whereas that of the plane wave remains a constant.

Example 3.6

Complete Table 3.6 that relates to spherical wave propagation from a point source to another point at a distance r (at which pressure and velocity are measured) from the source in a free-field condition. Assume the speed of airborne sound c_o as 340 mm/s and nominal density of air ρ_o as 1.2 kg/m³.

100 HZ

From Equation 3.42, the wave number of the sound wave is given by

$$k_o = \frac{\omega}{c_o} = \frac{2\pi \times 100}{340} = 1.85 \text{ rad/m} \qquad (1)$$

TABLE 3.6

Sound Parameters at Various Frequencies (Example 3.6)

Frequency, Hz	100	200	300
Dynamic pressure (rms), Pa	2	1	
Phase angle between dynamic pressure and particle velocity, ϕ_o	15°		
Particle velocity (rms), mm/s			3
Distance from the source, m			
Sound intensity (rms), mW/m²			
Sound power, mW			
L_p (sound pressure level, dB)			
L_W (sound power level, dB)			
L_I (sound intensity level, dB)			

From Equation 3.106, the distance from which the measurement point is located from the source is given by

$$r = \frac{1}{k_o \tan \phi_o} = \frac{1}{1.85 \tan 15°} = 2.02 \text{ m} \tag{2}$$

From Equation 3.104, the rms particle velocity due to the 100 Hz wave at a distance of 2.02 m is given by

$$u_{rms} = \frac{p_{rms}\sqrt{1+(k_o r)^2}}{\rho_o c_o k_o r} = \frac{2\sqrt{1+(1.85 \times 2.02)^2}}{1.2 \times 340 \times 1.85 \times 2.02} = 5.1 \text{ mm/s} \tag{3}$$

From Equation 3.117, the sound intensity at a distance of 2.02 m from the source is given by

$$< I > = \frac{p_{rms}^2}{\rho_o c_o} = \frac{2^2}{1.2 \times 340} = 9.8 \text{ mW/m}^2 \tag{4}$$

From Equation 3.76, the sound power of the source is given by

$$W = 4I_{rms}\pi r^2 = 9.8 \times 10^{-3} \times 4\pi \times 2.02^2 = 502.5 \text{ mW} \tag{5}$$

From Equation 3.57, the sound pressure level at $r = 2.02$ m is given by

$$L_p = 10\log\frac{2^2}{(20 \times 10^{-6})^2} = 100 \text{ dB} \tag{6}$$

From Equation 3.75, the sound power level is given by

$$L_W = 10\log\frac{W}{W_{ref}} = 10\log\frac{0.5025}{10^{-12}} = 117 \text{ dB} \tag{7}$$

From Equation 3.77, the sound intensity level is given by

$$L_I = 10\log\frac{I}{I_{ref}} = 10\log\frac{9.8 \times 10^{-3}}{10^{-12}} = 100 \text{ dB} \tag{8}$$

200 HZ

From Equation 3.42, the wave number of the sound wave at 200 Hz is given by

$$k_o = \frac{\omega}{c_o} = \frac{2\pi \times 200}{340} = 3.7 \text{ rad/m} \tag{9}$$

From Equation 3.106, the phase angle between dynamic pressure and particle velocity of the sound wave at 200 Hz is given by

$$\phi_o = \tan^{-1}\frac{1}{k_o r} = \tan^{-1}\frac{1}{3.7\times2.02} = 7.6° \tag{10}$$

From Equation 3.104, the rms particle velocity due to the 200 Hz wave at a distance of 2.02 m is given by

$$u_{rms} = \frac{p_{rms}\sqrt{1+(k_o r)^2}}{\rho_o c_o k_o r} = \frac{1\sqrt{1+(3.7\times2.02)^2}}{1.2\times340\times3.7\times2.02} = 2.45 \text{ mm/s} \tag{11}$$

From Equation 3.117, the sound intensity at a distance of 2.02 m from the source is given by

$$<I> = \frac{p_{rms}^2}{\rho_o c_o} = \frac{1^2}{1.2\times340} = 2.5 \text{ mW/m}^2 \tag{12}$$

From Equation 3.76, the sound power of the source is given by

$$W = 4I_{rms}\pi r^2 = 2.5\times10^{-3}\times4\pi\times2.02^2 = 125.6 \text{ mW} \tag{13}$$

From Equation 3.57, the sound pressure level at $r = 2.02$ m is given by

$$L_p = 10\log\frac{1^2}{(20\times10^{-6})^2} = 94 \text{ dB} \tag{14}$$

From Equation 3.75, the sound power level is given by

$$L_W = 10\log\frac{W}{W_{ref}} = 10\log\frac{0.1256}{10^{-12}} = 111 \text{ dB} \tag{15}$$

From Equation 3.77, the sound intensity level is given by

$$L_I = 10\log\frac{I}{I_{ref}} = 10\log\frac{2.5\times10^{-3}}{10^{-12}} = 94 \text{ dB} \tag{16}$$

300 HZ

From Equation 3.42, the wave number of the sound wave at 300 Hz is given by

$$k_o = \frac{\omega}{c_o} = \frac{2\pi\times300}{340} = 5.54 \text{ rad/m} \tag{17}$$

From Equation 3.106, the phase angle between dynamic pressure and particle velocity of the sound wave at 300 Hz is given by

$$\phi_o = \tan^{-1}\frac{1}{k_o r} = \tan^{-1}\frac{1}{5.54 \times 2.02} = 5.1° \tag{18}$$

From Equation 3.104, the dynamic pressure of the sound wave at 300 Hz is given by

$$p_{rms} = \frac{u_{rms}\rho_o c_o k_o r}{\sqrt{1+(k_o r)^2}} = \frac{3\times 10^{-3} \times 1.2 \times 340 \times 5.54 \times 2.02}{\sqrt{1+(5.54\times 2.02)^2}} = 1.22\ \text{Pa} \tag{19}$$

From Equation 3.117, the sound intensity at a distance of 2.02 m from the source is given by

$$<I> = \frac{p_{rms}^2}{\rho_o c_o} = \frac{1.22^2}{1.2 \times 340} = 3.6\ \text{mW/m}^2 \tag{20}$$

From Equation 3.76, the sound power of the source is given by

$$W = 4I_{rms}\pi r^2 = 3.6\times 10^{-3} \times 4\pi \times 2.02^2 = 186.7\ \text{mW} \tag{21}$$

From Equation 3.57, the sound pressure level at $r = 2.02$ m is given by

$$L_p = 10\log\frac{1.22^2}{(20\times 10^{-6})^2} = 96\ \text{dB} \tag{22}$$

From Equation 3.75, the sound power level is given by

$$L_W = 10\log\frac{W}{W_{ref}} = 10\log\frac{0.1867}{10^{-12}} = 113\ \text{dB} \tag{23}$$

From Equation 3.77, the sound intensity level is given by

$$L_I = 10\log\frac{I}{I_{ref}} = 10\log\frac{3.6\times 10^{-3}}{10^{-12}} = 96\ \text{dB} \tag{24}$$

Table 3.7 is the completed table.

TABLE 3.7

Sound Parameters at Various Frequencies (Example 3.6)

Frequency, Hz	100	200	300
Dynamic pressure (rms), Pa	2	1	1.22
Phase angle between dynamic pressure and particle velocity	15°	7.6	5.1
Particle velocity (rms), mm/s	5.1	2.47	3
Distance from the source, m	2.02		
Sound intensity (rms), mW/m²	9.8	2.5	3.6
Sound power, mW	502.5	125.6	186.7
L_p (sound pressure level, dB)	100	94	96
L_W (sound power level, dB)	117	111	113
L_I (sound intensity level, dB)	100	94	96

Note: Sound intensity and sound pressure levels are the same in a free field.

3.10 Frequency Analysis of Airborne Sound Signals

A typical sound signal can have a combination of several frequencies in the audible frequency range. By using the rms value of the dynamic pressure, we can compute the overall sound pressure level using Equation 3.57. However, this information is not of much value due to the following reasons. First, without knowing the frequency content of the dynamic pressure, it would be difficult to characterize the source of sound. Second, since our hearing sensitivity varies widely for various frequencies, the magnitude of sound pressure levels as measured by the microphone and that perceived by our ears will not match. This not only leads to confusion but defeats the very purpose of reducing sound to prevent annoyance and noise induced hearing loss to humans; therefore, frequency analysis helps in arriving at certain corrections to compensate for our hearing sensitivity. Third, the sound attenuation measures for noisy machines need acoustic materials that have frequency dependent properties. Therefore, the average sound pressure levels in various frequency bands will be useful in designing sound reduction measures. Fourth, to reduce sound radiation at the source, it is necessary to know about the frequency content of the sound field produced by it. Finally, our hearing mechanism is capable of identifying various frequencies. Therefore, frequency analysis helps in identifying various types of sounds.

From Equation 3.42, it is seen that the wavenumber is the ratio of frequency and speed of airborne sound. Since the wavenumber is given by $\dfrac{2\pi}{\lambda}$, the wavelength is the ratio of speed of sound to its frequency. Since the speed of sound remains constant for all frequencies, the wavelength is inversely proportional to frequency. Figure 3.15 and Table 3.8 show the wavelength of sound having speed of 330 m/s at various frequencies; the wavelength is very large at low frequencies and very small at high frequencies. Hence wavelength can be correlated with physical dimensions of noise sources.

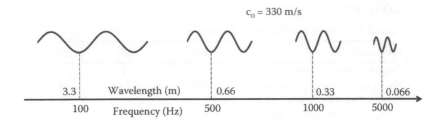

FIGURE 3.15
Frequency and wavelength of sound signals.

TABLE 3.8

Wavelength of Various Frequencies of Sound at c_o = 330 m/s

Frequency (Hz)	Wavelength (m)	Wavelength (mm)
16	20.625	20625
100	3.300	3300
300	1.100	1100
500	0.660	660
700	0.471	471
800	0.413	413
1000	0.330	330
1500	0.220	220
2000	0.165	165
3000	0.110	110
4000	0.083	83
5000	0.066	66
6000	0.055	55
7000	0.047	47
8000	0.041	41
16000	0.021	21

3.10.1 Octave and 1/3 Octave Bands

Table 3.9 shows seven musical notes of Indian and Western classical music. The ascending notes increase in pitch and after the seventh tone, the eighth tone will return to the beginning of the tone series but will be at twice the frequency when it started with the first series. This is the basis for defining an octave band that has the highest frequency twice the lower frequency. Since music has influenced acoustics, octave bands are used for airborne sound measurement. Hence, sound-measuring equipment that can identify frequency content is equipped with a number of band-pass filters at which sound pressure levels are measured. Such equipment can give the frequency content of sound pressure levels that would help identify the source as well design appropriate attenuation measures using acoustic materials that have frequency-dependent properties.

The frequency response of a typical octave band filter is shown in Figure 3.16. The general relationship between the upper and lower cutoff frequencies is given by

$$f_2 = 2^a f_1 \qquad\qquad (3.118)$$

TABLE 3.9

Musical Notes of Indian and Western Classical Music

Note Number	Indian	Western
1	Sa	Do
2	Re	Re
3	Ga	Mi
4	Ma	Fa
5	Pa	So
6	Da	La
7	Ni	Ti
8	Sa	Do

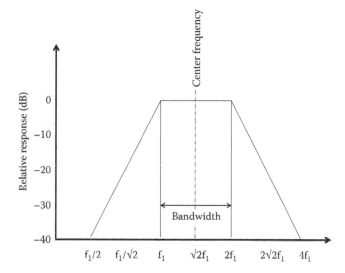

FIGURE 3.16
Octave band filter.

where a is a constant. The value of $a = 1$ corresponds to an octave band filter, where the upper frequency limit is twice the lower frequency. The value of $a = 1/3$ corresponds to a 1/3 octave band filter, where the upper frequency is $2^{1/3}$ times the lower frequency. The difference between an octave band filter and 1/3 octave band filter is that the 1/3 octave band filters analyze sound levels in a much narrower band. Digital filters in recent times can measure in much narrower bands. For specifying the sound power level of most machines it is sufficient to use the octave band filters in the measuring equipment. However, with the advent of digital filters with miniaturized electronics, handheld sound level meters that can measure 1/3 octave bands have become very common.

The upper and lower cutoff frequencies for octave and 1/3 octave band filters cannot be arbitrarily chosen. These have been standardized by the ISO, as shown in Table 3.10, with center frequencies from 16 Hz to 16 kHz that cover the audible range. The center frequencies corresponding to the octave band filter or 1/3 octave band filter are used in reporting the corresponding sound pressure levels. You might have also noticed the center frequencies of octave bands marked on the frequency equalizer of many sound production

TABLE 3.10

Frequencies for Octave and 1/3 Octave Band Filters

Octave Band			1/3 Octave Band		
Lower Frequency, f_1 (Hz)	Center Frequency, f_c (Hz)	Higher Frequency, f_2 (Hz)	Lower Frequency, f_1 (Hz)	Center Frequency, f_c (Hz)	Higher Frequency, f_2 (Hz)
11	16	22	14.1	16.0	17.8
			17.8	20.0	22.4
			22.4	25.0	28.2
22	31.5	44	28.2	31.5	35.5
			35.5	40.0	44.7
			44.7	50.0	56.2
44	63	88	56.2	63.0	70.8
			70.8	80.0	89.1
			89.1	100.0	112.0
88	125	177	112	125	141
			141	160	178
			178	200	224
177	250	355	224	250	282
			282	315	355
			355	400	447
355	500	710	447	500	562
			562	630	708
			708	800	891
710	1000	1420	891	1000	1122
			1122	1250	1413
			1413	1600	1778
1420	2000	2840	1778	2000	2239
			2239	2500	2818
			2818	3150	3548
2840	4000	5680	3548	4000	4467
			4467	5000	5623
			5623	6300	7079
5680	8000	11360	7079	8000	8913
			8913	10000	11220
			11220	12500	14130
11360	16000	22720	14130	16000	17780

systems. Generally, center frequencies from 63 Hz to 4000 Hz are used in most applications, but they should be carefully chosen for each application by ensuring that all the dominant frequencies are accounted for. Figure 3.17 shows the range of frequencies for various sound sources and the speech frequency range for humans. It can be seen that males and females have slightly different ranges of speech frequencies. Therefore, males with hearing loss that generally occurs in the high frequencies will have difficulty in recognizing sounds produced by females, which is also in the high frequency range.

The process of obtaining octave band sound pressure levels by measurement is illustrated in Figure 3.18. The output of a microphone is passed through several octave band filters and the output of each filter represents the sound pressure levels in the respective

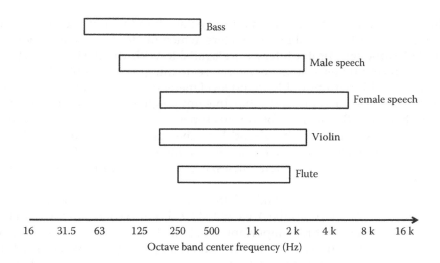

FIGURE 3.17
Frequency ranges of various signals.

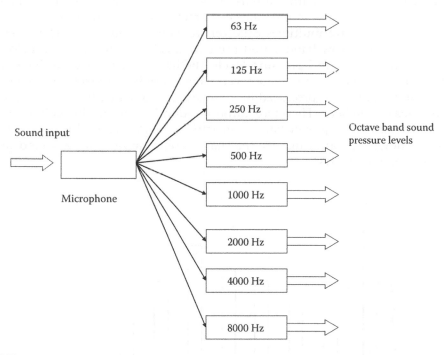

FIGURE 3.18
Octave band analysis.

octave bands. Although octave band filters in the entire audible range are available, only those between 63 Hz and 8000 Hz are shown in the figure. In addition, it may not be necessary to obtain information on octave band sound pressure levels in all such bands in the audible frequency range; if the sound pressure levels are less than 10 dB to that measured in the nearest band, it can be ignored. Therefore, depending on the source of sound, one could neglect octave bands in the lower and higher ends of the audible frequency range.

The filters that were used early in obtaining octave band sound pressure levels were of the analog type for the portable type meters, which made them heavy; they weighed around 10 kg! Although digital filters were available with the desktop spectrum analyzers, only in recent times, due to very large-scale integration (VLSI) electronics, have they become available with portable sound level meters; even 1/3 octave band handheld sound level meters are currently available. In summary, collecting sound pressure data in 1/3 octave bands, with a facility for various types of time averaging, has become very easy in recent times using portable sound level meters that weigh around 1 kg. Much simpler sound level meters that give only A-weighted and C-weighted (no weighting except at very high and very low frequencies) are available for quick assessment of airborne sound.

The sound pressure level data obtained from an octave band sound level meter are shown in Figure 3.19. Many times the A-weighted level is also shown along with the other octave band information. But it is important to note that sound pressure levels in each of the octave bands must be shown as obtained without subjecting to A-weighting. Therefore, knowing the actual sound pressure levels of a machine along with its A-weighted value is a standard practice.

Figure 3.20 shows octave band sound pressure levels that contain harmonics, assuming that all frequencies are within a narrow band of the respective octave band center frequencies. An airborne sound can be characterized as containing harmonics, if the higher frequencies are multiples of lower frequencies. Since this characterization is much easier with narrow band sound pressure levels, when the consecutive octave bands have significant sound pressure levels, the harmonic nature can be easily recognized. Harmonic sounds can be observed when sounds are produced due to longitudinal vibration and those produced by stringed instruments. In addition, harmonic sounds are pleasant to hear. Therefore, airborne sound when produced in appropriate proportions in various frequency bands can actually be pleasant. This has led to a systemic study known as sound quality that has become important for sound pressure levels from consumer goods.

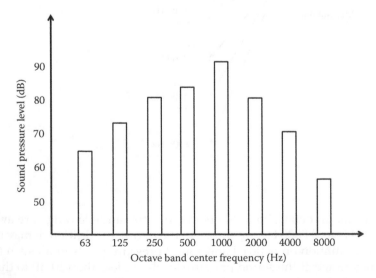

FIGURE 3.19
Displaying octave band information.

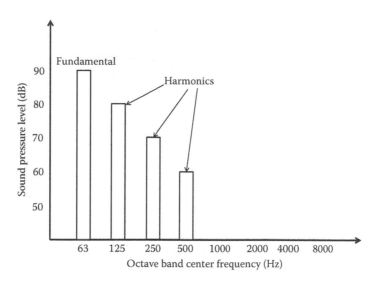

FIGURE 3.20
Harmonics of a sound signal.

3.10.2 A-Weighting of Sound Pressure Levels

The sound pressure at various octave bands can be measured using a microphone and associated frequency filters. This cannot give an objective assessment of sound as perceived by the listener, since our ears have a nonlinearly varying sensitivity to sound at different frequencies. One of the first attempts to account for this was to define loudness in the form of phons. A phon was defined as the sound pressure level of a sound at 1000 Hz. Corresponding to each phon, a phon contour was drawn at various frequencies. This was done by subjective assessment of the sound pressure level corresponding the phon at 1000 Hz, to the sound pressure level at other frequencies, which have the same loudness as that of a sound at 1000 Hz.

From the earlier discussion, it is clear that our ears have a varying sensitivity, for frequency as well as amplitude of sound. The quantity measured by a calibrated microphone and amplifier is modified (weighted) so that the frequency response follows approximately the equal loudness curve of any of the phon contours. The A-weighting contour is shown in Figure 3.21.

An A-weighted sound pressure level is the quantity measured by a calibrated microphone and amplifier of a sound level meter, modified so the frequency response follows approximately the equal loudness curve of 40 phons. This is known as the A-weighted sound level, dBA, and it has been found to correlate extremely well with subjective response and shows up consistently well in comparisons with other sound scales. This fact, along with the ease with which measurements can be made using a sound level meter has led this frequency weighting to be adopted internationally. Because, individual response varies so much and sound scales are highly correlated with each other, it is often argued that the A-weighting measurement is as good as any other method of subjective rating of sound.

A-weighting basically consists of subtracting or adding a certain number of decibels to sound pressure levels, measured at each of the octave or 1/3 octave frequency bands, and then computing an overall value of the sound pressure level. A-weighting for the octave band frequencies from 63 Hz to 8000 Hz are shown in Table 3.11; they are also available in 1/3 octave bands.

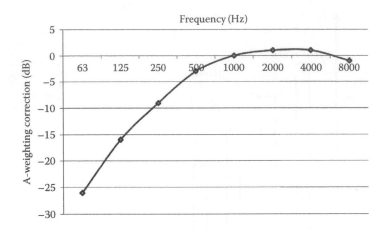

FIGURE 3.21
Weighting curves.

TABLE 3.11

A-Weighting Correction of Octave Band Sound Pressure Levels

Octave center frequency (Hz)	63	125	250	500	1000	2000	4000	8000
A-weighting correction (dB)	−26	−16	−9	−3	0	1	1	−1

The A-weighted sound level is widely used to state design goals as a single number and its usefulness is limited because it does not give any information on the spectrum content; it is very similar to a stock market index that gives a single number based on the performance of a specific set of stocks. It can be measured using a sound level meter, which gives specific weighting to the frequency content, according to the A-weighting scale, as per the human ear sensitivity established by means of extensive measurements. The A-weighted sound level thus has the advantage of identifying the desirable level as a single valued number and it correlates well with the human judgment of relative loudness. It does not, however, correlate well with the human judgment regarding the subjective quality of sound. A-weighted level comparisons are best suited for comparing sound levels that are alike in the frequency spectrum but differ in level.

The other scales that have been standardized are the B, C, and D weightings. The B network follows the 70 phon contour and the C follows the 100 phon contour. C-weighting is flat except for very low and very high frequencies. The D network follows a contour of perceived noisiness and is used for single event aircraft sound measurement.

In addition to these networks, some of the sound level meters have a linear or "Lin" mode, which allows the signal to pass through, unmodified, and does not weight the signal. Out of A, B, and C, the most widely used is the A network.

From this discussion it is clear that all the airborne sound frequencies are not equally perceived by our ears. This is the basis for the new digital storage formats of music audios such as MP3. In this format, only frequencies in the audible range are used and sounds that are less significant (less than 10 dB) are removed. In addition, repetitive sounds are recorded once and the instant at which they are to be played back are noted by the player.

TABLE 3.12

Sound Pressure Data (Example 3.7)

Octave Frequency (Hz)	63	125	250	500	1000	2000	4000	8000
Lp (dB)	91	95	101	100	96	90	84	75
A-weighting corrections (dB)	−26	−16	−8	−3	0	1	1	−1
A-weighted (dB)	65	79	93	97	96	91	85	74

These measures significantly reduce the size of these audio files and in addition remove unwanted sounds.

Example 3.7

The data in Table 3.12 refers to the octave band measurements carried out near a machine. Determine the overall sound pressure level and the A-weighted level.

$$L_p(\text{overall}) = 10\log 10(10^{91/10} + 10^{95/10} + 10^{101/10} + 10^{100/10} + 10^{96/10} + 10^{90/10} + 10^{84/10} + 10^{75/10})$$

$$= 105\,\text{dB}$$

$$L_{p,A} = 10\log 10(10^{65/10} + 10^{79/10} + 10^{93/10} + 10^{97/10} + 10^{96/10} + 10^{91/10} + 10^{85/10} + 10^{74/10}) = 101\,\text{dBA}$$

Note that $L_{p,A}$ is less than L_p due to weighting and the difference between them is −5 (dB) that falls between 250 to 500 Hz A-weighting correction as per Table 3.11. Hence we can sometimes guess the dominant frequency of airborne sound using the difference of $L_{p,A}$ and L_p.

3.11 Transducers for Sound Measurement

Measurement of sound is the first step in working toward its reduction. The first step in this process is the use of microphones to convert the dynamic pressure of sound into voltage signals that can be further processed for obtaining their frequency information. We are generally concerned about the free-field response of microphones, which is defined as the relation between microphone output voltage and the sound pressure that existed at the microphone location before the microphone was introduced into the sound field. Microphones are also classified based on whether they are used in a free field (where there are no reflections) or diffuse field (where the sound field is uniform due to multiple reflections) environment and they should be chosen accordingly.

3.11.1 Condenser Microphones

The schematic diagram of a condenser microphone is shown in Figure 3.22. It consists of a thin metallic diaphragm and a rigid back plate. The diaphragm and back plate are electrically insulated from each other and connected to a stabilized direct voltage polarization

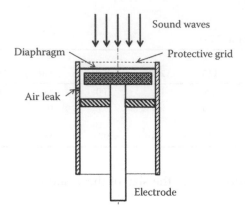

FIGURE 3.22
Condenser microphone.

source, forming a parallel-plate capacitor. When a microphone is exposed to a sound pressure wave, its thin diaphragm moves to and fro relative to the rigid back plate, and this movement causes an alternating charge in the capacitance between the diaphragm and the back plate. This charge produces a corresponding voltage signal across the output terminals. There is an additional circuit that provides voltage output at low impedance. Since the diaphragm is balanced on both sides by the atmospheric pressure, the condenser microphone is sensitive only to the dynamic sound pressure.

3.11.2 Sound Level Meter

The most commonly used instrument for sound measurements is a sound-level meter as shown in Figure 3.23. Initially, the dynamic sound pressure, p, is converted to voltage by

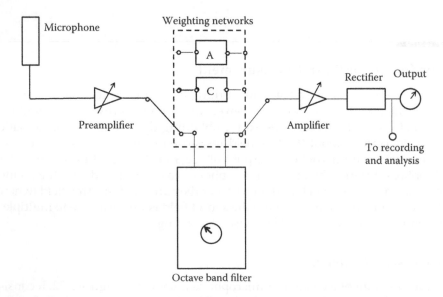

FIGURE 3.23
Sound level meter.

means of a condenser microphone that was discussed earlier. The microphone output is amplified using a preamplifier and then passed through either the weighting networks or octave filters that are analog in nature. The weighting networks such A-weighting account for the frequency-related hearing characteristics of humans; the C-weighting gives less importance to very low and very high frequencies but allows the rest of the frequencies through it. The octave band filters give the sound pressure level data for each octave band. The voltage output of these filters is then passed through an amplifier and is available for possible analysis using external spectrum analyzers and for storage. The same signal is passed through a rectifier to give an output on the meter. Depending on the time-averaged variation of the sound pressure, fast or slow options can be used to obtain a steady reading on the meter.

The sound level meter shown in Figure 3.23 is an analog instrument that was used earlier, long before the advent of electronics. Such sound level meters were very heavy and one had to manually switch to various octave bands to obtain sound pressure levels in the corresponding band. Therefore, it took a long time to cover all the octave bands and if one had to measure at many points around a machine, it was a rather tedious exercise. Now, with the advent of electronics, digital filters can be used to filter the microphone output to obtain the desired octave or 1/3 octave band output in a fraction of second without any manual intervention. The sound pressure signals can also be time-averaged and stored for possible downloading to the computer. Therefore, 1/3 octave band sound pressure measurement has become easier than ever before.

According to the standards, the sound level meters are divided into three categories: Type 0, Type 1, and Type 2. These differ in the tolerances of accuracy allowed. The tolerances broaden as the number increases and differ for various types to a degree, which significantly affects the manufacturing cost. The Type 0 meter is intended as a laboratory standard reference and is also the most expensive. The Type 1 meter is a precision portable meter intended for accurate field measurements. The Type 2 meter is intended for general-purpose field measurements.

Sound level meters can be calibrated by using a standard sound pressure source that produces 94 dB at 1000 Hz. If the meter does not read 94 dB, a screw adjustment for a potentiometer is given to adjust the output levels to 94 dB. Now, with the recent digital electronic devices-based sound level meters, there is a facility for direct calibration. Sound level meters can also be calibrated in various bands by using a pink noise calibrator.

3.11.3 Sound Intensity Measurement

Sound intensity, at any point in a sound field, is the rate of flow of sound energy in a specified direction through a unit area normal to this direction at the point under consideration. The unit of sound intensity is watts per square meter (W/m^2) and is denoted by the symbol *I*. Sound intensity is a vector quantity having both magnitude and direction, whereas sound power and sound pressure level are scalar quantities. Unlike sound power, which is indirectly estimated through sound pressure measurement, sound intensity can be directly measured. Sound intensity fluctuates with time, but its time average is measured (see Harris 1991).

Sound intensity measurements are advantageous for the following reasons:

1. Measurements can be made in situ in almost any field situation.
2. Measurements can be made without the use of special facilities such as anechoic and reverberation rooms.

3. Measurements can be made in the presence of steady moderate or high background sound.

4. Measurements can be used to locate sources of sound because sound intensity gives a measure of the direction of sound energy flow as well as its magnitude.

However, sound intensity measurements are often cumbersome, involving a lot of time and effort.

The following sections deal with the basics of sound intensity and the transducer for its measurement.

3.11.3.1 Sound Fields

Sound intensity at a given point measured in a direction normal to a specified unit area through which the sound energy is flowing is given by

$$p \times u \tag{3.119}$$

where p is the sound pressure at that point and u is the particle velocity in that direction.

Consider a source in a free field as shown in Figure 3.11. All the sound energy radiated by the source must pass through an imaginary surface, which need not be always spherical as shown in the figure, enclosing the source. The sound power of the source can be obtained by measuring the sound intensity averaged over the imaginary surface enclosing the area and multiplying it by the area. The sound intensity diminishes with distance in a free-field following the inverse square law. Therefore, doubling the distance from the source decreases sound intensity by 4 times. This is equivalent to a decrease of 6 dB for every doubling of the distance from the source. However, since the sound power of a source is a constant, the decrease in the sound intensity due to increase in the distance from the source is offset by the increase in the area of the imaginary enclosing surface.

In a free field, the relationship between the mean square sound pressure, p, and the sound intensity, I, is

$$I = \frac{p^2}{\rho_o c_o} \tag{3.120}$$

where ρ_o is the density of air and c_o is the speed of airborne sound. Sound intensity is measured in the region in which the particle velocity and sound pressure are in phase with each other. This generally occurs at a point about 0.6 m away from the sound source, or about 1.5 times the typical dimension of the source, whichever is greater. In this region, sound intensity fluctuates at a frequency twice that of the sound wave, has a direction normal to the wave front, and is associated with the sound energy flow.

3.11.3.2 Measurement

In order to determine sound intensity, it is necessary to measure the sound pressure and particle velocity simultaneously at the same point in the sound field. Commercially available sound intensity probes use two closely spaced sound pressure transducers (microphones) in various configurations for measuring sound pressure and particle velocity. Particle velocity is related to the pressure gradient from Equation 3.98 as

$$u = -\frac{1}{\rho_o} \int \frac{\partial p}{\partial r} dt \tag{3.121}$$

where ρ_o is the density of air. The sound pressure gradient $\frac{\partial p}{\partial r}$ can be approximated by measuring the pressure at each microphone and dividing by the distance between them:

$$\frac{\partial p}{\partial r} \cong \frac{p_B - p_A}{\Delta r} = \frac{\Delta p}{\Delta r} \tag{3.122}$$

where p_A is the sound pressure at microphone A, p_B is the sound pressure at microphone B, and Δr is the distance between the microphones. This is the finite-difference approximation from using which the particle velocity thus becomes

$$u = -\frac{1}{\rho_o} \int \frac{p_B - p_A}{\Delta r} dt \tag{3.123}$$

and is the particle velocity at the acoustic center of the probe between the two microphones. At this point, pressure is determined by taking the average pressure of the two microphones:

$$p = \frac{p_A + p_B}{2} \tag{3.124}$$

The preceding equations for particle velocity and pressure can be used for computing the time-averaged sound intensity as follows:

$$< I >_t = \frac{1}{T} \int p u \, dt \tag{3.125}$$

But we have already used an extension of Equation 1.96 to compute the sound intensity of plane and spherical waves. Equation 3.125 can be practically implemented for measuring sound intensity around a machine by measuring the cross spectrum of the sound pressure measured by the two microphones, which can be related to sound intensity in the frequency domain.

3.11.3.3 Sound Intensity Probes

There are many types of sound intensity probes depending upon the configuration of the microphones. A typical sound intensity probe is shown in Figure 3.24, which is a face-face probe. Sound intensity can be related to Equation 3.125 or an equivalent formulation relating time-averaged sound intensity to the imaginary part of the cross spectrum of the two microphone signals. It may be recalled here that our hearing mechanism also computes the cross spectrum of sound signals in the two ears to detect the direction of sound.

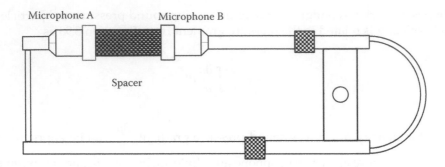

FIGURE 3.24
Sound intensity probe.

The sound intensity probe has directivity characteristics and it has maximum sensitivity in the direction in which the intensity is being measured. At an incidence angle of 90°, no intensity is sensed because of zero pressure difference in the signals. At other incident angles, the sound intensity will be less than that sensed along the probe axis. This is given by $10 \log \cos \theta$.

Based on the directivity characteristics of a sound intensity probe, the direction of a sound source can be detected. This is illustrated in Figure 3.25. There is a sound source, and sound intensity measurements are conducted by pointing the sound intensity axis perpendicular to a measurement surface and then moving the probe vertically down in discrete steps. The sound intensity measured by the probe will be at maximum when its axis is along the source. Therefore, by conducting a series of sound intensity measurements one can detect the exact location of a source.

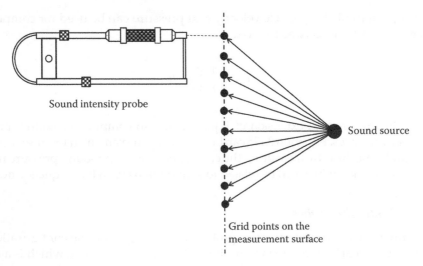

FIGURE 3.25
Direction of sound source from sound intensity measurement.

In addition to locating a source, sound intensity measurement can also be used for sound power measurement. By scanning the sound intensity probe around an imaginary surface of the machine covering all the noise radiating directions and multiplying the average quantity by the corresponding area, the net sound power of the machine can be obtained. The international standards for the measurement of sound intensity are covered under ISO 9614-1 and IEC 1043.

Sound intensity probes are calibrated by calibrating each of the microphones using a standard calibrator and also matching the phase between the microphones. This can be calibrated using a standard sound intensity calibrator.

3.12 Conclusions

This chapter has covered most characteristics of airborne sound. The wave equation for airborne sound is very similar to that derived for longitudinal vibration. In addition, the characteristic impedance discussed in this chapter has many important applications. Furthermore, the same equation that was derived for power input to a vibratory system was used to determine the basic equation for sound intensity of airborne sound. Sound pressure levels computed using the dynamic pressure of airborne sound provide the basis for their quantification by measuring with a microphone.

Sound pressure level measurement will be extremely useful in identifying the source machinery sound, which would further help in designing sound reduction measures. In addition, it helps to compute sound power, which is discussed in Chapter 8. Sound power has become an international standard for objective quantification of machinery sound. If sound power data of machinery are available, it is possible to predict sound pressure levels, even before the machine is installed. In addition, one could calculate the extra investment in terms of acoustic treatment that would be necessary for bringing down sound levels to acceptable levels. Thus it is very useful in evaluating the purchase of machinery from the viewpoint of sound.

Sound intensity presented in this chapter is not only useful for measuring sound power but can also be used to identify sources of sound in complex machinery. Using similar principles, the number of microphones can be increased to obtain an acoustic camera, for precise identification of sound sources within complex machinery. Therefore, sound intensity is very useful for reducing the main source of sound in machinery.

The material presented in this chapter on airborne sound will be useful in later chapters to discuss quantification of acoustic sources and to relate how vibration will result in sound.

PROBLEMS

3.1. The fundamental frequency of a note played from an instrument was 250 Hz with an rms dynamic pressure of 5 mPa. Determine the following:

a. Wavelength of the tone in air

b. Effective or rms value

c. Sound pressure level in decibels

3.2. The following octave-band sound pressure level measurements were obtained using a microphone and sound level meter system. Fill in the missing numbers and locate the dominant octave band frequency of the machine.

Octave center frequency (Hz)	63	125	250	500	1k	2k	4k	Overall Levels
Lp (dB)	95		80		90		85	
A-weighting correction	−26	−16	−9	−3	0	1	1	
$L_{p,A}$ (A-weighting) (dB)		85		75		85		

3.3. The sound pressure levels of a machine A in various octave bands are shown next. It is proposed to add another machine B so the resulting sound pressure levels decrease at the rate of 5 dB per octave, beginning from 63 Hz, and the combined sound pressure level should be 105 dB at 250 Hz. Determine the following:

a. Sound pressure levels of machine B that would achieve this

b. A-weighted level of when only machine A is operating

c. A-weighted levels when both machines are operating

Octave frequency (Hz)	63	125	250	500	1000	2000	4000	8000
Lp (dB) (Machine A)	91	95	102	100	95	90	85	75
Lp (dB) (Machine B)								
Lp (dB) (Machine A + B)			105					

3.4. The overall sound pressure level of a machine is 90 dB (linear unweighted) and the A-weighted sound pressure level is 83 dBA. Determine the dominant frequency of this machine. Can you expect your answer to be always correct? Explain.

3.5. The following table shows the sound power level data of four machines in various octave bands. Compute the sound power level in octave bands if all the machines are operated at the same time. In addition, determine the overall sound power level and A-weighted sound power level.

	Octave Center Frequency (Hz)	63	125	250	500	1k	2k	4k	8k
M1	Lw (dB)	70	75	80	85	90	95	80	60
M2	Lw (dB)	75	80	85	90	95	100	75	65
M3	Lw (dB)	80	75	75	85	90	95	85	75
M4	Lw (dB)	85	80	75	70	70	75	65	65

3.6. Two incoherent sound signals produce a sound pressure level at the receiver of 75 dBA. When the first signal was removed, a sound pressure level of 70 dBA results. Determine the sound level of the first signal.

3.7. The sound pressure levels measured at a particular location in a noisy factory, with and without one particular machine operating, are given in the following table in octave bands. Calculate the noise level in dBA due to the machine alone.

Octave Band Center Frequency (Hz)	63	125	250	500	1k	2k	4k	Overall Sound Pressure Level
$L_{p,machine+background}$ (dB)	98	94	90	90	87	84	76	
$L_{p,background}$ (dB)	97	90	85	88	83	80	73	

3.8. The following data refers to A-weighted overall sound pressure levels in a noisy workshop for the duration noted. Determine whether the total dose is acceptable. If each hour of work results in a productivity of $25,000, calculate the total loss per day due to noisy machines that do not allow for the 8-hour shift.

No.	Sound Pressure Level (dBA)	Time (hours), C_i
1	85	5
2	87	2
3	89	0.35
4	91	0.25
5	93	0.2
6	95	0.15
7	97	0.05
Total time		8

3.9. The overall A-weighted sound pressure level of machine A is 70 dBA and that of machine B is 75 dBA. If there are five machines of type B, determine the number of machines of type A that can be operated together along with the five machines of type B, without exceeding an overall A-weighted level of 85 dBA.

3.10. There are two machines, A and B, that have overall sound pressure levels (A-weighted) of 75 dBA and 70 dBA, respectively.

a. During the day (6 am to 10 pm) both of them will have to work together for 10 hours and machine B alone has to work for 2 hours. How many hours can machine A work alone so the L_{eq} does not exceed 75 dBA? Assume a background level of 60 dBA during the day.

b. If the two machines are to be simultaneously used overnight (10 pm to 6 am) and the L_{eq} cannot exceed 65 dBA, how many hours can they work together? Assume a background level of 50 dBA during the night.

3.11. The local regulations at a certain place prescribe that the sound pressure levels (from all sources) should not exceed 45 dBA during night and 50 dBA during the day. The background levels, when measured without any machine operating, are 38 dBA during the night and 46 dBA during the day. If a certain machine has a sound pressure level of 41 dBA, how many such similar machines can operate during the day and how many during the night without violating the local regulations?

3.12. The following dBA readings were taken inside a hospital room over an 8-hour period during night. The readings were taken every 2 minutes and the number of times they occurred were also recorded. Determine the L_{eq} value in dBA. If the allowable L_{eq} is 40 dBA during night in the hospital zone, is there a cause for complaint to the authorities?

dBA	No. of Times
60	30
58	30
56	25
54	40
52	30
50	30
48	50
46	0
44	5

3.13. The local regulations at a certain place prescribe that the sound pressure levels (from all sources) should not exceed 45 dBA during night and 50 dBA during the day. The background levels, when measured without any machine operating, are 40 dBA during the night and 45 dBA during the day. If a certain machine has a sound pressure level of 40 dBA, how many such similar machines can operate during the day and how many during the night without violating the local regulations?

$$L_{p,allowable,day} = 50 \text{ dB}; L_{p,allowable,night} = 45 \text{ dB}; L_{p,machine} = 40 \text{ dB}$$

$$L_{p,background,day} = 45 \text{ dB}; L_{p,background,night} = 40 \text{ dB}$$

3.14. The dynamic pressure measured from a point source is given by the following equation: $\tilde{p} = e^{j(2\pi 100t)}$ Pa. The particle velocity is given by $\tilde{u} = 2.5e^{j(2\pi 100t - 15.14°)}$ mm/s. Determine the following:

   a. Magnitude and phase angle of specific acoustic impedance
   b. Distance from the source

3.15. At some location, the dynamic pressure peak amplitude and particle velocity of a 100 Hz sound wave in air are measured to be 2 Pa and 10 mm/s. Assuming this as a spherical wave, find the distance from the source and the phase angle between dynamic pressure and particle velocity.

3.16 A simple spherical source in free field radiating at a frequency of 100 Hz produces a dynamic pressure (rms) of 6 Pa at 1 m and 0.6 Pa at 10 m. Evaluate the amplitude and phase (relative to the acoustic pressure) of the acoustic particle velocity at each of the two locations.

3.17. A spherical point source of sound radiates into air at 25°C and 100 kPa. The frequency of the sound wave is 200 Hz. At a radial distance of 1 m from the source, if $p_{rms} = 1$ Pa, determine the sound pressure level, rms acoustic particle velocity, rms particle displacement, phase difference between dynamic pressure and particle velocity, and dynamic pressure and particle displacement. At 2 m from the source, determine all the above values.

3.18. Dynamic pressure and particle velocity measurements were conducted at a certain distance from a source radiating sound in air and they are, respectively, as follows:

$$\tilde{p} = (1.3289 + j0.4837)e^{j250\pi t} \text{ Pa}$$

$$\tilde{u} = (3.5 + j0.7)e^{j250\pi t} \text{ mm/s}$$

$\rho_o = 1.16 \text{ kg/m}^3$; $c_o = 344 \text{ m/s}$

a. Do they represent plane waves or spherical waves at this location?

b. Determine the distance of this measurement location from the source.

c. What is the peak value of dynamic density at this location?

d. What is the sound pressure level at (dB and dBA) at this location?

e. Determine its wavelength.

f. At what distance from the source do they become plane waves?

3.19. Complete the following table that relates to spherical wave propagation from a point source to another point at a distance r (at which pressure and velocity are measured) from the source in a free-field condition. Let the characteristic impedance be $\rho_o c_o = 400$ rayl and the velocity of sound be $c_o = 344 \text{ m/s}$.

Frequency, Hz	100	200	300
Dynamic pressure (rms) (Pa)	2	1	
Particle velocity (rms) (mm/s)	10		5
Phase angle between dynamic pressure and particle velocity (degrees)			
Sound intensity (rms) (mW/m²)			
Sound power (mW)			
L_p (sound pressure level, dB)			
L_W (sound power level, dB)			
L_I (sound intensity level, dB)			

3.20. The sound pressure level measured at a distance of 1 m from a spherical source is 94 dB for a 1 kHz frequency wave. Assume $c_o = 340 \text{ m/s}$ and $\rho_o = 1.2 \text{ kg/m}^3$.

a. Determine the rms value of particle velocity and its phase difference with the dynamic pressure.

b. Write the general equation for dynamic pressure and particle velocity if there is only an outgoing wave.

4

Random Vibration

Random vibration is the one that is not deterministic; it will generally be arbitrary and cannot be expressed as a function of time of known characteristics. Much of the vibration arising from nature falls into this category, for example, earthquake, wind loading, and turbulence, etc. So, new techniques will have to be devised to analyze the response of structures to such excitations. Fortunately, much of random vibration has a certain statistical regularity and therefore statistics can be used effectively for such a study.

The main objective of introducing this chapter in the study of vibro-acoustics is to give a general systematic approach to the study of arbitrary signals and to show that they do have a statistical regularity that easily falls into place due to frequency analysis. In addition, the signals that we analyze in this chapter are not very different from those we normally encounter in our day-to-day activity. Broadband excitation is one such excitation that is discussed in this chapter. The concept of broadband excitation introduced in this chapter is perhaps the most important takeaway from this chapter and is the driving force of many vibro-acoustic systems. Furthermore, the correlation between two or more signals arising out of vibration and sound from a system is a very important aspect of identification of source excitation that will be introduced in this chapter; such a correlation is also very important for the functioning of our hearing mechanism in identifying the direction of sound.

The organization of this chapter is as follows. Initially, a deterministic time function is studied to compute parameters like mean, mean square value, variance, and autocorrelation. Then, the Fourier transform of a deterministic signal is presented, which later plays a major role in determining the response of structures subjected to random vibration. Then we begin with the study of random time-independent events by discussing the properties of discrete (tossing a coin or dice) and continuous random variables (diameter of a batch of machined shafts) like probability distribution and probability density functions are studied next, which are of course not dependent on time. Some important statistical distributions that represent probability density functions are also introduced. The expectation operation is introduced, which enables computation of mean and mean square value through the probability density function. The concept of covariance and correlation coefficients are also introduced. Time-independent random events are much easier to study and these concepts can be later extended for studying time-dependent random events.

A time-dependent random variable is then studied by introducing the concept of ensemble averages. Stationary random vibration, which has statistical regularity, is introduced with the help of these ensemble averages. Since ensemble averages of stationary signals are independent of the instant of time at which they are measured, their study once again closely resembles the study of time-independent random variables. Autocorrelation and cross correlation are defined for stationary signals and their frequency counterparts; they are respectively related to spectral density and cross spectral density through Fourier transform. By using spectral density, the response of structures of known system transfer function can be analyzed. This would help determine the mean square response of structures to stationary excitation of known spectral density, the most popular one being white noise excitation.

Study of random vibration is also closely related to the measurement and analysis of vibration and acoustic signals. Computing the auto spectrum, cross spectrum, autocorrelation, cross correlation, system transfer function, coherence, and sound intensity requires basic knowledge of random vibration. In addition, testing of structures for random excitation is also very important. The concept of level crossing helps us determine the exact number of cycles a structure can function without failure. Thus this chapter provides the basis for introducing many concepts that are useful for many applications in addition to vibro-acoustics.

4.1 Time Averages of a Function

Consider a single function $x(t)$ of period T (in theory T could be very large or could even tend to infinity). However, it is possible to define an average with respect to time. The word *temporal* is many times used to emphasize that the averaging carried out in time domain.

4.1.1 Mean

The temporal mean of $x(t)$ is given by

$$\left\langle x(t) \right\rangle \equiv \mu_x = \lim_{T \to \infty} \frac{1}{T} \int_{-T/2}^{T/2} x(t)\, dt \tag{4.1}$$

A new function called *autocorrelation* can be defined for a time-dependent function. Physically, it is a measure of how a function correlates with itself after a certain time shift, τ.

4.1.2 Autocorrelation

The temporal autocorrelation function is defined as

$$= R_x(\tau) \equiv \lim_{T \to \infty} \frac{1}{T} \int_{-T/2}^{T/2} x(t)\, x(t + \tau)\, dt \tag{4.2}$$

which is also sometimes represented by $\Phi(\tau)$.

4.1.3 Mean Square Values

The mean square value of a time-dependent function provides a measure of energy associated with the motion described by that variable and it is defined by

$$\Psi_x^2 = \lim_{T \to \infty} \frac{1}{T} \int_{-T/2}^{T/2} x^2(t)\, dt \tag{4.3}$$

The positive square root of this value is known as the root mean square or the rms value.

The mean value μ_x is constant and is the static component of $x(t)$, and $x(t) - \mu_x$ as the dynamic component. In many applications, the mean square value of the dynamic component is important. This quantity is the mean square value about the mean and is known as variance. The expression for variance is given by

$$\sigma_x^2 = \lim_{T \to \infty} \frac{1}{T} \int_{-T/2}^{T/2} [x(t) - \mu_x]^2 \, dt \tag{4.4}$$

The positive square root of variance is known as the standard deviation.

Expanding Equation 4.4, we obtain

$$\sigma_x^2 = \lim_{T \to \infty} \frac{1}{T} \int_{-T/2}^{T/2} x^2(t) \, dt - 2\mu_x \lim_{T \to \infty} \frac{1}{T} \int_{-T/2}^{T/2} x(t) \, dt + \mu_x^2 \tag{4.5}$$

From Equations 4.1, 4.3, and 4.5

$$\sigma_x^2 = \Psi_x^2 - \mu_x^2 \tag{4.6}$$

Therefore, variance is equal to the mean square value minus the square of the mean value.

Let us now look at some periodic signals to compute the preceding basic parameters.

Example 4.1

Consider the square pulse as shown in Figure 4.1. Determine the mean, mean square value, and variance of $x(t)$ using the time waveform; and obtain the autocorrelation function of $x(t)$.

From Equation 4.1, the mean value of the given time waveform is given by

$$\mu_x = \lim_{T \to \infty} \frac{1}{T} \int_{-T/2}^{T/2} x(t) \, dt = \frac{1}{T} \int_{-T/4}^{T/4} dt = \frac{1}{T} \left[\frac{T}{4} - \left(-\frac{T}{4} \right) \right] = \frac{1}{2} \tag{1}$$

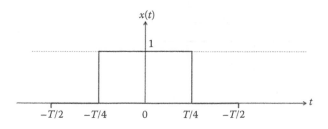

FIGURE 4.1
Square pulse of Example 4.1.

From Equation (4.3), the mean square value is given by

$$\Psi_x^2 = \lim_{T \to \infty} \frac{1}{T} \int_{-T/2}^{T/2} x^2(t)dt = \frac{1}{T} \int_{-T/4}^{T/4} dt = \frac{1}{2} \tag{2}$$

From Equation 4.6, the variance is given by

$$\sigma_x^2 = \Psi_x^2 - \mu_x^2 = \frac{1}{2} - \left(\frac{1}{2}\right)^2 = \frac{1}{4} \tag{3}$$

To determine the autocorrelation function of a time waveform, it is necessary to consider two such similar overlapping waveforms and then apply Equation 4.2. Figure 4.2 shows two such square pulses similar to the one shown in Figure 4.1 that are separated by time τ, which is bound by $0 < \tau < T/2$, and it is used to compute the autocorrelation of the given pulse in this interval.

From Equation 4.2, the autocorrelation function in the interval $0 < \tau < T/2$ is given by

$$R_x(\tau) \equiv \lim_{T \to \infty} \frac{1}{T} \int_{-T/2}^{T/2} x(t)x(t+\tau)dt = \frac{1}{T} \int_{-T/4}^{T/4-\tau} dt$$

$$, 0 < \tau < T/2 \tag{4}$$

$$= \frac{1}{2} - \frac{\tau}{T}$$

The autocorrelation for the period $T/2 < \tau < T$ can be obtained by either replacing τ with $T - \tau$ in Equation 4 or by drawing a figure similar to Figure 4.2 that represents two overlapping signals within $T/2 < \tau < T$; the easier option is the first one from which the autocorrelation for this time shift is given by

$$R_x(\tau) = -\frac{1}{2} + \frac{\tau}{T} , T/2 < \tau < T \tag{5}$$

The autocorrelation function for the negative time shift, since it is a symmetric function, can be obtained by replacing τ with $-\tau$ in Equations 4 and 5 due to its symmetric properties. The complete autocorrelation function is shown in Figure 4.3.

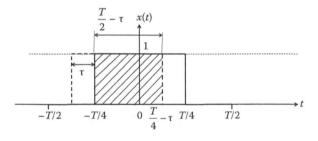

FIGURE 4.2
Overlapping of two signals of Figure 4.1 separated by τ.

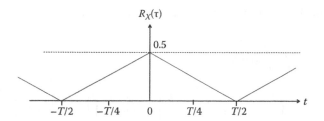

FIGURE 4.3
Autocorrelation function of the waveform shown in Figure 4.1.

The square pulse of Example 4.2 has constant values within each period. Thus, integration was easy and the autocorrelation function was a linear function. Next we shall look at a triangular wave that has linear variation with respect to time.

Example 4.2

Determine the mean, mean square value, and variance of the sawtooth wave shown in Figure 4.4 using its time waveform.

The time-domain function of the sawtooth wave shown in Figure 4.4 can be expressed as

$$
\begin{aligned}
x(t) &= A\left(\frac{4t}{T} + 2\right) && -\frac{T}{2} \le t \le -\frac{T}{4} \\
&= \frac{4A}{T}t && -\frac{T}{4} \le t \le \frac{T}{4} \\
&= A\left(-\frac{4t}{T} + 2\right) && \frac{T}{4} \le t \le \frac{T}{2}
\end{aligned} \tag{1}
$$

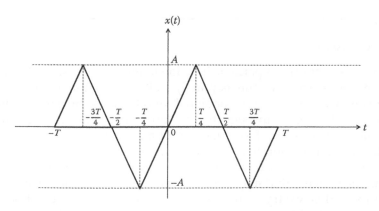

FIGURE 4.4
Sawtooth wave of Example 4.2.

From Equation 4.1, the mean value of the given time waveform is given by

$$\mu_x = \lim_{T \to \infty} \frac{1}{T} \int_{-T/2}^{T/2} x(t)\,dt$$

$$= \frac{1}{T}\left\{ \int_{-T/2}^{-T/4}\left[-A\left(\frac{4t}{T}+2\right)\right]dt + \int_{-T/4}^{T/4}\left[\frac{4A}{T}t\right]dt + \int_{T/4}^{T/2}\left[A\left(-\frac{4t}{T}+2\right)\right]dt \right\} \tag{2}$$

$$= 0$$

From Equation 4.3, the mean square value is given by

$$\Psi_x^2 = \lim_{T \to \infty} \frac{1}{T} \int_{-T/2}^{T/2} x^2(t)\,dt$$

$$= \frac{1}{T}\left\{ \int_{-T/2}^{-T/4}\left[-A\left(\frac{4t}{T}+2\right)\right]^2 dt + \int_{-T/4}^{T/4}\left[\frac{4A}{T}t\right]^2 dt + \int_{T/4}^{T/2}\left[A\left(-\frac{4t}{T}+2\right)\right]^2 dt \right\} \tag{3}$$

$$= \frac{A^2}{3}$$

From Equation 4.6, the variance is given by

$$\sigma_x^2 = \Psi_x^2 - \mu_x^2 = \frac{A^2}{3} \tag{4}$$

The autocorrelation of the given time waveform is left to Problem 4.2.

4.2 Fourier Analysis

Fourier series representation of a periodic function in terms of the sum of a number of frequencies is an important step in frequency analysis. It is well known that frequency analysis is an easier way of analyzing dynamic signals and hence is the natural choice of many sensor organs that developed through the evolutionary process. Therefore, frequency analysis of signals is also the preferred method of analyzing dynamic signals measured through transducers. Since Fourier series representation is restricted only to periodic signals, it cannot be, therefore, used for a general class of signals; there are many signals arising in dynamic systems that may not be periodic, although they may have statistical regularity. In this section, we try to modify the Fourier series representation to obtain the Fourier transform pair that form the basis for analyzing a general class of signals; in addition this Fourier transform pair also provides for moving from time to frequency domain and vice versa, which is very advantageous in signal processing. The Fourier integrals can be further extended to define the concept of spectral density and autocorrelation that are useful in obtaining statistical regularity of a general class of signals and in identifying the source of vibration and sound. Hence, Fourier analysis is useful in the study of vibro-acoustics.

4.2.1 Need for Frequency Analysis

The output signals from transducers, used in both vibration and sound measurement, are analog electrical signals. They can be either periodic or nonperiodic. This output signal can be seen on a cathode ray oscilloscope (CRO). However, such a display of the time variation of the signal will not give useful information in most cases. The signal that is displayed on the CRO will be a combination of signals of several frequencies. This combined signal that is displayed on the CRO does not reveal the frequency composition of the signal. Therefore, special efforts are required to separate the individual frequencies that comprise the combined signal. Once the individual frequencies are known, then it will be easier to relate them to their source of origin, depending upon the characteristics of the machine, such as rotational speed or number of teeth in the gearbox. Therefore, determining the frequency content of a vibration or sound signal is a very important step in the measurement process. The Fourier transform pair helps in determining the frequency content and is therefore discussed in detail in this section. In addition to the practical aspect of the importance of frequency-domain information, frequency-domain functions are very useful in analyzing random signals. Therefore, the Fourier transform plays a very important role in studying vibration and sound.

First, we shall discuss the importance of frequency analysis through some simple examples. Figure 4.5 shows the time-domain representation of a sinusoidal signal of dynamic pressure; only one wavelength of the signal is shown. Such signals, known as pure tones, may arise out of, for example, the vibration of a tuning fork or vibration of a transformer. A is the peak amplitude of the signal and the period of the signal is given by T. For a simple signal like this, it is very easy to compute its frequency, which is the reciprocal of the time period T, that is $f = 1/T$.

Figure 4.6 shows the frequency-domain representation of a sinusoidal signal of Figure 4.5. The peak amplitude of the time signal gives the height of the signal and it has a frequency f, which is a reciprocal of the time period.

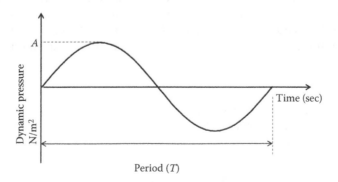

FIGURE 4.5
Time-domain representation of a sinusoid.

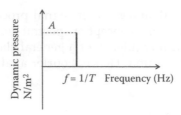

FIGURE 4.6
Frequency-domain representation of a sinusoid.

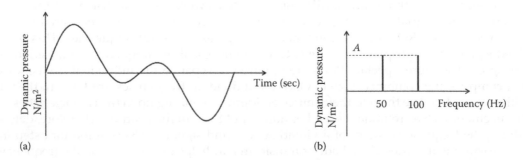

FIGURE 4.7
Addition of two signals.

If all signals that are normally encountered in vibration and noise studies are sinusoidal signals, as shown in Figure 4.5, then there will not be any need for frequency analysis. However, in actual practice, most signals that are available from vibration and noise measurements will be more complex our example, with a number of frequencies and each frequency component will have a different amplitude. Therefore, sophisticated instrumentation is necessary to determine the frequencies and the respective amplitudes that are contained in a signal.

Figure 4.7a shows that even adding two sinusoids of two frequencies 50 and 100 Hz can result in a complex signal that cannot be easily analyzed for its frequencies. However, the frequency-domain representation of Figure 4.7b presents a clear picture about the frequency content of the combined signal.

By now the reason for frequency analysis of a signal must be very clear. Although we have stated the frequency-domain representation of a signal, we have not yet explained the manner in which that is obtained. First, let us try to obtain the Fourier transform pair that is the basis for obtaining frequency-domain information of a time signal.

4.2.2 Fourier Transform

Since the Fourier series expansion of a periodic signal is well known, we shall first start with a periodic function and expand it in the Fourier series. If a function $f(t)$ is a periodic function of period T, it can be represented by a Fourier series as

$$f(t) = a_o + \sum_{k=1}^{\infty} \left(a_k \cos \frac{2\pi kt}{T} + b_k \sin \frac{2\pi kt}{T} \right) \tag{4.7}$$

where the constants a_k and b_k are defined as

$$a_o = \frac{1}{T} \int\limits_{-T/2}^{T/2} f(t)\,dt$$

$$a_k = \frac{2}{T} \int\limits_{-T/2}^{T/2} f(t)\cos\frac{2\pi kt}{T}\,dt \qquad (4.8)$$

$$b_k = \frac{2}{T} \int\limits_{-T/2}^{T/2} f(t)\sin\frac{2\pi kt}{T}\,dt$$

Since we are generally interested in the dynamic component of signals, $a_o = 0$.

The values of a_k and b_k can be computed from Equation 4.8. They are represented for various frequencies as shown in Figure 4.8 and the spacing between these frequency values is given by

$$\Delta\omega = \frac{2\pi}{T} \qquad (4.9)$$

Using Equation 4.8 in Equation 4.7, the periodic function $f(t)$ becomes

$$f(t) = \sum_{k=1}^{\infty}\left\{\frac{2}{T}\int\limits_{-T/2}^{T/2} f(t)\cos\frac{2\pi kt}{T}\,dt\right\}\cos\frac{2\pi kt}{T}$$

$$+ \sum_{k=1}^{\infty}\left\{\frac{2}{T}\int\limits_{-T/2}^{T/2} f(t)\sin\frac{2\pi kt}{T}\,dt\right\}\sin\frac{2\pi kt}{T} \qquad (4.10)$$

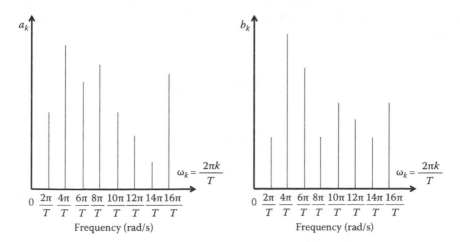

FIGURE 4.8
The value of coefficients a_k and b_k of Equation 4.8.

Using the frequency spacing of Equation 4.9, and replacing $\dfrac{2\pi k}{T}$ by ω_k, Equation 4.10 becomes

$$f(t) = \sum_{k=1}^{\infty} \left\{ \frac{\Delta\omega}{\pi} \int_{-T/2}^{T/2} f(t)\cos\omega_k t \, dt \right\} \cos\omega_k t$$
$$+ \sum_{k=1}^{\infty} \left\{ \frac{\Delta\omega}{\pi} \int_{-T/2}^{T/2} f(t)\sin\omega_k t \, dt \right\} \sin\omega_k t \tag{4.11}$$

When the time period T becomes extremely large, the frequency spacing becomes smaller and smaller, $\Delta\omega \to d\omega$, and the summation in Equation 4.11 becomes an integral as follows:

$$f(t) = \int_{\omega=0}^{\infty} \frac{d\omega}{\pi} \left\{ \int_{-\infty}^{\infty} f(t)\cos\omega t \, dt \right\} \cos\omega t$$
$$+ \int_{\omega=0}^{\infty} \frac{d\omega}{\pi} \left\{ \int_{-\infty}^{\infty} f(t)\sin\omega t \, dt \right\} \sin\omega t \tag{4.12}$$

The concept of a time-dependent function with a very large period helps in extending the Fourier series concept, which is only applicable to periodic functions, and to nonperiodic functions also by considering the entire signal as a single periodic function of a large period. A single time history can, therefore, be considered as a single periodic function that does not repeat itself.

Equation 4.12 can be further simplified to obtain a pair of Fourier integrals that would help convert the time-domain function into the frequency domain and vice versa. This is carried out as follows. Based on Equation 4.12 let us define the following frequency-dependent functions:

$$A(\omega) = \frac{1}{2\pi} \int_{-\infty}^{\infty} f(t)\cos\omega t \, dt$$
$$B(\omega) = \frac{1}{2\pi} \int_{-\infty}^{\infty} f(t)\sin\omega t \, dt \tag{4.13}$$

By substituting the functions of Equation 4.13 in Equation 4.12, the function $f(t)$ becomes

$$f(t) = 2\int_{0}^{\infty} A(\omega)\cos\omega t \, d\omega + 2\int_{0}^{\infty} B(\omega)\sin\omega t \, d\omega \tag{4.14}$$

Using Equation 4.13, the following function can be defined

$$F(\omega) = A(\omega) - jB(\omega) \tag{4.15}$$

By using the values of $A(\omega)$ and $B(\omega)$ in Equation 4.13, Equation 4.15 becomes

$$
\begin{aligned}
F(\omega) &= \frac{1}{2\pi} \int_{-\infty}^{\infty} f(t) \cos \omega t \, dt - \frac{j}{2\pi} \int_{-\infty}^{\infty} f(t) \sin \omega t \, dt \\
&= \frac{1}{2\pi} \int_{-\infty}^{\infty} f(t)[\cos \omega t - j \sin \omega t] \, dt
\end{aligned}
\tag{4.16}
$$

By using the exponential function, the function within the parenthesis of the integral of Equation 4.16 becomes

$$F(\omega) = \frac{1}{2\pi} \int_{-\infty}^{\infty} f(t) e^{-j\omega t} \, dt \tag{4.17}$$

Equation 4.17 is known as the Fourier integral, which transforms a time-domain signal into a frequency-domain signal.

Both integrals of Equation 4.14 are even functions. Therefore, the limits of these integrals can be changed as $-\infty$ to ∞ as follows

$$f(t) = \int_{-\infty}^{\infty} A(\omega) \cos \omega t \, d\omega + \int_{-\infty}^{\infty} B(\omega) \sin \omega t \, d\omega \tag{4.18}$$

Since $j \int_{-\infty}^{\infty} A(\omega) \sin \omega t \, d\omega$ and $j \int_{-\infty}^{\infty} B(\omega) \cos \omega t \, d\omega$ are zero due to the odd functions of the respective integrands, they can be added to the Equation 4.18 as follows:

$$
\left.
\begin{aligned}
f(t) &= \int_{-\infty}^{\infty} A(\omega) \cos \omega t \, d\omega + j \int_{-\infty}^{\infty} A(\omega) \sin \omega t \, d\omega \\
&\quad + \int_{-\infty}^{\infty} B(\omega) \sin \omega t \, d\omega - j \int_{-\infty}^{\infty} B(\omega) \cos \omega t \, d\omega
\end{aligned}
\right\}
\tag{4.19}
$$

By regrouping terms, Equation 4.19 becomes

$$f(t) = \int_{-\infty}^{\infty} [A(\omega) - jB(\omega)][\cos \omega t + j \sin \omega t] \, d\omega \tag{4.20}$$

By using Equation 4.15 and the exponential function, Equation 4.20 becomes

$$f(t) = \int_{-\infty}^{\infty} F(\omega)e^{j\omega t}\, d\omega \tag{4.21}$$

Equation 4.21 is the inverse Fourier transform of $F(\omega)$. Equations 4.17 and 4.21 form the Fourier transform pair that is useful in converting time-domain signals into frequency domain and vice versa.

Example 4.3

Determine the Fourier transform of the function $f(t) = ae^{-bt^2}$.
From Equation 4.17, the frequency-domain function of $f(t)$ is given by

$$F(\omega) = \frac{1}{2\pi} \int_{-\infty}^{\infty} ae^{-bt^2} e^{-j\omega t}\, dt = \frac{a}{\pi} \int_{0}^{\infty} e^{-(bt^2 + j\omega t)}\, dt \tag{1}$$

$bt^2 + j\omega t$ can be written as follows:

$$bt^2 + j\omega t = \left(\sqrt{b}t + \frac{j\omega}{2\sqrt{b}}\right)^2 + \frac{\omega^2}{4b} \tag{2}$$

Substituting Equation 2 in 1

$$F(\omega) = \frac{ae^{-\frac{\omega^2}{4b}}}{\pi} \int_{0}^{\infty} e^{-\left(\sqrt{b}t + \frac{j\omega}{2\sqrt{b}}\right)^2}\, dt \tag{3}$$

Defining the following variable

$$z = \sqrt{b}t + \frac{j\omega}{2\sqrt{b}} \tag{4}$$

Then the differential of Equation 4 becomes

$$dz = \sqrt{b}dt \tag{5}$$

From Equations 3 through 5

$$F(\omega) = \frac{ae^{-\frac{\omega^2}{4b}}}{\pi\sqrt{b}} \int_{0}^{\infty} e^{-z^2}\, dz \tag{6}$$

The integral portion of Equation 6 is an error function of value $\dfrac{\sqrt{\pi}}{2}$ and from this the Fourier transform becomes

$$F(\omega) = \frac{ae^{-\frac{\omega^2}{4b}}}{2\sqrt{\pi b}} \tag{7}$$

which is the frequency-domain Fourier transform of the given equation. Incidentally, both the time- and frequency-domain functions of this example decay exponentially with the square of the respective time or frequency variable that is not the general case.

4.2.3 Spectrum Analyzers

The task of separating different frequencies of a signal obtained from a transducer has been made easy with the advent of electronic instrumentation in the form of spectrum analyzers that have incorporated advances in digital signal processing. The signals obtained from the transducers are in a continuous form known as analog signals. By using digital electronic circuits, this analog signal is first converted into a digital signal, which can be easily processed with the digital instrumentation. This section covers details of signal processing and instrumentation related to spectrum analyzers. From the data resulting from the frequency analysis, a number of parameters can be evaluated, which are very useful in vibration and noise analysis. Since Fourier transform forms the basis of operation of spectrum analyzers, they are discussed here.

Apart from converting the time-domain information obtained from transducers to frequency domain, the spectrum analyzers can perform many signal processing actions that can throw light on finer aspects on the signals from various viewpoints. This would enhance identification of important characteristics of the signal that will be useful in identifying the source of noise and vibration.

Now we will discuss a specific spectrum analyzer, fast Fourier transform (FFT) analyzers. They are the most important instrumentation for the analyses of vibration and noise signals. They derive the name FFT from the name of the algorithm, which is used for converting time-domain information into frequency domain.

Since Fourier transform can be applied on a single, long, periodic signal, a suitable record length has to be decided depending on the number of data points that can be stored in the RAM (random-access memory). This is known as record length and the sampling frequency used for digital conversion decides the range of frequencies this data can provide. The number of data points in a record length is known as the block of data.

4.2.3.1 Fast Fourier Transform Algorithm

The basis for obtaining frequency-domain information from time-domain data is the FFT algorithm and it is derived from the Fourier integral that is expressed in the discrete form.

The Fourier transform integral pair available from Equations 4.17 and 4.21 is rewritten as follows:

$$F(\omega) = \frac{1}{2\pi} \int_{-\infty}^{\infty} f(t)e^{-j\omega t}\, dt \tag{4.22}$$

and

$$f(t) = \int_{-\infty}^{\infty} F(\omega)e^{j\omega t}\, dt \tag{4.23}$$

Equations 4.22 and 4.23 give the relation between the continuous time-domain signal and the corresponding frequency-domain signal. Since the microprocessor can handle discrete data, Equations 4.22 and 4.23 are modified as follows to handle such data:

$$F(k) = \frac{1}{N} \sum_{n=0}^{N-1} f(n)e^{-j2\pi nk/N} \tag{4.24}$$

and

$$f(n) = \sum_{k=0}^{N-1} F(k)e^{j2\pi nk/N} \tag{4.25}$$

Equations 4.24 and 4.25 form the discrete Fourier transform, or DFT, pair. $f(n)$ represents the discrete N samples of a time function $f(t)$ at N different instants evenly spaced from zero to the record length time T, and $F(k)$ represents the discrete spectrum at N different frequencies equally distributed from zero to Fs, the sampling frequency. The transformation from time domain to frequency domain of the discrete data can be directly carried out using Equation 4.24. The direct computation takes about N^2 complex multiplications. Since the fast Fourier algorithm can carry out the same operation with $N \log_2 N$ complex computations, it results in considerable savings of computation time. The savings of computation is obtained by noting that since the product of indices $n \times k$ in Equations 4.24 and 4.25 is same as $k \times n$, it need not be computed again.

The basic operation of an FFT analyzer can be explained as follows. Consider the analog signal of Figure 4.9. This may be a typical vibration or noise signal. This signal, when fed into an FFT analyzer, is first sampled to convert into a discrete or digital signal. Although there is a small difference between a digital signal and discrete signal, we will use both of them without distinguishing between them. For the current discussion, it is sufficient to know that a discrete or digital signal is defined only at discrete intervals of time.

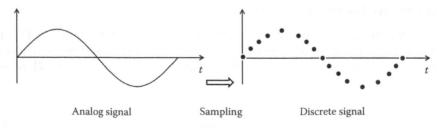

Analog signal   Sampling   Discrete signal

FIGURE 4.9
Sampling of an analog signal.

Now the signal, which is in the form of discrete numbers, has to be suitably manipulated to obtain the frequency-domain information. From Equations 4.24 and 4.25, it is clear that the FFT algorithm is applied on a block of data and therefore the number of points in each block must be first decided. Since the signal from transducers is continuously transmitted, the corresponding digitized signal is continuously available. During the early stages of development of the FFT analyzer, the RAM of the processors was about 1024 bytes. Since each byte can store one sample of the discrete signal, 1024 points of the discrete signal was treated as a block and transferred to the RAM and further processed. At the present time the RAM size available with the processors is much higher and therefore instrumentation with a higher block size is now available.

When the FFT algorithm is applied to a block of discrete data (1024 points) in the time domain, we get a block of discrete data (1024 points) in the frequency domain from Equation 4.24. Due to the nature of the algorithm and other properties of the sampling process, the frequency information so obtained is symmetric. Therefore, there are only 512 points of useful information. Even among the 512 points, only the first 400 points are used to prevent some errors in the sampling process, which get accrued in the higher frequency end of the frequency-domain signal. These 400 points of information can now be displayed on the analyzer. Therefore, earlier analyzers were called 400 line analyzers. It basically means that they are displaying frequency-domain information containing 400 points. The frequency spacing of these points can be controlled in the sampling process, which will be discussed later; this will also help us control the upper limit of frequency. If we keep the bandwidth very large, the spacing of the points displaying frequency-domain information also becomes very large. Therefore, it is possible that we might miss some important information that may be available between the points. Hence, the frequency range of interest must be carefully selected.

In practice, as explained earlier, a commercial FFT frequency analyzer usually takes 1024 digital samples of a time function into a memory as a data block with a sampling frequency, *Fs*. It then carries out the FFT computation using these 1024 samples and obtains a frequency spectrum at 512 different frequencies evenly distributed from 0 to *Fs*/2 for display on a screen. The DFT transform works on finite time records of length *T*. The time signal is sampled at $n\Delta t$, where Δt is the sampling time; the reciprocal of sampling time is the sampling frequency. The sampling frequency refers to the rate at which the incoming analog signal is checked and the corresponding value recorded. The Nyquist sampling theorem states that this sampling frequency must be at least twice the frequency of the analog signal, in order that the digital signal retains complete information about the original signal. The DFT results in the Fourier spectrum of finite time record sampled at $k\Delta f$; $\Delta f = 1/T$ (resolution). The analyzer will then use the next block of 1024 samples to generate a new spectrum.

From Table 4.1 it can be seen that the sampling frequency is about 2.5 times the highest frequency of the range. It takes longer time to sample a signal in the frequency range 0 to 10 Hz than in the frequency range 0 to 1000 Hz. Time taken is given by the product of time period of the highest frequency and the number of FFT lines. Since only 400 lines of frequency points are used, the maximum frequency will therefore be less than *Fs*/2, as seen in Table 4.1. The 400-line analyzers were popular in the early 1980s. But now, since the RAM size can be very large, FFT analyzers having 6000 lines are available.

In practice, however, several block lengths will have to be considered by the spectrum analyzer. Therefore, the resulting spectrum of several blocks must be averaged to eliminate signal noise and other inaccuracies that creep in spectral analysis, because the actual signal from a machine cannot exactly fit inside a block. In addition, continuous signals are

TABLE 4.1

Characteristics of a 400-Line FFT Analyzer

Range (Hz)	Sampling Frequency (Hz)	Frequency Spacing, f (Hz)	Time Taken (sec)
0–10	25	0.025	40
0–20	51	0.05	20
0–50	128	0.125	8
0–100	256	0.25	4
0–200	512	0.5	2
0–500	1280	1.25	0.8
0–1000	2560	2.5	0.4

arbitrarily chopped off at the beginning and end of blocks. The error arising out of this can be reduced by using window functions that smoothly reduce the signal to zero at the beginning and end of the blocks. One can also synchronize the start of measurement with respect to a specific time reference from the machine. For example, the start of measurement can be synchronized with the instant at which a gearbox tooth passes a reference point. Such a measurement will help trace defects in a particular tooth. This is known as synchronous averaging and is used in machinery diagnostics.

Although one might choose a sampling frequency depending on the frequencies of interest, the original signal might contain frequencies higher than the upper limit of frequency based on sampling frequency. This results in aliasing that represents a higher frequency signal as a lower frequency signal, which can be avoided by prefiltering the input signal to the spectral analyzers to block frequencies higher than those of interest.

4.3 Statistics, Probability, and Probability Density Function

The following definitions are necessary before discussing the theory of random variables:

Probability—Classical definition of probability is the ratio of the favorable outcomes to the total number of outcomes.

Trial—A trial is defined as a single real or conceptual experiment, whose possible outcomes are assumed to be known.

Sample space (Ω)—A sample space is a set containing all possible outcomes of a trial.

Sample point (ω)—Each element of the sample space is known as the sample point.

If a sample space contains only one sample point, then there is only one possible outcome and the trial is therefore deterministic. Table 4.2 shows some examples of discrete space time-independent random variables.

4.3.1 Theory of Random Variable

A *real random variable* $X(\omega)$, $\omega \in \Omega$ is a set function defined on Ω such that for every real number x, there exists the probability $P(X \leq x)$, which defines the probability of the random variable X being less than x, where x represents the range of values that the random

TABLE 4.2

Examples of Finite Sample Space

Trial	Sample Space
Coin toss	Heads, tails
Dice throw	1, 2, 3, 4, 5, 6

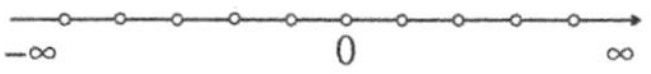

FIGURE 4.10
Discrete sample space.

$$-9 \quad -7 \qquad 3 \qquad 8$$

FIGURE 4.11
Continuous sample space.

variable X can possibly take. Note that the uppercase letter always represents the random variable and the lower case letter denotes the corresponding values.

The sample space can be along a real line R from $-\infty$ to ∞. A discrete random time-independent variable will take a sequence of discrete values on R (Figure 4.10 and Table 4.2), where as a continuous random variable can take any value on one or several intervals on R (Figure 4.11). For example, the diameter of a workpiece manufactured from a machine is an example of a continuous random variable that is also time-independent.

An *event* is a subset of a sample space to which a consistent probability can be assigned. For example, heads or tails can be considered as events of the coin toss experiment.

4.3.2 Relative Frequency

Relative frequency is the ratio of the number of times the event occurs (in an experiment) to the total number of times the experiment was performed, which eventually defines statistical regularity.

Consider a trial with discrete sample space Ω: $(x_1, x_2, \ldots, x_n)$. Let N be the total number of trials during which the sample points $x_1, x_2, \ldots, x_i, \ldots, x_n$ are observed $N_1, N_2, \ldots, N_i, \ldots, N_n$ times, respectively (i.e., x_1 occurs N_1 times, x_2 occurs N_2 times, x_i occurs N_i times, and so on). Then clearly

$$\sum_{i=1}^{n} N_i = N \tag{4.26}$$

The relative frequency of occurrence of x_i is defined by

$$r_N(x_i) = N_i/N \tag{4.27}$$

As an example, relative frequency that is dependent on the number of trials for a coin toss is as shown in Figure 4.12. One can observe that as the number of trials, N, becomes very large, the relative frequency has converged to 0.5.

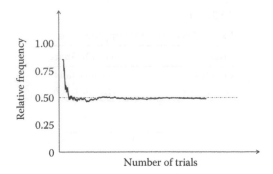

FIGURE 4.12
Relative frequency versus number of trials of a coin toss experiment.

It is clear that the value of relative frequency is bounded between zero and one:

$$0 \leq r_N(x_i) \leq 1 \tag{4.28}$$

The sum of relative frequencies of all the events will be equal to one:

$$\sum_{i=1}^{n} r_N(x_i) = \frac{1}{N} \sum_{i=1}^{n} N_i = 1 \tag{4.29}$$

A mathematical statement of the limiting process in which the relative frequency defines probability of the event is given by

$$\lim_{N \to \infty} P\left(\left| r_N(x_i) - p \right| > \varepsilon \right) = 0 \tag{4.30}$$

where P() denotes the probability of the event contained in the parenthesis, $p = P(x_i)$ is the probability of occurrence of x_i, and ε is an arbitrary small number.

Equation 4.30 is referred to as Bernoulli's law of large numbers and provides the basic definition of probability. Incidentally, this concept can be applied to dynamic systems of vibro-acoustics (see Chapter 10) that require averaging of information over a large number of modes and we find that with larger numbers of modes, the averaged information gets better. Similarly, the concept of block averaging of FFT data reduces random noise in the measured data.

4.3.3 Probability Distribution Function

For a complete description of a random variable, it is necessary to specify its distribution. This helps in classifying various types of random variables. Consider the event $(X \leq x)$, which is the probability of a random variable being less than a specified value of x. This event defines the probability distribution function $F(x) = P(X \leq x)$, $x \in R$.

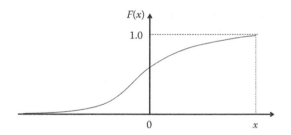

FIGURE 4.13
Probability distribution function $F(x)$.

A typical probability distribution function is shown in Figure 4.13. First, the extreme values can be easily recognized. The probability of a random variable being less than a very large positive value of x is one and the probability of the same random variable being less than a very large negative of value of x is zero. This can be expressed as

$$F(\infty) = P(X \le \infty) = 1$$
$$F(-\infty) = P(X \le -\infty) = 0$$

(4.31)

The variation of $F(x)$ between the extreme values of one and zero will depend on the nature of random variable; therefore, this variation has to be determined for each type of random variable. However, $F(x)$ is monotonically increasing with $F(-\infty) = 0$ and $F(\infty) = 1$.

4.3.4 Probability Density Function

The probability structure can also be defined by the derivative of the probability distribution function, called the *probability density function*, $p(x)$.

$$p(x) = \frac{dF(x)}{dx} = \lim_{dx \to 0} \frac{F(x+dx) - F(x)}{dx}$$

(4.32)

From the definition of $F(x)$,

$$F(x+dx) - F(x) = P(x < X \le x + dx)$$

(4.33)

From Equations 4.32 and 4.33

$$p(x)dx = P(x < X \le x + dx)$$

(4.34)

Therefore, $p(x)dx$ represents the probability of random variable X lying in the interval $(x, x + dx)$. Then, from Equation 4.32

$$F(x) = \int_{-\infty}^{x} p(\xi)\,d\xi \tag{4.35}$$

For any interval R_1: (x_1, x_2)

$$P(x_1 < X \le x_2) = \int_{x_1}^{x_2} p(\xi)\,d\xi \tag{4.36}$$

Since $F(\infty) = 1$,

$$P(-\infty < X \le \infty) = \int_{-\infty}^{\infty} p(\xi)\,d\xi = 1 \tag{4.37}$$

Probability density functions of many random processes converge to some of the well-known distributions, whose properties are well established, including normal (Gaussian) distribution, binomial distribution, and Poisson distribution. In addition, probability distribution functions can be constructed based on the measured values of a signal; many spectrum analyzers have this capability.

4.3.5 Probability Description of Several Random Variables

It is very common in many practical applications and naturally occurring events to have several random variables that are dependent on each other. The joint probability distribution function between these variables helps in establishing a definite relationship between them. Therefore, in many applications, it is necessary to know the joint behavior of two or more random variables. This is described by the *joint probability functions*.

Let us consider the joint behavior of two random variables X_1 and $X_2 \in R$ (see Figure 4.14). Let X_1 and X_2 be defined along two Cartesian axes. Plane $X_1 - X_2$ is now the sample space. The joint probability distribution function of X_1 and X_2 is given by

$$F(x_1, x_2) = P\{(X_1 \le x_1) \cap (X_2 \le x_2)\} \tag{4.38}$$

The event $\{(X_1 \le x_1) \cap (X_2 \le x_2)\}$ is the shaded area in Figure 4.14. The probability distribution function satisfies the following conditions:

$$F(-\infty, x_2) = F(x_1, -\infty) = F(-\infty, -\infty) = 0 \tag{4.39}$$

$$F(\infty, \infty) = 1 \tag{4.40}$$

$$F(x_1, \infty) = F(x_1), \quad F(\infty, x_2) = F(x_2) \tag{4.41}$$

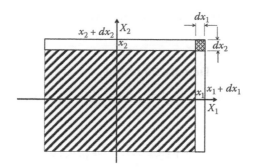

FIGURE 4.14
Joint probability distribution of two random variables X_1 and X_2.

The joint probability density function of X_1 and X_2 is defined as the mixed partial derivative of $\Gamma(x_1, x_2)$; that is,

$$p(x_1, x_2) = \frac{\partial^2}{\partial x_1 \partial x_2} F(x_1, x_2) \tag{4.42}$$

Then it follows that

$$p(x_1, x_2) dx_1 dx_2 = P\{(x_1 < X_1 \le x_1 + dx_1) \cap (x_2 < X_2 \le x_2 + dx_2)\} \tag{4.43}$$

which is the probability of both X_1 and X_2 lying simultaneously in the elemental area $dx_1 dx_2$, as shown in the figure. From Equation 4.42, we obtain

$$F(x_1, x_2) = \int\limits_{-\infty}^{x_1} \int\limits_{\infty}^{x_2} p(\xi_1, \xi_2) d\xi_1 \, d\xi_2 \tag{4.44}$$

Also, from Equations 4.39 through 4.41

$$\int\limits_{-\infty}^{\infty} \int\limits_{-\infty}^{\infty} p(x_1, x_2) dx_1 \, dx_2 = 1 \tag{4.45}$$

$$F(x_1) = \int\limits_{-\infty}^{\infty} d\xi_2 \int\limits_{-\infty}^{x_1} p(\xi_1, \xi_2) d\xi_1 \tag{4.46}$$

Therefore

$$p(x_1) = \frac{\partial}{\partial x_1} F(x_1) = \int\limits_{-\infty}^{\infty} p(\xi_1, \xi_2) d\xi_2 \tag{4.47}$$

Thus the probability density function for X_1 alone can be determined from the joint probability density function.

4.3.6 Conditional Probability Density Function

Condition probability $p(x_1/x_2)$ defines the probability of X_1 occurring subject to X_2 also occurring and is given by

$$p(x_1/x_2) = \frac{p(x_1, x_2)}{p(x_2)} \quad \text{if } p(x_2) \neq 0 \tag{4.48}$$

and $p(x_1/x_2)$ is assumed to be zero if $p(x_2) = 0$. If X_1 and X_2 are independent, then

$$\left.\begin{array}{l} p(x_1/x_2) = p(x_1) \\[2mm] \text{and } p(x_1, x_2) = p(x_1)p(x_2)(x_2) \end{array}\right\} \tag{4.49}$$

Thus if the joint probability function can be expressed as a product of several functions, each function being a function of a variable, then the random variable are uncorrelated.

The theory of joint distribution of two variables can be generalized to n variables directly. Consider n random variables X_1, X_2, X_3, ..., $X_n \in R$. The n-dimensional space formed by $x_1, x_2, ..., x_n$ comprises the sample space:

$$p(x_1, x_2, ..., x_n) = \frac{\partial^n}{\partial x_1, \partial x_2, ... \partial x_n} F(x_1, x_2, ..., x_n) \tag{4.50}$$

Similarly, other properties can be defined (see Problem 4.5).

4.4 Expected Values, Moments, and Characteristic Functions

Although many dynamic variables vary arbitrarily with time, their statistical averages can be established by computing some type of averaging. We have computed such averages before for deterministic functions based on their time waveforms. But similar averages can be obtained for random variables based on their probability density functions. In order to compute such averages, a compact operator known as the expectation operator, E, is defined and it plays a major role in analyzing dynamic signals. The type of average that will be obtained depends on the functions on which the expectation operator acts upon.

Let X be a discrete random variable with elements $x_1, x_2, ..., x_n$. Consider a function $f(X)$. Let N be the total number of trials and N_i the number of times $X = x_i$ occurs. The average value of $f(X)$ is defined as

$$\text{Average } [f(x)] = \sum_{i=1}^{n} f(x_i) \frac{N_i}{N} = \sum_{i=1}^{n} f(x_i) r_N(x_i) \tag{4.51}$$

(It is like a weighted sum of $f(x_i)$, where the weight is the relative frequency of the event.)
In the limit as $N \to \infty$, Equation 4.51 becomes

$$E[f(X)] = \lim_{N \to \infty} \text{Average}[f(X)] \tag{4.52}$$

$$E[f(X)] = \sum_{i=1}^{n} f(x_i) \lim_{N \to \infty} r_N(x_i) = \sum_{i=1}^{n} [f(x_i)] P(x_i) \tag{4.53}$$

If X is a continuous random variable, then

$$E[f(X)] = \int_{-\infty}^{\infty} [f(x)] p(x) \, dx \tag{4.54}$$

provided it is bounded. $E[f(X)]$ is called the *expected value* of $f(X)$.

We note that the expectation operator $E[]$ is a linear operator defined by

$$E[\ ] = \begin{cases} \sum_{i=1}^{n} [\ ] P(x_i) & \text{if } X \text{ is discrete} \\[2em] \int_{-\infty}^{\infty} [\ ] p(x) \, dx & \text{if } X \text{ is continous} \end{cases} \tag{4.55}$$

4.4.1 Moments

For a random variable X, let $f(X) = X^n$, where n is an integer and let

$$m_n = E[X^n] = \int_{-\infty}^{\infty} x^n p(x) \, dx \tag{4.56}$$

m_n is called the nth moment of the random variable X.

The first moment or mean is the expected value of X given by

$$m_1 \equiv \mu = E[X] = \int_{-\infty}^{\infty} x p(x) \, dx \tag{4.57}$$

The second moment or the mean square value is the expected value of X^2 given by

$$m_2 \equiv E[X^2] = \int_{-\infty}^{\infty} x^2 p(x) \, dx \tag{4.58}$$

$\sqrt{m_2}$ is the root mean square (rms) value of X.

4.4.2 Central Moments

For dynamic variables, the variation around the mean is more important than the absolute variation. Therefore, central moments are defined that lead to the computation of variance and standard deviation that are similar to those computed earlier for deterministic functions. But they are defined in terms of their mean and probability density function, and the mean itself is determined using the probability density function.

Let

$$f(X) = (X - \mu)^n \tag{4.59}$$

where n is an integer and μ is as defined by Equation 4.57.

K_n is called the nth central moment of random variable X given by

$$K_n = E[(X - \mu)^n] = \int_{-\infty}^{\infty} (x - \mu)^n p(x)\, dx \tag{4.60}$$

For $n = 2$, the second central moment, variance, is given by

$$K_2 = \sigma^2 = E[(X - \mu)^2] = \int_{-\infty}^{\infty} (x - \mu)^2 p(x)\, dx \tag{4.61}$$

Since E is a linear operator, expanding $E[(X - \mu)^2]$

$$\begin{aligned}
\sigma^2 &= E[(X - \mu)^2] = E[X^2 + \mu^2 - 2\mu X] \\
&= E[X^2] - 2\mu E[X] + \mu^2 = E[X^2] - \mu^2
\end{aligned} \tag{4.62}$$

Compare these results with Equation 4.6. They are the same. However, earlier the averages were time averages of a function, whereas the above are related to a continuous random variable of known probability density function.

Two related random variables can be found in many dynamic systems. For example, force and response of a vibratory system, sound signals from the two microphones of a sound intensity probe, the vibration signal of a structure, and the airborne sound signal produced by it are relevant examples. The correlation between these signals is an important aspect of identifying the relationship between them and to know about the dynamic system that produced them. Our sensory organs also use this important correlation property. For example, the correlation between our two ears helps in identifying the direction from which a sound is coming.

Consider two random variables X_i and X_j from a set of random variables $X_1, X_2, ..., X_n$.

Define

$$K_{ij} = E\left[(X_i - \mu_i)(X_j - \mu_j)\right] \tag{4.63}$$

where K_{ij} is known as the covariance of X_i and X_j.

In terms of the joint probability density function, the covariance can be defined as

$$K_{ij} = \int\limits_{-\infty}^{\infty} \int\limits_{-\infty}^{\infty} (x_i - \mu_i)(x_j - \mu_j) p(x_i, \mu_j) dx_i \, dx_j \tag{4.64}$$

By expanding Equation 4.64, the covariance can be obtained as

$$K_{ij} = E[X_i X_j] - \mu_i \mu_j \tag{4.65}$$

The random variables X_i and X_j are said to be uncorrelated (no relation between each other) if $K_{ij} = 0$. Then, from Equation 4.65

$$E[X_i X_j] = \mu_i \mu_j \tag{4.66}$$

If X_i and X_j are independent, then $p(x_i, x_j) = p(x_i) p(x_j)$,

$$
\begin{aligned}
E[X_i X_j] &= \int\limits_{-\infty}^{\infty} \int\limits_{-\infty}^{\infty} (x_i)(x_j) p(x_i, x_j) dx_i \, dx_j \\
&= \int\limits_{-\infty}^{\infty} (x_i) p(x_i) dx_i \int\limits_{-\infty}^{\infty} (x_j) p(x_j) dx_j = \mu_i \mu_j
\end{aligned}
\tag{4.67}
$$

which is same as Equation 4.65. Thus, if two random variables are independent, they are necessarily uncorrelated.

The normalized value of covariance, the correlation coefficient, is defined as

$$\rho_{ij} = Kij / \sigma_i \sigma_j \tag{4.68}$$

where σ_i and σ_j are, respectively, the standard deviations of the random variables X_i, and it can be shown that $|\rho ij| \le 1$.

Example 4.4

Consider the sawtooth wave of Figure 4.4 in Example 4.2.

a. Determine the probability density and probability distribution function for this waveform.
b. Determine the mean, mean square value, and variance using the probability density function.

ANSWERS

a. Figure 4.15 shows the probability distribution function of the sawtooth wave shown in Figure 4.4. It can be described by the following equation:

$$F(x) = 0, \quad x < -A$$

$$= \frac{x}{2A} + \frac{1}{2}, \quad -A < x < A \qquad (1)$$

$$= 1, \quad x > A$$

Now the probability density function can be obtained by differentiating the probability distribution function defined in Equation 1 and is given by

$$p(x) = \frac{1}{2A}, \quad -A < x < A$$

$$= 0, \quad x < -A \qquad (2)$$

$$= 0, \quad x > A$$

Figure 4.16 shows the probability density function of the waveform shown in Figure 4.4. Before we proceed further, it is important to check the basic property of a probability density function using Equation 4.37 as follows:

$$\int_{-\infty}^{\infty} p(\xi)\,d\xi = \frac{1}{2A} \int_{-A}^{A} dx = 1 \qquad (3)$$

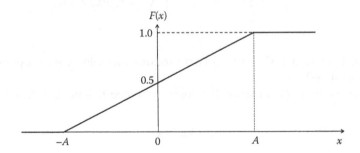

FIGURE 4.15
Probability distribution function of the sawtooth wave shown in Figure 4.4 (Example 4.4).

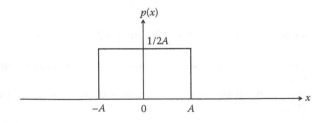

FIGURE 4.16
Probability density function of the waveform shown in Figure 4.4.

b. From Equation 4.57, the mean value of the square pulse using its probability density function is given by

$$\mu_X = E[X] = \frac{1}{2A} \int_{-A}^{A} x \, dx = 0 \tag{4}$$

From Equation 4.58, the mean square value of the sawtooth wave using its probability density function is given by

$$\psi_x^2 = E[X^2] = \frac{1}{2A} \int_{-A}^{A} x^2 \, dx = \frac{A^2}{3} \tag{5}$$

From Equation 4.61, the variance of the sawtooth wave using its probability density function is given by

$$\sigma_x^2 = E[(X - \mu)^2] = \int_{-\infty}^{\infty} (x - \mu)^2 p(x) \, dx$$

$$= \frac{1}{2A} \int_{-A}^{A} (x)^2 \, dx = \frac{A^2}{3} \tag{6}$$

The values of the mean, mean square, and variance computed in Equations 4 through 6 are the same as those computed directly from the waveform in Example 4.2. Therefore, we can see that probability density functions can be defined even for deterministic variables and hence deterministic variables can be considered as a special case of random variables.

4.4.3 Characteristic Function

Although we rarely compute moments beyond the second, it is important to understand that a complete description of a random variable requires computation of as many moments as possible. However, in most cases, it is not necessary to compute beyond the second moment. In addition, it is important to note that although the Fourier transform pair relates to time- and frequency-domain data, there could be other pairs of variables that can be related through a Fourier transform pair. A new variable known as the characteristic function of the random variable is related to its probability density function.

From Equation 4.54, let $f(X) = e^{j\theta X}$, where θ is a real number. Then the characteristic function $M(\theta)$ is defined as

$$M(\theta) \equiv E[e^{j\theta X}] = \int_{-\infty}^{\infty} e^{j\theta x} p(x) \, dx \tag{4.69}$$

It is clear that $M(\theta)$ is the Fourier transform of $p(x)$. Therefore, by using the inverse Fourier transform

$$p(x) = \frac{1}{2\pi} \int_{-\infty}^{\infty} M(\theta) e^{-j\theta x} \, d\theta \tag{4.70}$$

We see from the above equations that $M(\theta)$ and $p(x)$ form a Fourier transform pair. From Equation 4.64

$$M(\theta)\big|_{\theta=0} = \int_{-\infty}^{\infty} p(x)\,dx = 1 \tag{4.71}$$

Differentiating Equation 4.69 with respect to θ

$$\frac{dM(\theta)}{d\theta} = j \int_{-\infty}^{\infty} e^{j\theta x} x p(x)\,dx \tag{4.72}$$

And the nth differential is given by

$$\frac{d^n M(\theta)}{d\theta^n} = j^n \int_{-\infty}^{\infty} e^{j\theta x} x^n p(x)\,dx \tag{4.73}$$

Evaluating Equation 4.73 at $\theta = 0$

$$\frac{d^n M(\theta)}{d\theta^n}\bigg|_{\theta=0} = j^n \int_{-\infty}^{\infty} x^n p(x)\,dx \equiv j^n E[X^n] \tag{4.74}$$

Rewriting Equation 4.74

$$E[X^n] = \frac{1}{j^n} \frac{d^n M(\theta)}{d\theta^n}\bigg|_{\theta=0} \tag{4.75}$$

Therefore, the moment functions of a random variable can be expressed as the derivative of its characteristic function.

Consider the Taylor series expansion for $M(\theta)$

$$M(\theta) = M(\theta)\big|_{\theta=0} + \left(\sum_{n=1}^{\infty} \frac{d^n M(\theta)}{d\theta^n}\bigg|_{\theta=0} \right) \frac{\theta^n}{n!} \tag{4.76}$$

or

$$M(\theta) = 1 + \sum_{n=1}^{\infty} \frac{(j\theta)^n}{n!} E[X^n] \tag{4.77}$$

The right-hand side of the preceding equation contains the infinite number of moments of the random variable X. Hence, a complete description of the probability distribution of a random variable requires a complete set of moments.

4.5 Theory of Random Processes

Now let us look at continuous random variables that are time dependent. The definition of an ensemble provides the basis for establishing statistical regularity of random variables. However, the collection of ensemble data of the dynamic system is not arbitrary; they are collected under similar conditions to obtain a better measure of their statistical regularity. For example, if we collect a set of vibration data from a gearbox with respect to an arbitrary time reference, we may not be able to establish the relation between a specific defect and its resulting vibration. However, if each set of vibration data was collected with respect to a specific instant at which one of the gear sets passes through a reference marking, such data will be more useful in establishing their statistical regularity. Similar ensembles that are based on data collected under specific conditions will be studied in this section that has many practical applications. Similarly, data blocks of the time-domain signal of a certain record length that are used for applying the FFT algorithm in spectrum analyzers can also be considered as an ensemble.

Let us consider an experiment consisting of measuring the displacement of a package in a shipment container traveling on a rough road and denote it by $x_1(t)$ the time history corresponding to that displacement. If, at some other time, the same truck travels on the same road under similar conditions, then the associated time history $x_2(t)$ will in general be different from $x_1(t)$ because of various reasons such as the variation in tire pressure, the path traveled on the road, etc. Next, let us assume that the experiment is repeated a large number of times and plot the time histories $x_k(t)(k = 1,2,...)$ as shown in Figure 4.17. Similar time histories can be collected regarding vibration in machines. These time histories are different in general, as seen. The reasons for being different are very complex; perhaps all factors affecting the outcome are not completely understood. Hence this is a random phenomenon.

An individual time history $x_k(t)$ is called a *sample function* and the variable $x_k(t)$ itself is referred to as a *random variable*. The entire collection or *ensemble* of all possible time histories that might result in a given experiment ($k \to \infty$) is known as *random process* or *stochastic process* and is denoted by $\{x_k(t)\}$ or $x_k(t)$ or simply $x(t)$. (In practice, a reasonably large number of finite sample functions constitute an engineer's approximation to a true *random process*.) In this example, $\{x_k(t)\}$ plays the role of an excitation that the plane is subjected to random excitation. Because the input is not deterministic, the question arises as to how to compute the response of the system.

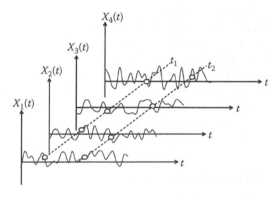

FIGURE 4.17
Ensemble of vibration data.

One approach is to find the response for each sample function. That would involve too much computation and is therefore not very practical from the viewpoint of handling the data. Moreover, each sample function and the response is a random function. Hence, such an approach is rejected. The other approach is to develop techniques based on statistical descriptions in terms of averages, as discussed next.

4.5.1 Ensemble Averages

Ensemble averages are conducted over the collection of sample functions. The mean value of a random process at a given time, t_1, is obtained by simply summing up the values corresponding to the time t_1 of all the individual sample functions in the ensemble and dividing the result by the number of sample functions. It is given by

$$\mu_x(t_1) = \lim_{n \to \infty} \frac{1}{n} \sum_{k=1}^{n} x_k(t_1) \tag{4.78}$$

A different type of ensemble average is obtained by summing the products of the instantaneous values of the sample functions at two instants of time $t = t_1$ and $t = t_1 + \tau$ (see Figure 4.17) and dividing the results by the number of sample functions. Such an average is called the autocorrelation function. The autocorrelation function, which was studied earlier during deterministic functions, in the present case is an indication of the relationship of the random variable at various instants of time. Between time instants t_1 and $t_1 + \tau$ it is given by

$$R_x(t_1, t_1 + \tau) = \lim_{n \to \infty} \frac{1}{n} \sum_{k=1}^{n} x_k(t_1) x_k(t_1 + \tau) \tag{4.79}$$

By fixing three or four times, such as t_1, $t_1 + \tau$, $t_1 + \sigma$, we can calculate the higher order averages. We discussed these higher order averages in Section 4.4.3 while defining a characteristic function. Such averages are seldom needed, however.

If $\mu_x(t_1)$ and $R_x(t_1, t_1 + \tau)$ do depend on t_1, then the random process $\{x_k(t)\}$ is said to be *nonstationary*. As a special case, when μ_x and R_x do not depend on t_1, the random process is called *weakly stationary*. Hence, for a weakly stationary random process, the mean value is constant, $\mu_x(t_1) = \mu_x = $ constant, and the autocorrelation function R_x depends on the time shift τ alone; that is, $R_x(t_1, t_1 + \tau) = R_x(\tau)$. When all possible averages over $\{x_k(t)\}$ are independent of t_1, the random process is said to be strongly stationary. In many practical cases, strong stationarity can be assumed if weak stationarity can be established. In view of this we shall not distinguish between the two processes, and refer to them as simply stationary. Fortunately, we generally deal with stationary signals in vibro-acoustics, as they are continuously produced by machines in a repetitive manner that results in statistical regularity of vibration or airborne sound signals. However, our speech signal is a nonstationary signal that requires a different analysis technique that is not discussed here.

4.5.2 Time Averages and Ergodic Random Processes

All the averages discussed so far have been ensemble averages. To evaluate them it is necessary to have the probability distribution of the samples or at least a large number of individual samples. Given a single sample $x_k(t)$ of duration T (in theory T is very large or $\to \infty$) it is, however, possible to define an average with respect to time along the sample.

Temporal mean of $x_k(t)$ is given by

$$\langle x_k(t) \rangle \equiv \mu_x(k) = \lim_{T \to \infty} \frac{1}{T} \int_{-T/2}^{T/2} x_k(t) \, dt \tag{4.80}$$

And temporal autocorrelation function is given by

$$R_x(k, \tau) \equiv \lim_{T \to \infty} \frac{1}{T} \int_{-T/2}^{T/2} x_k(t) x_k(t + \tau) \, dt \tag{4.81}$$

which is sometimes also represented by $\Phi(\tau)$.

If a random process is stationary and if $\mu_x(k)$ and $R_x(k, \tau)$ are independent of time history $x_k(t)$ (i.e., independent of k), then the process is said to be an *ergodic process*. The conditions for an ergodic process are given by

$$\mu_x = \mu_x(k)$$

$$\Phi(\tau) = R_x(k, \tau) = R(\tau) \tag{4.82}$$

From the preceding discussion it is clear that an ergodic process has to be a stationary process, whereas a stationary process need not be necessarily ergodic. The main usefulness of the definition of ergodicity is that it allows the use of a single sample function to compute the statistical averages of a random process rather than accounting for the entire ensemble. However, if the given random process is not ergodic, it would be inevitable to work with ensemble averages.

4.5.3 Description of Random Processes

A random process for a time-dependent random variable is defined as a set function of two arguments $\omega \in \Omega$ and $t \in T$, where Ω is the space of the family of random variables $X(., t)$ and T is the indexing set of the parameter t and can be written as $X(\omega, t)$. Generally, $X(\omega, t)$ is represented by $X(t)$ only.

A random process $X(t)$ reduces to a set of random variables at fixed instants of time $t = t_1, t_2,..., t_n$.

A hierarchy of joint probability density functions can define the probability structure.

$$p(x(t_1), t_1), \text{ for random variable } X_1$$
$$p(x(t_1), t_1; x(t_2), t_2), \text{ for random variables } X_1 \text{ and } X_2$$
$$\vdots$$

$$p(x(t_1), t_1) \, dx(t_1) = \text{Probability that } X(t) \text{ lies in the interval } [x(t_1), x(t_1) + dx(t_1)]$$
$$p(x(t_1), t_1; x(t_2), t_2) \, dx(t_1) \, dx(t_2) = \text{Joint probability that } X(t) \text{ lies in}$$
$$[x(t_1), x(t_1) + dx(t_1)] \text{ and } [x(t_2), x(t_2) + dx(t_2)] \text{ at time } t_1.$$

4.5.4 Stationary Process

An important aspect of machinery vibration and sound is that they have statistical regularity and hence make them easier to analyze. Therefore, a stationary process defined earlier for an ensemble is applicable to such signals. Hence, the averages computed for them will become time invariant and, therefore, similar equations that were used for time-independent random signals will become applicable through their probability density functions.

Probability structure is invariant under arbitrary translations of the indexing parameter t

$$p\big(x(t_1),t_1\big) = p(x) \ [\text{independent of time}]$$

$$p\big(x(t_1),t_1;x(t_2),t_2\big) = p(x_1,0;x_2,\tau) \quad \text{where } \tau = t_2 - t_1$$

4.5.5 Expected Values and Moments

These are the generalizations of definitions used before for a time-independent random variable. The average value of the function of a time-dependent random variable is given by

$$E\big[f(X(t))\big] = \int_{-\infty}^{\infty} [f(x)]p(x,t)\,dx \tag{4.83}$$

The average value of a time-dependent random variable is given by

$$\mu(t) = m_1(t) = E[X(t)] = \int_{-\infty}^{\infty} xp(x,t)\,dx \tag{4.84}$$

The mean square value of a time-dependent random variable is given by

$$m_2(t) = E[X^2(t)] = \int_{-\infty}^{\infty} x^2 p(x,t)\,dx \tag{4.85}$$

In Equations 4.83 through 4.85, $p(x,t)$ is the joint probability density function between the random variable and time.

The autocorrelation of a time-dependent random variable at time instants t_1 and t_2 is given by

$$\phi(t_1,t_2) = E[X(t_1)X(t_2)]$$
$$= \int_{-\infty}^{\infty}\int_{-\infty}^{\infty} x_1 x_2 p(x_1,t_1;x_2,t_2)\,dx_1\,dx_2 \tag{4.86}$$

The variance of the time-dependent random variable is given by

$$E[(X(t) - \mu(t))^2] = \sigma^2(t) \tag{4.87}$$

The autocovariance of the time-dependent random variable at two instants of time t_1 and t_2 is given by

$$K(t_1, t_2) = E\left[(X(t_1) - \mu(t_1))(X(t_2) - \mu(t_2)) \right] \tag{4.88}$$

The autocovariance of Equation 4.88 can be expressed in terms of autocorrelation function of Equation 4.81 by

$$K(t_1, t_2) = \Phi(t_1, t_2) - \mu(t_1)\mu(t_2) \tag{4.89}$$

When the two instants of time of the autocovariance function of the time-dependent random variable become the same, that is $t_1 = t_2 = t$, the autocovariance function becomes a measure of the covariance (Equation 4.89) given by

$$K(t, t) = \sigma^2(t) \tag{4.90}$$

For two time-dependent random processes $X(t)$, $Y(t)$, the cross-covariance at two instants of time t_1 and t_2 is given by

$$K_{XY}(t_1, t_2) = E\left[(X(t_1) - \mu_X(t_1))(Y(t_2) - \mu_Y(t_2)) \right] \tag{4.91}$$

In terms of cross-correlation function, Equation 4.91 becomes

$$K_{XY}(t_1, t_2) = \Phi_{XY}(t_1, t_2) - \mu_X(t_1)\mu_Y(t_2) \tag{4.92}$$

The correlation coefficient between two time-dependent random variables at two instants of time t_1 and t_2 is given by

$$\rho_{XY}(t_1, t_2) = \frac{K_{XY}(t_1, t_2)}{\sigma_X(t_1)\sigma_Y(t_2)} \tag{4.93}$$

For an autocorrelation function

$$\phi(t_1, t_2) = \phi(t_2, t_1) \tag{4.94}$$

For an autocovariance function

$$K(t_1, t_2) = K(t_2, t_1) \tag{4.95}$$

For a cross-covariance function

$$K_{XY}(t_1, t_2) = K_{YX}(t_2, t_1) \neq K_{XY}(t_2, t_1) \tag{4.96}$$

4.5.6 Stationary Random Processes

For a stationary process of a time-dependent random variable, as discussed before, the following relations hold:

$$\begin{aligned}
\mu(t) &= \mu \\
\sigma(t) &= \sigma \\
\phi(t_1, t_2) &= R(t_2 - t_1) = R_x(\tau) \quad t_2 = t_1 + \tau
\end{aligned} \tag{4.97}$$

Therefore, for a stationary process, Equations 4.83 through 4.94 yield simpler equations.

4.5.7 Properties of Autocorrelation Function

A typical autocorrelation function is shown in Figure 4.18. For a stationary process, the properties of an autocorrelation function are expressed as follows:

1. Autocorrelation function can be expressed in terms of the time shift τ.

$$R_X(\tau) = E\big[X(t)X(t+\tau)\big] \tag{4.98}$$

2. The autocorrelation function is a symmetric function of τ.

$$R_X(t_2 - t_1) = R_X(t_1 - t_2) \text{ or } R_X(\tau) = R_X(-\tau) \tag{4.99}$$

3. $R_X(\tau)$ is a nonnegative function.

$$R_X(\tau) > 0 \tag{4.100}$$

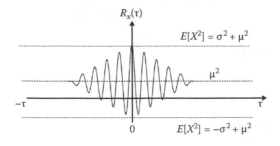

FIGURE 4.18
Autocorrelation function.

4. Since $\tau = 0$ is the reference value of time at which the random process is compared with later instants of time, the autocorrelation is a maximum at $\tau = 0$.

$$R_X(0) \geq |R_X(\tau)| \tag{4.101}$$

5. If a random process does not contain any periodic components, then $\lim_{T \to \infty} R(\tau) = 0$. This is very useful in separating periodic signals buried in noise.

6. The maximum value of $R_X(\tau) = R_X(0)$ is also the mean square value Ψ_x^2.

7. If $x(t)$ is periodic, then $R_X(t)$ is also periodic and the maximum value of $R_X(\tau)$ is obtained not only at $\tau = 0$ but also for value of $n\tau; n = 1, 2, 3, \ldots$.

4.5.8 Cross-Correlation

As discussed earlier, the correlation between two signals in both time and frequency domains is important. In the time domain, for stationary processes X and Y, the cross-correlation is given by

$$\begin{aligned} \phi_{XY}(t_1, t_2) = R_{XY}(\tau) &= E\big[X(t)Y(l+\tau)\big] \\ R_{YX}(\tau) &- E\big[Y(t)X(t+\tau)\big] \end{aligned} \tag{4.102}$$

Because of stationarity, it follows that

$$\begin{aligned} R_{XY}(\tau) &= E\big[X(t-\tau)Y(t)\big] = R_{YX}(-\tau) \\ R_{YX}(\tau) &= R_{XY}(-\tau) \end{aligned} \tag{4.103}$$

In general, $R_{XY}(\tau)$ and $R_{YX}(\tau)$ are not the same, unlike the autocorrelation function. Using Equations 4.92 and 4.93 we see that

$$R_{XY}(\tau) = K_{XY}(\tau) + \mu_X \mu_Y$$

$$\text{or } R_{XY}(\tau) = \rho_{XY}(\tau)\sigma_X\sigma_Y + \mu_X\mu_Y \tag{4.104}$$

$$\text{and } R_{YX}(\tau) = \rho_{YX}(\tau)\sigma_Y\sigma_X + \mu_Y\mu_X$$

A typical cross-correlation function is shown in Figure 4.19. Also, since $-1 < \rho_{XY} < 1$, we see that

$$-\sigma_X\sigma_Y + \mu_X\mu_Y \leq R_{XY}(\tau) \leq \sigma_X\sigma_Y + \mu_X\mu_Y \tag{4.105}$$

For most random processes, we expect that there will be no correlation between X and Y when the separation τ becomes large, that is $\rho_{XY} \to 0$,

$$R_{XY}\big|_{\tau \to \infty} = \mu_X\mu_Y = R_{YX}\big|_{\tau \to \infty} \tag{4.106}$$

Note that there is a maximum correlation at $\tau = \tau_0$.

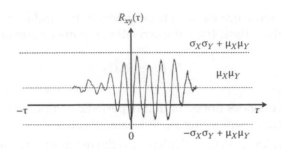

FIGURE 4.19
Cross-correlation function.

4.5.9 Spectral Density

In this section our attention is to find the frequency composition of naturally occurring random processes. The condition for using the Fourier integral is given by

$$\int_{-\infty}^{\infty} |X(t)| \, dt < \infty \tag{4.107}$$

And for a stationary process $x(t)$, the preceding condition may not be satisfied in general, so the classical approach of Fourier analysis cannot be applied to a sample function. If the preceding integral is not convergent, the Fourier transform of $F(\omega)$ does not exist. To avoid this difficulty, one can use autocorrelation function $R_X(\tau)$ instead of $x(t)$ itself.

If the zero value of the random process $x(t)$ is normalized (or adjusted) so the mean value of the process is $\mu = E[X] = 0$ and provided that $x(t)$ has no periodic components, $R_X(\tau)$ tends to zero for very large values of time shifts τ, and if the autocorrelation function has no periodic components, the following condition

$$\int_{-\infty}^{\infty} |R_X(\tau)| \, d\tau < \infty \tag{4.108}$$

is satisfied.

The frequency counterpart of autocorrelation function is the spectral density $S_X(\omega)$. Spectral density and autocorrelation functions are related through the Fourier transform integrals as follows. The Fourier transform of $R_X(\tau)$ is given by

$$R_X(\tau) = \int_{-\infty}^{\infty} S_X(\omega) e^{j\omega\tau} \, d\omega \tag{4.109}$$

and

$$S_X(\omega) = \frac{1}{2\pi} \int_{-\infty}^{\infty} R_X(\tau) e^{-j\omega\tau} \, d\tau \tag{4.110}$$

Note that the autocorrelation evaluated at the zero time shift, which is a measure of mean square value is given by

$$R_X(\tau = o) \equiv \psi_x^2 = \int_{-\infty}^{\infty} S_X(\omega) d\omega \qquad (4.111)$$

The mean square value of a stationary random process X is therefore given by the area under a graph of spectral density $S_X(\omega)$ and ω.

We also know that the complex Fourier transform can be expressed in terms of its real and imaginary parts. In this case

$$S_X(\omega) = A(\omega) - jB(\omega) \qquad (4.112)$$

Since $A(\omega)$ represents real part of the spectral density, from Equation 4.105 it is given by

$$A(\omega) = \frac{1}{2\pi} \int_{-\infty}^{\infty} R_X(\tau) \cos \omega \tau \, d\tau \qquad (4.113)$$

And since $B(\omega)$ represents the imaginary component of spectral density, it is given by

$$B(\omega) = \frac{1}{2\pi} \int_{-\infty}^{\infty} R_X(\tau) \sin \omega \tau \, d\tau \qquad (4.114)$$

$R_X(\tau)$ is an even function and $\sin \omega \tau$ is an odd function, and their product is an odd function. Therefore, in Equation (4.114), the value of integral between $-\infty$ and 0 is exactly of opposite sign to that between 0 and ∞ due to $B(\omega) \equiv 0$. Therefore

$$S_X(\omega) = A(\omega) \qquad (4.115)$$

which is a real and even function of ω.

In view of physical interpretation of spectral density as well as being defined as a measure of $E[X^2]$, we see that $S_X(\omega) \geq 0$ and a typical plot of spectral density is shown in Figure 4.20.

Example 4.5

Calculate the spectral density of a general periodic function.

Since $R_X(\tau)$ is periodic and even, we can expand it in terms of the following Fourier series:

$$R_X(\tau) = a_o + \sum_{k=1}^{\infty} a_k \cos \frac{2\pi k}{T} \tau \qquad (1)$$

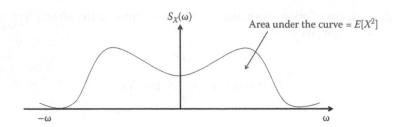

FIGURE 4.20
Spectral density function.

Using an exponential function, Equation 1 becomes

$$= a_o + \frac{1}{2}\sum_{k=1}^{\infty} a_k \left(e^{\frac{j2\pi k}{T}\tau} + e^{\frac{-j2\pi k}{T}\tau} \right) \tag{2}$$

where

$$a_o = \frac{2}{T}\int_0^{T/2} R_X(\tau)d\tau \text{ and } a_k = \frac{4}{T}\int_0^{T/2} R_X(\tau)\cos\frac{2\pi k}{T}d\tau \tag{3}$$

Spectral density and autocorrelation are related by

$$S_X(\omega) = \frac{1}{2\pi}\int_{-\infty}^{\infty} R_X(\tau)e^{-j\omega\tau}\,d\tau \tag{4}$$

Using Equation 2, the spectral density becomes

$$S_X(\omega) = \frac{1}{2\pi}\int_{-\infty}^{\infty} a_o e^{-j\omega\tau}\,d\tau + \frac{1}{2}\sum_{k=1}^{\infty}\frac{1}{2\pi}\int_{-\infty}^{\infty} a_k e^{\frac{j2\pi k}{T}\tau}e^{-j\omega\tau}d\tau$$

$$+ \frac{1}{2}\sum_{k=1}^{\infty}\frac{1}{2\pi}\int_{-\infty}^{\infty} a_k e^{\frac{-j2\pi k}{T}\tau}e^{-j\omega\tau}d\tau \tag{5}$$

The following integral is useful in simplifying Equation 5

$$\delta(\omega - a) = \frac{1}{2\pi}\int_{-\infty}^{\infty} e^{a\tau}e^{-j\omega\tau}\,d\tau \tag{6}$$

Now, by using Equation 6, Equation 5 becomes

$$S_X(\omega) = a_o\delta(\omega) + \frac{1}{2}\sum_{k=1}^{\infty} a_k\left[\delta\left(\omega + \frac{2\pi k}{T}\right) + \delta\left(\omega - \frac{2\pi k}{T}\right)\right] \tag{7}$$

Equation 7 can be used for any autocorrelation function and by substituting appropriate values of a_o and a_k obtained from Equation 3, the actual spectral density can be obtained.

4.5.10 Narrowband and Broadband Processes

For a narrow band process, the spectral density occupies only a narrow band of frequencies. Figure 4.21 shows the time waveform, spectral density, and autocorrelation of a narrow band signal of average frequency ω_o.

For a broadband process, the spectral density covers a broadband of frequencies. In the limit when the frequency band extends from $\omega_1 = 0$ to $\omega_2 = \infty$, the spectral density approaches an ideal white noise. The analogy to "white" is due to white light that contains all the colors and therefore all the frequencies of the visible spectrum. Figure 4.22

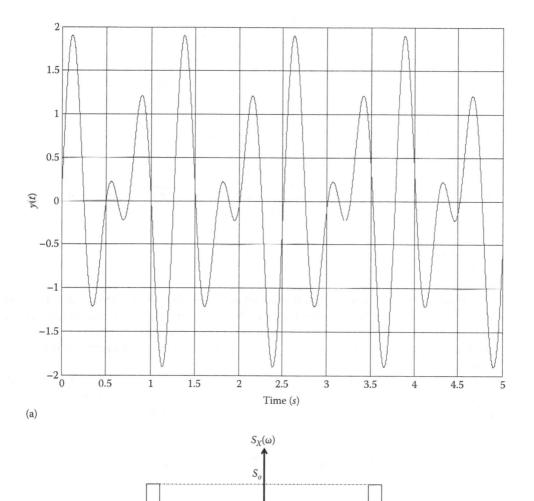

(a)

(b)

FIGURE 4.21
Narrow band excitation: (a) time waveform, (b) spectral density. *(Continued)*

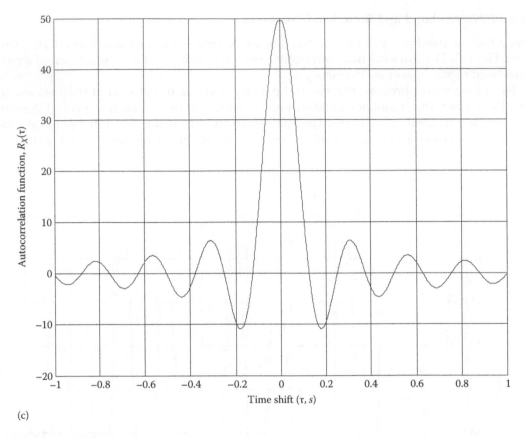

(c)

FIGURE 4.21 (CONTINUED)
Narrow band excitation: (c) autocorrelation.

shows the time waveform, spectral density, and autocorrelation of a broadband process. From this figure it can be clearly seen that a rather arbitrary time waveform has certain distinct features in terms of spectral density and autocorrelation functions. Therefore, spectral density and autocorrelation are useful in identifying and characterizing random signals.

Then, from Equation 4.111 it is seen that $E[X^2]$ or the mean square value for the white noise process must be infinite! Therefore, it implies that this is simply a theoretical concept.

From the example discussed earlier, we see that when $\omega_1 = 0$, $R_X(\tau)$ is

$$R_X(\tau) = \frac{4S_o}{\tau} \cos \frac{\omega_2 \tau}{2} \cdot \sin \frac{\omega_2 \tau}{2} = 2S_o \left(\frac{1}{\tau} \right) \cdot \sin \omega_2 \tau \tag{4.116}$$

In order to get the value of $R_X(\tau)$ at $\tau = 0 \Rightarrow$

$$\lim R_X(\tau \to 0) = \lim_{\omega_2 \tau \to 0} 2S_o \omega_2 \left(\frac{\sin \omega_2 \tau}{\omega_2 \tau} \right) = 2S_o \omega \tag{4.117}$$

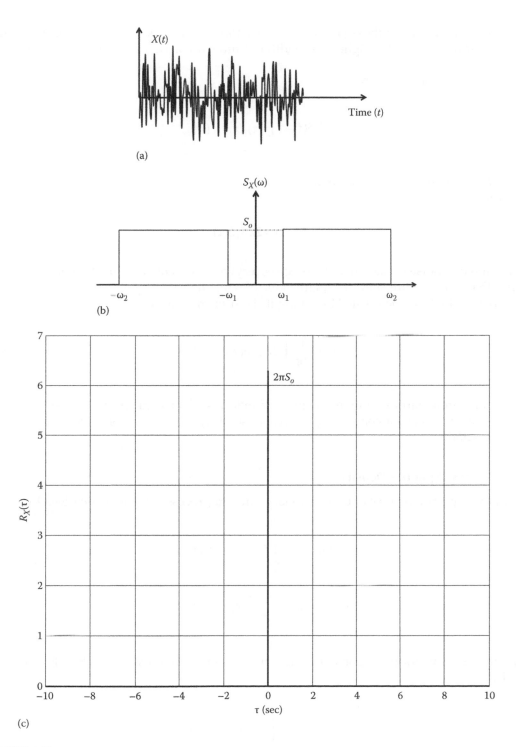

FIGURE 4.22
Broadband excitation: (a) time waveform, (b) spectral density, (c) autocorrelation.

As ω_2 becomes larger, the adjacent cycles pack together, and as $\omega_2 \to \infty$ they form a single vertical spike of infinite height, zero width but finite area. This area can be shown to be $2\pi S_o$.

For a Dirac delta function

$$\int_{-\infty}^{\infty} \delta(\tau - T) f(\tau) \, d\tau = f(T) \tag{4.118}$$

$\delta(\tau - T)$ is zero everywhere except at $\tau = T$.

Or for the limiting case

$$R_X(\tau) = 2\pi S_o \, \delta(\tau) \tag{4.119}$$

For a white noise excitation, $R_X(\tau)$ is zero everywhere except at $\tau = 0$ where it is infinite, since that is a measure of mean square value of white noise.

Equation 4.119 can be verified by taking its Fourier transform:

$$S_X(\omega) = \frac{1}{2\pi} \int_{-\infty}^{\infty} 2\pi S_o \, \delta(\tau) e^{-j\omega\tau} \, d\tau = S_o \tag{4.120}$$

Therefore, broadband excitation is expressed in terms of constant spectral density, S_o, and this makes it easier for computing the response of dynamic systems subjected to white noise excitation.

4.5.11 Cross-Spectral Density

The cross-spectral density of two stationary random processes X and Y are given by

$$S_{XY}(\omega) = \frac{1}{2\pi} \int_{-\infty}^{\infty} R_{XY}(\tau) e^{-j\omega\tau} \, d\tau$$

$$\text{and } S_{YX}(\omega) = \frac{1}{2\pi} \int_{-\infty}^{\infty} R_{YX}(\tau) e^{-j\omega\tau} \, d\tau \tag{4.121}$$

The inverse transform relations of Equation 4.121 represent cross-correlation between X and Y

$$R_{XY}(\tau) = \int_{-\infty}^{\infty} S_{XY}(\omega) e^{j\omega\tau} \, d\omega$$

$$R_{YX}(\tau) = \int_{-\infty}^{\infty} S_{YX}(\omega) e^{j\omega\tau} \, d\omega \tag{4.122}$$

Recall that according to the classical Fourier transform theory, we must require

$$\int_{-\infty}^{\infty} |R_{XY}(\tau)| d\tau < \infty \tag{4.123}$$

which implies that $x(t)$ and $y(t + \tau)$ must be uncorrelated when $\tau \to \infty$ and either of the means μ_X or μ_Y must be zero.

Cross-spectral density is useful in the following applications: (a) computing sound intensity and (b) determining the influence of one signal on another through a frequency-dependent parameter called *coherence*. For example, when a machine produces airborne sound due to many sources and the frequencies of airborne sound produced due to vibration can only be determined by computing the coherence between the vibration of a machine surface and the sound produced by it; the coherence will be the maximum (unity) at those frequencies in which the airborne sound is produced due to vibration. Another application using coherence is to identify resonant frequencies of a structure from the spectrum of transfer function obtained from two channel measurements of force and response; the resonant frequencies are those corresponding to frequencies where the coherence between force and response is a maximum (unity).

Example 4.6

The single-degree-of-freedom (SDOF) spring–mass–damper shown in Figure 4.23 is subjected to a stationary random excitation, $F(t)$. Determine the mean, autocorrelation, and mean square response using autocorrelation.

Consider the special case when $F(t)$ is a white noise. If the white noise excitation has zero mean, the response also has zero mean. The autocorrelation of the white noise excitation is given by

$$R_F(\tau) = 2\pi S_o \delta(\tau) \tag{1}$$

Equation 1 is plotted in Figure 4.24.

The autocorrelation of the response in terms of impulsive response function is given by

$$R_Y(\tau) = 2\pi S_o \int_{-\infty}^{\infty} \int_{-\infty}^{\infty} h(\eta_1) h(\eta_2) \delta(\tau + \eta_1 - \eta_2) d\eta_1 d\eta_2 \tag{2}$$

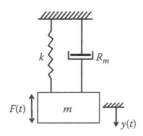

FIGURE 4.23
SDOF subjected to force excitation.

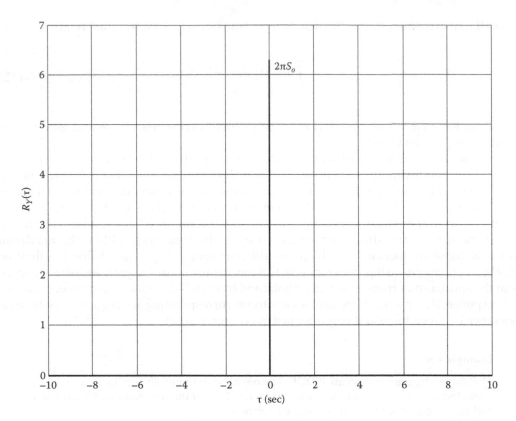

FIGURE 4.24
Autocorrelation of a white noise of spectral density S_o.

Substituting the respective impulse response functions, the autocorrelation of the response becomes

$$R_Y(\tau) = 2\pi S_o \int_{-\infty}^{\infty}\int_{-\infty}^{\infty} \frac{1}{\omega_d^2 m^2} e^{-\zeta\omega_n(\eta_1+\eta_2)}(\sin\omega_d\eta_1)\cdot(\sin\omega_d\eta_2)\cdot\delta(\tau+\eta_1-\eta_2)d\eta_1\,d\eta_2 \qquad (3)$$

Equation 3 can be regrouped as

$$R_Y(\tau) = 2\pi S_o \int_{-\infty}^{\infty} \frac{1}{\omega_d^2 m^2} e^{-\zeta\omega_n(\eta_1)}(\sin\omega_d\eta_1)\left\{ \int_{-\infty}^{\infty} e^{-\xi\omega_n(\eta_2)}(\sin\omega_d\eta_2)\cdot\delta(\tau+\eta_1-\eta_2)d\eta_2 \right\}d\eta_1 \qquad (4)$$

The general property of a Dirac delta function is given by

$$\int_{-\infty}^{\infty} \delta(t-T)f(t)dt = f(T) \qquad (5)$$

By using Equation 5 the bracketed portion of Equation 4 is given by

$$\int_{-\infty}^{\infty} e^{-\zeta\omega_n(\eta_2)}(\sin\omega_d\eta_2)\cdot\delta(\tau+\eta_1-\eta_2)d\eta_2 = e^{-\zeta\omega_n(\tau+\eta_1)}\sin\omega_d(\tau+\eta_1) \tag{6}$$

From Equations 4 and 6, the autocorrelation of the response becomes

$$R_Y(\tau) = \frac{2\pi S_o}{\omega_d^2 m^2}\int_{-\infty}^{\infty} e^{-\zeta\omega_n(\tau+2\eta_1)}(\sin\omega_d\eta_1)\cdot\left(\sin\omega_d(\tau+\eta_1)\right)d\eta_1 \tag{7}$$

Then, from the table of integrals in Equation 6,

$$R_Y(\tau) = \frac{\pi}{2}\frac{S_o}{\zeta m^2\omega_n^3}e^{-\zeta\omega_n\tau}\left\{\cos\omega_d\tau + \frac{\zeta\omega_n}{\omega_d}\sin\omega_d\tau\right\} \tag{8}$$

for $\tau > 0$.

The process may be repeated for $\tau < 0$; or we can simply use the fact that $R_X(\tau)$ is an even function of τ to write

$$R_Y(\tau) = \frac{\pi}{2}\frac{S_o}{\zeta m^2\omega_n^3}e^{-\zeta\omega_n\tau}\left\{\cos\omega_d\tau - \frac{\zeta\omega_n}{\omega_d}\sin\omega_d\tau\right\} \tag{9}$$

for $\tau < 0$.

The curve has continuous derivatives (first and second) at $\tau = 0$; however the third derivative is discontinuous. For small τ, the correlation looks like a sinusoid of frequency ω_n, but for large τ it falls to zero. This is typical of a narrowband process.

We can also combine Equations 7 and 8 as

$$R_Y(\tau) = \frac{\pi S_o}{2m^2\omega_n^3\zeta}e^{-\zeta\omega_n|\tau|}\left\{\cos\omega_d|\tau| + \frac{\omega_n\zeta}{\omega_d}\sin\omega_d|\tau|\right\} \tag{10}$$

If we set $\tau = 0$ in Equation 10 we obtain the mean square response:

$$\sigma_Y^2 \equiv E\left[Y^2(t)\right] = \frac{\pi S_o}{2\zeta m^2\omega_n^3} \tag{11}$$

4.6 Response due to an Arbitrary Excitation

Now we would like to compute the response of a spring–mass–damper system subjected to a forcing function given earlier.

Consider an SDOF subjected to a $F(t)$, and during a period of time $\Delta\eta$ (small) gives to the system an impulse $F(\eta)\Delta\eta$. Therefore, for a small time interval $\Delta\eta$, the response can be determined as

$$\Delta y(t) \cong F(\eta)\Delta\eta\, h(t-\eta) \tag{4.124}$$

Hence the total response (i.e., for the entire $F(t)$) can be expressed as

$$y(t) = \int_0^t F(\eta)h(t-\eta)d\eta \quad 0 \le t \le t_p \tag{4.125}$$

This integral is called Duhamel's integral or the superposition integral that was discussed in Chapter 1.

We can also write Equation 4.125 in a slightly different form given by

$$y(t) = \int_0^t F(t-\eta)h(\eta)d\eta \tag{4.126}$$

Sometimes, the form of Equation 4.126 may lead to some simplification.

Note that Equation 4.125 is valid for $t \le t_p$, where t_p indicates the duration of the force. For $t > t_p$

$$y(t) = \int_0^{t_p} F(\eta)h(t-\eta)d\eta \quad t \ge t_p \tag{4.127}$$

Equation 4.127 can be written as

$$y(t) = \int_0^{t_p} F(\eta)h(t-\eta)d\eta + \int_{t_p}^t F(\eta)h(t-\eta)d\eta \tag{4.128}$$

Since the force is zero for $t > t_p$, the second integral of Equation 4.123 is zero.

For some $F(\eta)$, integration of the Equation 4.128 in closed form is possible. Otherwise, we may have to use a numerical integration technique such as Simpson's rule.

It should be noted that η is the independent variable in the preceding equations. Accordingly, the integration is performed with respect to η, with t being treated as a constant.

4.6.1 System Transfer Function

Now let us derive the relationship between system transfer function and impulse response function.

We shall assume that $F(\omega)$, the Fourier transform of $F(t)$, exists. Then

$$F(t) = \int_{-\infty}^{\infty} e^{j\omega t} F(\omega) d\omega \tag{4.129}$$

Substituting this expression of $F(t)$ in Equation 4.125,

$$y(t) = \int_{0}^{t} h(t-\eta) \left[\int_{-\infty}^{\infty} e^{j\omega\eta} F(\omega) d\omega \right] d\eta \tag{4.130}$$

By interchanging the order of integration

$$y(t) = \int_{-\infty}^{\infty} \left\{ \int_{0}^{t} h(t-\eta) e^{j\omega\eta} d\eta \right\} F(\omega) d\omega \tag{4.131}$$

The response can be expressed in terms of the transient system transfer function

$$y(t) = \int_{-\infty}^{\infty} H(\omega, t) F(\omega) d\omega \tag{4.132}$$

The transient frequency response function can be expressed in terms of the impulse response function

$$H(\omega, t) = \int_{0}^{t} h(t-\eta) e^{j\omega\eta} d\eta \tag{4.133}$$

Since $h(t - \eta)$ is the unit impulse response, it vanishes for $\eta > t$ (or $t < \eta$). Also $h(t) = 0$ for $t < 0$, therefore the transient frequency response function becomes

$$H(\omega, t) = \int_{-\infty}^{\infty} h(t-\eta) e^{j\omega\eta} d\eta \tag{4.134}$$

From Equation 4.134 it is clear that the transient frequency response function is the Fourier transform of the impulse response function.

Equation 4.134 can also be written as

$$H(\omega, t) = e^{j\omega t} \int_{-\infty}^{\infty} h(\eta) e^{-j\omega\eta} d\eta \tag{4.135}$$

From Equation 4.135, the steady state frequency response function can be defined as

$$\bar{H}(\omega) = \int_{-\infty}^{\infty} h(\eta) e^{-j\omega \eta} \, d\eta \qquad (4.136)$$

Equation (4.136) can be used to obtain the natural frequencies of a dynamic system by giving it an impulse input and then converting the response into the frequency domain. If we are unable to measure the amplitude of input force, we get a scaled version of the transfer function from a single channel measurement of response; however, the scaled transfer function can still give the natural frequencies of the dynamic system.

4.6.2 Autocorrelation of the Response

The mean response of a single-degree-of-freedom system, neglecting initial conditions, is given by

$$\mu_Y(t) = E[Y(t)] \qquad (4.137)$$

The mean response in terms of impulse response function and the mean value of force excitation is given by

$$\mu_Y = \int_0^t h(t - \tau) E[F(\tau)] \, d\tau \qquad (4.138)$$

For very large values of time

$$\lim_{t \to \infty} \mu_Y(t) = \lim_{t \to \infty} \int_0^t h(t - \tau) \mu_F(\tau) \, d\tau \qquad (4.139)$$

If $F(t)$ is stationary, $\mu_F(t)$ = constant = $\mu_F(t)$ = constant = μ_F and Equation 4.139 reduces to

$$\lim_{t \to \infty} \mu_Y(t) = \mu_Y \int_0^{\infty} h(\eta) \, d\eta = \mu_Y = \text{const.} \qquad (4.140)$$

We shall now consider the steady-state response of a stationary random process. The response as time tends to a very large value is given by

$$\lim_{t \to \infty} y(t) = \int_{-\infty}^{\infty} h(t - \tau) F(\tau) \, d\tau = \int_{-\infty}^{\infty} h(\eta) F(t - \eta) \, d\eta \qquad (4.141)$$

Based on Equation 4.141, the response is written for two instants of time with a shift of τ:

$$\left.\begin{array}{c} y(t) = \int\limits_{-\infty}^{\infty} h(\eta_1)F(t-\eta_1)d\eta_1 \\[4mm] \text{and } y(t+\tau) = \int\limits_{-\infty}^{\infty} h(\eta_2)F(t+\tau-\eta_2)d\eta_2 \end{array}\right\} \qquad (4.142)$$

Then, the autocorrelation is given by

$$R_Y(\tau) = E\left[y(t)y(t+\tau)\right] \qquad (4.143)$$

From Equations 4.142 and 4.143, the autocorrelation of the response is given by

$$\begin{aligned} R_Y(\tau) &= E\left[\int\limits_{-\infty}^{\infty} h(\eta_1)F(t-\eta_1)d\eta_1 \int\limits_{-\infty}^{\infty} h(\eta_2)F(t+\tau-\eta_2)d\eta_2\right] \\[2mm] &= E\left[\int\limits_{-\infty}^{\infty}\int\limits_{-\infty}^{\infty} h(\eta_1)h(\eta_2)F(t-\eta_1)F(t+\tau-\eta_2)d\eta_1\,d\eta_2\right] \qquad (4.144)\\[2mm] &= \int\limits_{-\infty}^{\infty}\int\limits_{-\infty}^{\infty} h(\eta_1)h(\eta_2)E\left[F(t-\eta_1)F(t+\tau-\eta_2)\right]d\eta_1\,d\eta_2 \end{aligned}$$

Because $F(t)$ is assumed to be stationary, the autocorrelation of the force excitation becomes

$$\begin{aligned} E\left[F(t-\eta_1)F(t+\tau-\eta_2)\right] &= E\left[F(t)\,F(t+\tau+\eta_1-\eta_2)\right] \\[2mm] &= R_F(\tau+\eta_1-\eta_2) \end{aligned} \qquad (4.145)$$

Therefore, the autocorrelation of the response becomes

$$R_Y(\tau) = \int\limits_{-\infty}^{\infty}\int\limits_{-\infty}^{\infty} h(\eta_1)h(\eta_2)R_F(\tau+\eta_1-\eta_2)d\eta_1\,d\eta_2 \qquad (4.146)$$

Equation 4.146 implies that stationary excitation to a deterministic system will result in a stationary response.

The mean square $E[Y^2(t)]$ of the response process can be obtained either from the autocorrelation function $R_Y(\tau)$ or from the knowledge of spectral density $S(\omega)$ of the response function.

From Equation 4.141, the mean square response is given by

$$R_Y(0) = E[Y^2(t)] = \int\limits_{-\infty}^{\infty} \int\limits_{-\infty}^{\infty} h(\eta_1)h(\eta_2)R_F(\eta_1 - \eta_2)\,d\eta_1\,d\eta_2 \tag{4.147}$$

4.6.3 Spectral Density

Spectral density is given by

$$S_Y(\omega) = \frac{1}{2\pi} \int\limits_{-\infty}^{\infty} R_Y(\tau)e^{-j\omega\tau}\,d\tau \tag{4.148}$$

From Equations 4.147 and 4.148, spectral density is given by

$$S_Y(\omega) = \frac{1}{2\pi} \int\limits_{-\infty}^{\infty} e^{-j\omega\tau} \left[\int\limits_{-\infty}^{\infty} \int\limits_{-\infty}^{\infty} h(\eta_1)h(\eta_2)R_F(\tau + \eta_1 - \eta_2)\,d\eta_1\,d\eta_2 \right] d\tau \tag{4.149}$$

But $R_F(\tau + \eta_1 - \eta_2)$ can be expressed as the inverse Fourier transform of the spectral density of the input excitation given by

$$R_F(\tau + \eta_1 - \eta_2) = \int\limits_{-\infty}^{\infty} S_F(\omega)e^{j\omega(\tau+\eta_1-\eta_2)}\,d\omega \tag{4.150}$$

From Equations 4.149 and 4.150

$$S_Y(\omega) = \frac{1}{2\pi} \int\limits_{-\infty}^{\infty} e^{-j\omega\tau} \left\{ \int\limits_{-\infty}^{\infty} S_F(\omega) \left[\int\limits_{-\infty}^{\infty} h(\eta_1)e^{j\omega\eta_1}\,d\eta_1 \cdot \int\limits_{-\infty}^{\infty} h(\eta_2)e^{j\omega\eta_2}\,d\eta_2 \right] e^{j\omega\tau}\,d\omega \right\} d\tau \tag{4.151}$$

which simplifies to

$$= \frac{1}{2\pi} \int\limits_{-\infty}^{\infty} e^{-j\omega\tau} \left[\int\limits_{-\infty}^{\infty} S_F(\omega)\bar{H}(\omega)\bar{H}^*(\omega)e^{j\omega\tau}\,d\omega \right] d\tau \tag{4.152}$$

where $\bar{H}^*(\omega)$ stands for the complex conjugate of $H(\omega)$. Then

$$S_Y(\omega) = \frac{1}{2\pi} \int\limits_{-\infty}^{\infty} e^{-j\omega\tau} \left[\int\limits_{-\infty}^{\infty} S_F(\omega)|\bar{H}(\omega)|^2 e^{j\omega\tau}\,d\omega \right] d\tau \tag{4.153}$$

From the basic definition of spectral density and autocorrelation pair, the autocorrelation of the response is given by

$$R_Y(\tau) = \int_{-\infty}^{\infty} |\bar{H}(\omega)|^2 S_F(\omega) e^{j\omega\tau} d\omega \qquad (4.154)$$

Autocorrelation and spectral density of the response are related by

$$R_Y(\tau) = \int_{-\infty}^{\infty} S_Y(\omega) e^{j\omega\tau} d\omega \qquad (4.155)$$

By comparing Equations 4.154 and 4.155, the spectral density of the response, steady-state frequency response function, and spectral density of the force are given by

$$S_X(\omega) = |\bar{H}(\omega)^2| S_F(\omega) \qquad (4.156)$$

4.6.4 Mean Square Response

From Equation 4.154, the mean square response is given by

$$E[x^2(t)] = R_X(0) = \int_{-\infty}^{\infty} S_X(\omega) d\omega \qquad (4.157)$$

From Equations 4.156 and 4.157, the mean square response is given by

$$E[x^2(t)] = \int_{-\infty}^{\infty} |\bar{H}(\omega)|^2 S_F(\omega) d\omega \qquad (4.158)$$

Example 4.7

The SDOF spring–mass–damper shown in Figure 4.25 is subjected to a stationary random excitation $F(t)$ of spectral density S_o. Determine the spectral density of response and its mean square value.

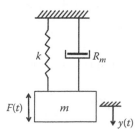

FIGURE 4.25
SDOF subjected to a force excitation.

The spectral density of the response in terms of system transfer function and spectral density of the force excitation is given by

$$S_Y(\omega) = \left|\bar{H}(\omega)\right|^2 S_F(\omega) \tag{1}$$

For an SDOF, the system transfer function is given by

$$\bar{H}(\omega) = \frac{1}{m\left[\left(\omega_n^2 - \omega^2\right) + j2\zeta\omega\omega_n\right]} \tag{2}$$

Therefore, from Equations 1 and 2, the spectral density of the response is given by

$$S_Y(\omega) = \frac{S_o}{m^2\left[\left(\omega_n^2 - \omega^2\right)^2 + 4\zeta^2\omega_n^2\omega^2\right]} \tag{3}$$

From Equation 3, the mean square response can be obtained as

$$E[Y^2(t)] = R_Y(0) = \int_{-\infty}^{\infty} S_Y(\omega)\,d\omega \tag{4}$$

Taking S_o outside the integral, the mean square response becomes

$$E[Y^2(t)] = \frac{S_o}{m^2} \int_{-\infty}^{\infty} \frac{d\omega}{\left|\left(\omega_n^2 - \omega^2\right) + i2\zeta\omega_n\omega\right|^2} \tag{5}$$

Equation 5 can be integrated using the results of a standard integral as follows:

$$I_2 = \int_{-\infty}^{\infty} \left|\frac{B_0 + i\omega B_1}{A_0 + j\omega A_1 - \omega^2 A_2}\right|^2 d\omega = \frac{\pi\left\{A_0 B_1^2 + A_2 B_0^2\right\}}{A_0 A_2 A_1} \tag{6}$$

Comparing Equations 5 and 6, the following relationships can be obtained:

$$B_0 = 1, \quad B_1 = 0$$
$$A_0 = \omega_n^2, \quad A_1 = 2\zeta\omega_n \tag{7}$$
$$A_2 = 1$$

From Equations 6 and 7, the mean square response is given by

$$E[Y^2(t)] = \frac{\pi S_o}{2\zeta m^2 \omega_n^3} \tag{8}$$

The mean square value obtained in Equation 8 is the same as that obtained in Example 4.6 using autocorrelation, but the spectral density method of this example is much easier.

4.7 Equivalent Bandwidth of a Single Degree of Freedom Subjected to White Noise Excitation

Consider an SDOF system consisting of a spring–mass-damper system shown in Figure 4.25. It is useful in many applications to determine the equivalent bandwidth of such a system when it is subjected to white noise excitation. It can be obtained as follows.

From Equation 1.112 (see Chapter 1), the mobility (admittance) of an SDOF subjected to complex harmonic excitation is given by

$$\bar{Y}_m = \frac{\tilde{v}}{\tilde{F}} = \frac{1}{\bar{Z}_m} = \frac{1}{R_m + j(\omega m - k/\omega)} \tag{4.159}$$

By using $2\zeta = \eta$, the damping coefficient can be expressed in terms of loss factor as

$$R_m = 2\zeta m\omega_n = \eta m\omega_n \tag{4.160}$$

From the preceding equations, admittance is given by

$$\bar{Y}_m = \frac{1}{\eta m\omega_n \left(1 + j\left(\dfrac{\omega m}{\eta m\omega_n} - \dfrac{k}{\eta m\omega\omega_n} \right) \right)} \tag{4.161}$$

Equation 4.161 can be further simplified as

$$\bar{Y}_m = \frac{1}{\eta m\omega_n \left(1 + \dfrac{j\left(\omega^2 - \omega_n^2\right)}{\eta\omega\omega_n} \right)} \tag{4.162}$$

When this SDOF is subjected to a white noise excitation of spectral density S_o (N^2/Hz), the mean square velocity response in terms of admittance is given by

$$\langle v^2 \rangle = S_o \int_0^\infty \left| \bar{Y}_m \right|^2 df \tag{4.163}$$

From Equations 4.162 and 4.163, the mean square response is given by

$$\langle v^2 \rangle = \frac{S_o}{2\pi} \int_0^\infty \frac{1}{\left| \eta m\omega_n \left(1 + \dfrac{j\left(\omega^2 - \omega_n^2\right)}{\eta\omega\omega_n} \right) \right|^2} d\omega \tag{4.164}$$

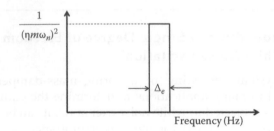

FIGURE 4.26
Equivalent bandwidth of an SDOF subjected to white noise excitation.

The solution of the standard integral is given by

$$I_2 = \int_{-\infty}^{\infty} \left| \frac{B_0 + i\omega B_1}{A_0 + j\omega A_1 - \omega^2 A_2} \right|^2 d\omega = \frac{\pi \left\{ A_0 B_1^2 + A_2 B_0^2 \right\}}{A_0 A_2 A_1} \tag{4.165}$$

By comparing Equations 4.164 and 4.165

$$B_1 = \eta\omega_n; \ B_0 = 0; \ A_0 = \omega_n^2; \ A_1 = \eta\omega_n; \ A_2 = 1 \tag{4.166}$$

From Equations 4.164 through 4.166, the mean square response is given by

$$\langle v^2 \rangle = \frac{S_o}{(\eta m\omega_n)^2} \left(\frac{\eta\pi f_n}{2} \right) \tag{4.167}$$

From Equation 4.167, the equivalent bandwidth for the mean square response of the SDOF to white noise excitation is given by

$$\Delta_e = \left(\frac{\eta\pi f_n}{2} \right) \text{Hz} \tag{4.168}$$

From Equations 4.167 and 4.168, the mean square velocity is given by

$$\langle v^2 \rangle = \frac{S_o \Delta_e}{(\eta m\omega_n)^2} \tag{4.169}$$

where the spectral density is in N^2/Hz and the equivalent bandwidth is in Hz. It is as shown in Figure 4.26.

Example 4.8

A single-degree-of-freedom spring–mass–damper shown in Figure 4.27 is subjected to a stationary random excitation $F(t)$ of spectral density S_o ($N^2/rad/s$). The resonant frequency of the SDOF is 150 Hz and the damping factor is 0.005 and its mass is 1 kg.

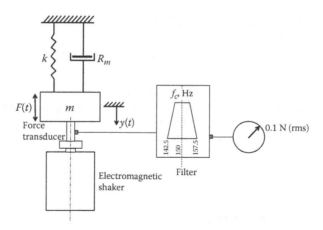

FIGURE 4.27
SDOF subjected to a band-limited force excitation.

The applied force is measured with a piezoelectric force transducer and filtered using a bandwidth filter of 15 Hz that is centered around 150 Hz. If the rms value of the measured force is 0.1 N, determine the rms velocity and power input to the system. Is it practical to neglect the force measurement at nonresonant frequencies?

The lower frequency limit of the filter is 150 − (15/2) − 142.5 Hz and the upper limit 150 + (15/2) = 157.5 Hz. The loss factor is $\eta = 2\zeta = 2 \times 0.005 = 0.01$.

The force measurement accounts for spectral density in the range of 142.5 to 157.5 Hz, and it is assumed that the spectral density is constant in this region.

From Equation 4.157, the mean square value of force between frequencies ω_1 and ω_2 is given by

$$E[f^2(t)] = R_F(0) = \int_{\omega_1}^{\omega_2} S_o(\omega)\,d\omega \tag{1}$$

If the spectral density remains constant in the given bandwidth, it can be related to the mean square value as

$$S_o = \frac{E[f^2(t)]}{\omega_2 - \omega_1} = \frac{0.1^2}{2\pi(157.5 - 142.5)} = 1.061 \times 10^{-4}\,\frac{N^2}{rad/s} \tag{2}$$

From Equation 4.168, the effective bandwidth (Hz) is given by

$$\Delta_e = \left(\frac{\eta\pi f_n}{2}\right) = \frac{0.01 \times \pi \times 150}{2} = 2.36\,Hz\;(14.80\,rad/s) \tag{3}$$

The effective mean square value of force is given by

$$E\left[f_e^2(t)\right] = S_o\Delta_e = 1.061 \times 10^{-4} \times 14.80 = 0.0016\,N^2 \tag{4}$$

(spectral density in $N^2/rad/s$ and effective bandwidth in rad/s).

The rms value of the effective force is given by

$$f_e(\text{rms}) = \sqrt{E\left[f_e^2(t)\right]} = \sqrt{0.0016} = 0.0396\,\text{N} \tag{5}$$

From Equation 5 it is clear that although the input force was measured as 0.1 N (rms) within a bandwidth of 15 Hz, the effective force on the SDOF is only 0.0396 (rms) within the effective bandwidth of 2.36 Hz (14.80 rad/s).

From Equation 4.160, the mechanical resistance is given by

$$R_m = 2\zeta m\omega_n = \eta m\omega_n = 0.01 \times 1 \times 2\pi \times 150 = 9.425\,\frac{\text{N}-\text{s}}{\text{m}} \tag{6}$$

The rms velocity of response is given by

$$v(\text{rms}) = \frac{f_e(\text{rms})}{R_m} = \frac{0.0396}{9.425} = 4.2\,\text{mm/s} \tag{7}$$

The rms value of displacement is given by

$$y(\text{rms}) = \frac{v(\text{rms})}{2\pi f_n} = \frac{4.2 \times 10^{-3}}{2\pi \times 150} = 4.46\,\mu\text{m} \tag{8}$$

The rms value of acceleration is given by

$$a(\text{rms}) = 2\pi f_n v(\text{rms}) = 2\pi \times 150 \times 4.2 \times 10^{-3} = 3.96\,\text{m/s}^2\,(\approx 0.4\ g) \tag{9}$$

From Equation 1.86, the power input to the system is given by

$$\langle W \rangle = \frac{f_e^2(rms)}{R_m} = \frac{0.0016}{9.425} = 0.16\ \text{mW} \tag{10}$$

Yes, it is justified to neglect force measurement outside the equivalent filter bandwidth of the SDOF, as the response of the system is negligible in this region.

4.8 Conclusions

A detailed discussion on random vibration was presented in this chapter. The discussion that initially started with time-independent discrete variables eventually ended up with time-dependent continuous variables. Although only the response of an SDOF for arbitrary random excitation was presented here, the same procedure can be extended to MDOF (multidegree-of-freedom) systems and continuous systems. The main objective of presenting this chapter was to explain the basic concept of spectral density and broadband excitation that is very important in most vibro-acoustic problems. In addition, much of the discussion presented in this chapter can also be used to understand the principle of noise and vibration measurement using spectral analyzers.

PROBLEMS

4.1. Calculate the mean, mean square value, rms value, variance, standard deviation and autocorrelation of the waveform shown in the following figure.

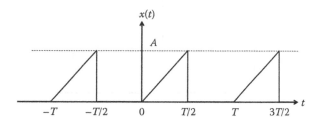

4.2. Determine the autocorrelation function of the sawtooth wave of Example 4.2 (see Figure 4.4).

4.3. A random variable X is distributed over 0 and 1 so that $p(x) = 1$ for $0 \leq x \leq 1$ and $p(x) = 0$ for $x < 0$ and $x > 1$. Determine $E[X]$, $E[X^2]$, and σ_X.

4.4. Two random variables X and Y have mean values $E[X]$ and $E[Y]$, respectively. If X and Y are statistically independent, and if $Z = X + Y$, first show that $E[Z^2] = E[X^2] + E[Y^2] + 2E[X]E[Y]$, then find

$$E[Z^2] \text{ and } \sigma_Z \text{ if}$$

$$p(x) = \begin{cases} 1 \text{ for } 0 \leq x \leq 4 \\ 0 \text{ for } x < 0 \text{ and } x > 4 \end{cases}$$

$$p(y) = \begin{cases} 1 \text{ for } 0 \leq y \leq 4 \\ 0 \text{ for } y < 0 \text{ and } y > 4 \end{cases}$$

4.5. Two random variables X and Y have the joint probability density function

$$p(x,y) = \frac{1}{\pi^2} \left[\frac{1}{1 + x^2 + y^2 + x^2 y^2} \right]$$

Prove that the random variable X and Y are statistically independent, and then compute $E[X]$, $E[Y]$, $E[X^2]$ and $E[Y^2]$.

4.6. A random variable, x, has the following probability density function as shown in the following figure.

$$p(x) = 0.25 \text{ for } 0 < x \leq 0.25$$
$$= p_{max} \text{ for } 0.25 < x \leq 1$$
$$= 0.25 \text{ for } 1 < x < 2$$

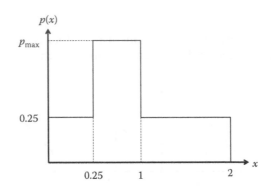

where p_{max} represents the maximum probability density of the signal.

a. Estimate p_{max}, mean value, mean square value, variance, and standard deviation of the random variable.

b. Determine and sketch the probability distribution function.

4.7. Consider the following half-sine pulse: $A\left|\sin\dfrac{2\pi t}{T}\right|$

a. Determine the mean and mean square value using the given half-sine pulse.

b. Determine the autocorrelation equation.

c. Determine the probability density function.

d. Use the probability density function of part c to determine the mean and mean square value.

4.8. If θ is a random variable uniformly distributed between $-180°$ and $+180°$ and the random variable Y is related to θ by $Y = a \tan \theta$, determine the probability density function and characteristic function for Y.

4.9. Random variables X and Y are independent and exponentially distributed. Following random variables are defined in terms of X and Y: $U = X + Y$ and $V = Y/(X + Y)$. Determine the joint probability density function between U and V and check whether they are independent.

4.10. A random variable has the following probability density function expressed in terms of unit step functions:

$$p(x) = \frac{1}{3}[u(x) - u(x - 3)]$$

If another random variable Y is defined as $Y = X^3$, determine the mean, mean square, and variance of the random variable Y.

4.11. A single-degree-of-freedom system shown in the following figure with a dashpot, spring, and negligible mass is subjected to a force excitation $F(t)$ that can be modeled as a white noise excitation of spectral density S_o.

a. Determine the autocorrelation of the response from which to determine the mean square value of response.

b. Determine the spectral density of the response, and then using this determine the mean square value of response.

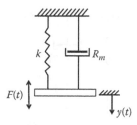

4.12. The dynamic system of negligible mass shown in the following figure is subjected to a random excitation of spectral density $S_f(\omega) = \dfrac{S_o}{\left|1 + j\dfrac{\omega}{\omega_n}\right|^2}$, where ω_n is a constant. Determine the mean square value of the response.

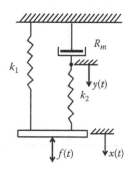

4.13. A vehicle of negligible mass that moves along a straight line carries a spring–mass–damper system as shown in the following figure; the coordinates are indicated as shown. Determine the transfer function between the relative displacement of the mass and the input acceleration $\ddot{x}$ of the vehicle. If the acceleration of the vehicle is a white noise excitation of spectral density S_o, determine the mean square response of the relative response of the mass, m.

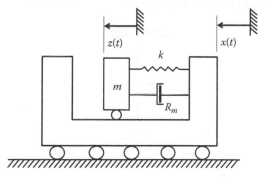

4.14. The dynamic system shown in the following figure is subjected to a white noise random excitation $f(t)$ of spectral density S_o.

a. Determine the mean square value of response of the mass, if $m = 0.1$ kg, $k_1 = k_2 = 5$ kN/m, and $R_m = 10$ N-s/m and $S_o = 10$ N²/Hz.

b. Determine the mean square value of the force in the spring k_2 for the same parameters of part a.

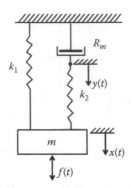

4.15 Determine the mean squared value and autocorrelation function for the stationary random process $X(t)$ whose spectral density is as shown in the following figure.

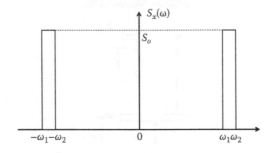

4.16 Consider the following autocorrelation function: $R_X(\tau) = A^2 e^{-\beta|\tau|} \cos \Omega\tau$. Determine the spectral density $S_x(\omega)$.

4.17 The following figure shows a stylus tracing a displacement curve $X(t)$ and the stylus has a viscous damper of damping coefficient R_m. The spectral density of the displacement curve is given by $S_X(\omega) = \dfrac{a^2}{(\omega^2 + b^2)^2}$. Determine the mean square value of the force in the damper.

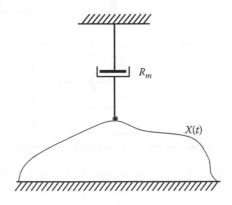

4.18 Consider a single-degree-of-freedom spring–mass–damper system that is subjected to a number of stationary random excitations: $F_1, F_2, ..., F_n$. Prove that the spectral density of the response is given by the following equation:

$$S_X(\omega) = \sum_{i=1}^{n} \sum_{j=1}^{n} \bar{H}_i^*(\omega) \bar{H}_j(\omega) S_{F_i F_j}(\omega)$$

where $S_{F_i F_j}(\omega)$ is the spectral density between various forces.

4.19. A stationary random process has a spectral density of the form $S_X(\omega) = \dfrac{50}{\omega^2 + 9}$ V^2/Hz. Determine the mean square value of the process in various octave bands in the audible range.

4.20. The following joint stationary random processes have time waveforms of the form $X(t) = 2 \sin(10t + \theta)$ and $Y(t) = 20 \cos(10t + \theta)$. Compute the cross correlation functions $R_{XY}(\tau)$ and $R_{YX}(\tau)$.

5

Flexural Vibration of Beams

Continuous systems like beams, plates, and shells are the most commonly used elements in the construction of any structure or machine. Flexural vibration in these elements, also known as bending vibration, is the vibration perpendicular to the beam axis or plate surface, and is the most commonly encountered wave type. In addition, flexural vibration very easily couples with the surrounding air and is thus the most dominant source of airborne noise from machines. Although plates are better candidates for producing airborne sound due to their flexural vibration, flexural vibration of beams is studied first since it is much easier to extend the earlier knowledge of longitudinal vibration in bars to the study of flexural vibration of beams. A fundamental difference between longitudinal waves and flexural waves is that the wave speeds due to flexural waves are frequency dependent, and, in addition, resonant frequencies of flexural vibration are not harmonic in nature; resonant frequencies of longitudinal vibration are harmonic and their wave speeds are frequency independent. Therefore, flexural waves produce unpleasant sound at high frequencies in the audible range.

The general differential equation for flexural vibration of beams is derived first. Then equations for natural frequencies and their corresponding mode shapes are derived. These are used to derive the general equations for resonant frequencies and mode shapes of cantilever and simple supported beams. Orthogonality conditions are derived and used to obtain uncoupled equations in terms of generalized coordinates, generalized masses, and forces to obtain the response. Then the wave approach is applied to flexural vibration to derive equations for bending wave speeds and bending wave numbers. The coincident condition of flexural waves with sound waves is presented that forms the basis for the definition of critical frequency. Phase velocity, group velocity, and modal density equations are obtained that are useful in statistical energy analysis (SEA) discussed in Chapter 10.

5.1 General Equation

The *Euler-Bernoulli beam theory* will be discussed in this section, which neglects the effect of shear deformation and rotary inertia. This theory can be applied to slender beams with a length-to-depth ratio greater than 100. The effect of shear and rotary inertia are accounted for in the *Timoshenko beam theory*. A slender beam that is subjected to a load $f(x,t)$ and a free body diagram of the cross section are shown in Figure 5.1. EI is *flexural rigidity*, where E is the Young's modulus of elasticity of the beam material and $I = I_{zz}$ the moment of inertia of the beam cross section about the z-z axis; ρ_l is the mass per unit length. I and ρ_l in general are not constant throughout the length of the beam; they are constant only for a beam of

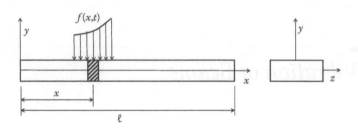

FIGURE 5.1
Beam subjected to a distributed excitation.

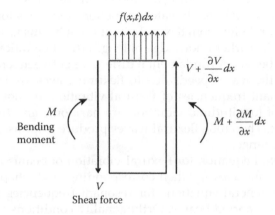

FIGURE 5.2
Shear forces and bending moments of the elemental area of Figure 5.1.

uniform cross section. The shear forces and bending moments of the elemental area of Figure 5.1 are shown in Figure 5.2. Summing the vertical forces

$$V + \frac{\partial V}{\partial X}dx - V + f(x,t)dx = \rho_l dx \frac{\partial^2 y}{\partial t^2} \tag{5.1}$$

Summing moments we obtain

$$V = \frac{\partial M}{\partial x} \tag{5.2}$$

where M is the bending moment. The bending moment is related to the beam curvature as follows:

$$M = -EI\psi \tag{5.3}$$

where Ψ is the curvature of the beam and is given by

$$\Psi = \frac{\dfrac{\partial^2 y}{\partial x^2}}{\left[1+\left(\dfrac{\partial y}{\partial x}\right)^2\right]^{1.5}} \approx \frac{\partial^2 y}{\partial x^2} \qquad (5.4)$$

With the preceding assumption, the partial (for very small slopes) differential equation corresponding to a beam of uniform cross section is given by

$$EI\frac{\partial^4 y}{\partial x^4}+\rho_l\frac{\partial^2 y}{\partial t^2}=f(x,t) \qquad (5.5)$$

5.2 Free Vibrations

For free vibration, $f(x,t) = 0$ and Equation 5.5 becomes

$$EI\frac{\partial^4 y}{\partial x^4}+\rho_l\frac{\partial^2 y}{\partial t^2}=0 \qquad (5.6)$$

Since Equation 5.6 is a linear partial differential equation, it can be written in terms of two linear ordinary differential equations by the method of separation of variables as follows:

$$y(x,t)=\phi(x)q(t) \qquad (5.7)$$

Substituting Equation 5.7 in Equation 5.6, the following equation is obtained:

$$-\frac{EI}{\rho}\frac{\dfrac{d^4\phi}{dx^4}}{\phi(x)}=\frac{\dfrac{d^2q}{dt^2}}{q(t)}=-\omega^2 \qquad (5.8)$$

From Equation 5.8 the following linear ordinary differential equations are obtained:

$$\phi^{iv}(x)-\frac{\omega^2\rho_l}{EI}\phi(x)=0 \qquad (5.9)$$

and

$$\ddot{q}(t)+\omega^2 q(t)=0 \qquad (5.10)$$

By defining a variable $k_b^4 = \dfrac{\omega^2 \rho_l}{EI}$,* Equation 5.9 can be written as follows:

$$\phi^{iv}(x) - k_b^4 \phi(x) = 0 \tag{5.11}$$

A general solution to Equation 5.11 is given by

$$\phi(x) = A \sinh k_b x + B \cosh k_b x + C \sin k_b x + D \cos k_b x \tag{5.12}$$

Equation 5.12 can be used to determine the natural frequencies and mode shapes of beams with various boundary conditions.

5.2.1 Cantilever Beam

The boundary conditions for the cantilever beam shown in Figure 5.3 are (a) deflection and slope are zero at the fixed end, and (b) the bending moment and shear forces are zero at the free end. These conditions are expressed by the following equations:

$$y(0) = 0; \left.\frac{dy}{dx}\right|_{x(0)} = 0 \tag{5.13}$$

$$\frac{d^2 y}{dx^2} = \left.\frac{d^3 y}{dx^3}\right|_{x=\ell} = 0 \tag{5.14}$$

Since Equations 5.13 and 5.14 are dependent on the space variable x, they can be expressed in terms of the function $\phi(x)$ as follows:

$$\phi(0) = \phi'(0) = \phi''(\ell) = \phi'''(\ell) = 0 \tag{5.15}$$

From Equations 5.15 and 5.12, two homogeneous linear equations can be obtained, which can be expressed in the matrix form as follows:

$$\begin{bmatrix} \sinh k_b \ell + \sin k_b \ell & \cosh k_b \ell + \cos k_b \ell \\ \cosh k_b \ell + \cos k_b \ell & \sinh k_b \ell - \sin k_b \ell \end{bmatrix} \begin{bmatrix} A \\ B \end{bmatrix} = 0 \tag{5.16}$$

The determinant of the matrix in Equation 5.16 equated to zero results in the following frequency equation:

$$\cosh k_b \ell \cos k_b \ell + 1 = 0 \tag{5.17}$$

Equation 5.17 is a *transcendental* equation, which has many roots. These roots can be obtained by using numerical methods, and are shown in Table 5.1, corresponding to the

---

* k_b will be interpreted as k_B the bending wave number in the wave approach.

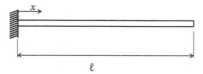

FIGURE 5.3
Cantilever beam.

TABLE 5.1

Roots of the Frequency Equation 5.17
(Cantilever Beam)

Mode No.	$k_b \ell$
1	1.87510406871200
2	4.69409113297420
3	7.85475743823760
4	10.99554073487500
5	14.13716839104600

first five natural frequencies. Each root corresponds to a natural frequency of the system. Corresponding to each natural frequency ω_n is a function $\phi_n(x)$. The constants A_n and B_n for each natural frequency cannot be explicitly determined; they can only be expressed in terms of each other. Assuming $B_n = 1$, the value of A_n is given by

$$A_n = -\frac{\cosh k_b \ell + \cos k_b \ell}{\sinh k_b \ell + \sin k_b \ell} = \alpha_n \tag{5.18}$$

Equations 5.12 and 5.18 can be combined as follows:

$$\phi_n(x) = \alpha_n [\sin k_b x - \sinh k_b x] + [\cosh k_b x - \cos k_b x] \tag{5.19}$$

Equation 5.19 is known as the *eigenfunction* or the *mode shape* for the nth natural frequency. It represents the relative displacement of the beam at different modes of flexural vibration. The mode shapes for the first four natural frequencies of a cantilever beam are shown in Figure 5.4.

5.2.2 Simply Supported Beam

The boundary conditions for a simple supported beam shown in Figure 5.5 are displacement and bending moment are zero at locations of the support. This can be mathematically represented as follows:

$$\phi(0) = 0, \; \phi(\ell) = 0, \; \phi''(0) = 0, \; \phi''(\ell) = 0 \tag{5.20}$$

Using Equations 5.12 and 5.20, the following frequency equation can be obtained:

$$\sin k_b \ell = 0 \tag{5.21}$$

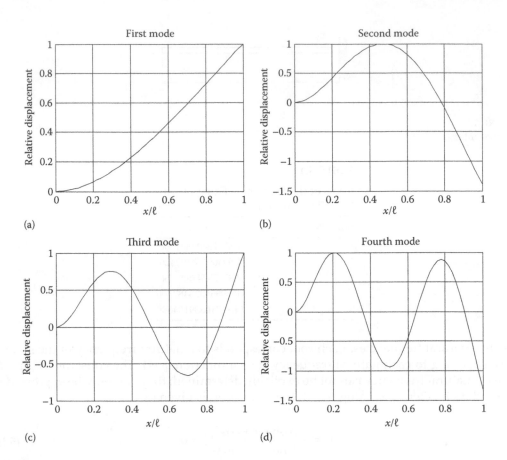

FIGURE 5.4
Mode shapes of a cantilever beam.

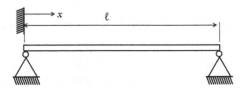

FIGURE 5.5
Simply supported beam.

Therefore, $k_b \ell = \dfrac{n\pi}{\ell}$, for $n = 1, 2, 3, \ldots$, satisfies Equation 5.21 and corresponds the natural frequencies of the simply supported beam. The mode shapes of the simply supported beam are given by

$$\phi_n(x) = D_n \sin \frac{n\pi x}{\ell} \tag{5.22}$$

where D_n is an arbitrary constant. The mode shapes of a simply supported beam are as shown in Figure 5.6.

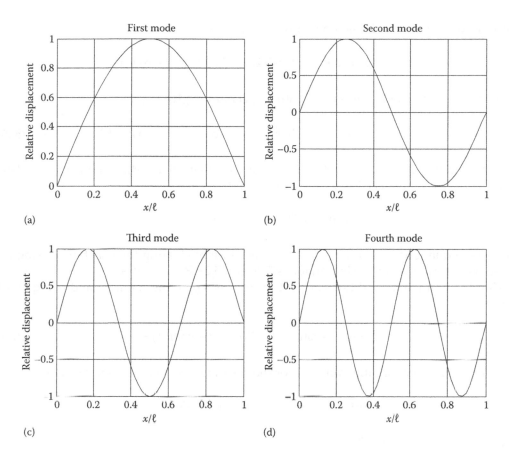

FIGURE 5.6
Mode shapes of a simply supported beam.

5.2.3 Free-Free Beam

A free-free beam is often used as a model for space structures and the boundary conditions are that the bending moment and shear force are zero at the free ends. These boundary conditions can be obtained in practice by supporting the free ends by elastic ropes.

From Equation 5.11, the general equation for the mode shape of a beam in flexural vibration is given by

$$\phi^{iv}(x) - k_b^4 \phi(x) = 0 \tag{5.23}$$

The solution of Equation 5.23 is given by

$$\phi(x) = A \sinh k_b x + B \cosh k_b x + C \sin k_b x + D \cos k_b x \tag{5.24}$$

A, B, C, and D are constants that can be solved from boundary conditions.

Before applying boundary conditions, let us obtain the first three derivatives of Equation 5.24 with respect to the spatial variable x. The first derivative of Equation 5.24 with respect to x is given by

$$\phi'(x) = A \cosh k_b x + B \sinh k_b x + C \cos k_b x - D \sin k_b x \tag{5.25}$$

The second derivative of Equation 5.24 with respect to x is given by

$$\phi''(x) = (A \sinh k_b x + B \cosh k_b x - C \sin k_b x - D \cos k_b x)k_b^2 \qquad (5.26)$$

The third derivative of Equation 5.24 with respect to x is given by

$$\phi'''(x) = (A \cosh k_b x + B \sinh k_b x - C \cos k_b x + D \sin k_b x)k_b^3 \qquad (5.27)$$

The boundary conditions of a free-free beam are that the bending moment and shear force are zero at both ends. They can be expressed as

$$\phi''(0) = \phi'''(0) = \phi''(\ell) = \phi'''(\ell) = 0 \qquad (5.28)$$

Therefore, Equations 5.26 and 5.27 are to be used for evaluating the constants. Using the boundary conditions at $x = 0$ in Equations 5.26 and 5.27, $B = D$ and $A = C$.

Using the boundary conditions at $x = \ell$ for the zero bending moment and using $D = B$ and $C = A$

$$A(\sinh k_b \ell - \sin k_b \ell) + B(\cosh k_b \ell - \cos k_b \ell) = 0 \qquad (5.29)$$

Using the boundary conditions at $x = \ell$ for the zero shear force and using $D = B$ and $C = A$

$$A(\cosh k_b \ell - \cos k_b \ell) + B(\sinh k_b \ell + \sin k_b \ell) = 0 \qquad (5.30)$$

From Equations 5.29 and 5.30

$$\begin{bmatrix} \sinh k_b \ell - \sin k_b \ell & \cosh k_b \ell - \cos k_b \ell \\ \cosh k_b \ell - \cos k_b \ell & \sinh k_b \ell + \sin k_b \ell \end{bmatrix} \begin{Bmatrix} A \\ B \end{Bmatrix} = 0 \qquad (5.31)$$

Equation 5.31 represents a system of linear homogenous equations and hence the determinant is equal to zero

$$\begin{vmatrix} \sinh k_b \ell - \sin k_b \ell & \cosh k_b \ell - \cos k_b \ell \\ \cosh k_b \ell - \cos k_b \ell & \sinh k_b \ell + \sin k_b \ell \end{vmatrix} = 0 \qquad (5.32)$$

Expanding the determinant of Equation 5.32

$$\cosh k_b \ell \cos k_b \ell = 1 \qquad (5.33)$$

Solving Equation 5.33 using the Newton-Raphson method or bisection method, the possible values of $k_b \ell$ are 0, 4.73, 7.85, 10.99, and so on. (0 corresponds to the rigid body mode.)

For every $k_b \ell$ satisfying Equation 5.33, Equation 5.29 becomes

$$A_n \left(\cosh k_{b_n} \ell - \cos k_{b_n} \ell \right) + B_n \left(\sinh k_{b_n} \ell + \sin k_{b_n} \ell \right) = 0 \qquad (5.34)$$

Assuming $B_n = 1$,

$$A_n = -\frac{\left(\cosh k_{b_n}\ell - \cos k_{b_n}\ell\right)}{\left(\sinh k_{b_n}\ell + \sin k_{b_n}\ell\right)} \tag{5.35}$$

From Equations 5.24 and 5.35 the nth mode shape of a free-free beam is given by

$$\phi_n(x) = A_n\left(\cosh k_{b_n}\ell - \cos k_{b_n}\ell\right) + \left(\sinh k_{b_n}\ell + \sin k_{b_n}\ell\right) \tag{5.36}$$

5.3 Orthogonality Conditions

Applying boundary conditions to Equation 5.12, the natural frequencies and the corresponding mode shapes can be obtained. The index of these natural frequencies and mode shapes is defined by n, which can take values 1, 2, ..., ∞.

The mode shape and wave number of a beam under flexural vibration is given by

$$\phi^{iv}(x) = k_B^4\phi(x) \tag{5.37}$$

where k_B is the bending wave number.

The response of the beam system to any general excitation can be expressed as the linear combination of its mode shapes. This can be expressed as

$$y(x,t) = \sum_{n=1}^{\infty} q_n(t)\phi_n(x) \tag{5.38}$$

$q_n(t)$ are the time-dependent linear combination parameters that link mode shapes with the response, and they are dependent on the force excitation and characteristics of the dynamic system; they are also known as generalized coordinates.

Differentiating Equation 5.38 four times with respect to the space variable x,

$$\frac{\partial^4 y}{\partial x^4} = \sum_{n=1}^{\infty} q_n(t)\phi_n^{iv}(x) \tag{5.39}$$

The second derivative of Equation 5.38 with respect to time is given by

$$\frac{\partial^2 y}{\partial t^2} = \sum_{n=1}^{\infty} \ddot{q}_n(t)\phi_n(x) \tag{5.40}$$

The beam equation can be recalled again:

$$EI\frac{\partial^4 y}{\partial x^4} + \rho_\ell \frac{\partial^2 y}{\partial t^2} = f(x,t) \tag{5.41}$$

Substituting Equations 5.39 and 5.40 in Equation 5.41

$$EI\sum_{n=1}^{\infty} q_n(t)\phi_n^{iv}(x) + \rho_\ell \sum_{n=1}^{\infty} \ddot{q}_n(t)\phi_n(x) = f(x,t) \tag{5.42}$$

Using Equation 5.37, Equation 5.42 becomes

$$EI\sum_{n=1}^{\infty} q_n(t)k_{B_n}^4\phi_n(x) + \rho_\ell \sum_{n=1}^{\infty} \ddot{q}_n(t)\phi_n(x) = f(x,t) \tag{5.43}$$

Using the relationship between bending wave number and natural frequency, Equation 5.43 becomes

$$EI\sum_{n=1}^{\infty} q_n(t)\frac{\omega_n^2\rho_\ell}{EI}\phi_n(x) + \rho_\ell \sum_{n=1}^{\infty} \ddot{q}_n(t)\phi_n(x) = f(x,t) \tag{5.44}$$

On simplification, Equation 5.44 becomes

$$\sum_{n=1}^{\infty} \phi_n(x)\left[\omega_n^2 q_n(t) + \ddot{q}_n(t)\right] = \frac{f(x,t)}{\rho_\ell} \tag{5.45}$$

Multiplying both sides of Equation 5.45 with a mode shape of index m and integrating along the beam length, it becomes

$$\sum_{n=1}^{\infty}\left[\int_0^\ell \phi_n(x)\phi_m(x)\,dx\right]\left[\omega_n^2 q_n(t) + \ddot{q}_n(t)\right] = \frac{1}{\rho_\ell}\int_0^\ell \phi_m(x)f(x,t)\,dx \tag{5.46}$$

Due to a property of the mode shapes expressed as

$$\int_0^\ell \phi_n(x)\phi_m(x)\,dx = 0 \text{ for } n \neq m \tag{5.47}$$

Equation 5.46 has a nonzero value only for $n = m$. Therefore, only one term of the summation of Equation 5.46 will be left.

For $m = n$, Equation 5.46 becomes

$$\ddot{q}_n(t) + \omega_n^2 q_n(t) = \frac{\displaystyle\int_0^\ell \phi_n(x)f(x,t)dx}{\displaystyle\int_0^\ell \rho_\ell \phi_n^2(x)dx} \tag{5.48}$$

The denominator of the right-hand side of Equation 5.48 is known as the generalized mass, given by

$$m_n = \int_0^\ell \rho_\ell \phi_n^2(x)dx \tag{5.49}$$

The generalized force is given by

$$Q_n(t) = \frac{\displaystyle\int_0^\ell \phi_n(x)f(x,t)dx}{m_n} \tag{5.50}$$

The uncoupled equations of motion for the beam in terms of generalized coordinates and forces are given by

$$\ddot{q}_n(t) + \omega_n^2 q_n(t) = Q_n(t) \tag{5.51}$$

The uncoupled equations of Equation 5.51 can be solved for each mode and then the total solution can be obtained from the equation. Therefore, this general procedure can be used for any arbitrary excitation.

Example 5.1

A simply supported beam of rectangular cross section (width = 40 mm and thickness = 5 m) and length $\ell = 0.5$ m is subjected to a ramp input at its center that reaches unit force in 10 sec (see Figure 5.7). The material of the beam is steel, $E = 200$ GPa, and the density is 7800 kg/m³. Determine the beam response at $t = 5$ sec by accounting for the first five modes.

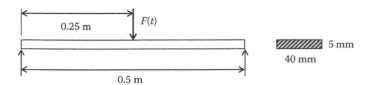

FIGURE 5.7
Simply supported beam of Example 5.1.

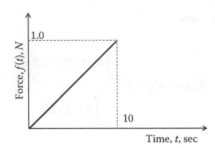

FIGURE 5.8
Time history of the force acting on the simply supported beam of Figure 5.7.

The force acting at the center, as shown in Figure 5.8, can be defined by

$$f(x,t) = 0.1t\delta\left\langle x - \frac{\ell}{2}\right\rangle \tag{1}$$

From Equation 5.22, the mode shape for a simple supported beam is given by

$$\phi_n(x) = \sin\left(\frac{n\pi x}{\ell}\right), n = 1,2,3... \tag{2}$$

From Equation 5.49, the generalized mass is given by

$$m_n = \int_0^\ell \rho A \phi_n^2(x)\,dx$$

$$= \rho A \int_0^\ell \sin^2\left(\frac{n\pi x}{\ell}\right)dx = \frac{\rho A \ell}{2} = 0.39 \tag{3}$$

From Equation 5.50, the generalized force is given by

$$Q_n(t) = \frac{1}{m_n}\int_0^\ell f(x,t)\phi_n(x)\,dx$$

$$= \frac{0.1}{0.39}\int_0^\ell t\delta\left\langle x - \frac{\ell}{2}\right\rangle \sin\left(\frac{n\pi x}{\ell}\right)dx \tag{4}$$

$$= 0.2564t\sin\left(\frac{n\pi}{2}\right), n = 1,2,3...$$

$$= 0.2564t(-1)^{\frac{n-1}{2}}, \quad n = 1,3,5...$$

From Equation 5.51, the uncoupled equations for the beam are given by

$$\ddot{q}_n + \omega_n^2 q_n = 0.2564t(-1)^{\frac{n-1}{2}}, n = 1,3,5... \tag{5}$$

From Equation 1.51 (see Chapter 1), the impulse response function of the uncoupled equations of the beam is given by

$$h_n(t) = \frac{\sin \omega_n t}{\omega_n} \qquad (6)$$

From Equations 5 and 6, and Equation 1.55 (see Chapter 1), the beam response in principal coordinates is given by

$$q_n(t) = \int_0^t h_n(\tau) Q_n(t-\tau) d\tau$$

$$= \frac{0.2564(-1)^{\frac{n-1}{2}}}{\omega_n} \int_0^t \sin \omega_n \tau (t-\tau) d\tau \qquad (7)$$

$$= \frac{0.2564(-1)^{\frac{n-1}{2}} (\omega_n t - \sin \omega_n t)}{\omega_n^3} \qquad n = 1,3,5\ldots$$

From Equation 5.38 and Equation 7, the response at any location of the beam is given by

$$y(x,t) = \sum_{n=1,3,5}^{\infty} \frac{0.2564(-1)^{\frac{n-1}{2}} (\omega_n t - \sin \omega_n t) \sin \frac{n\pi x}{\ell}}{\omega_n^3} \qquad n = 1,3,5\ldots \qquad (8)$$

From Equation 8, the displacement at the center of the beam is given by

$$y\left(\frac{\ell}{2},t\right) = 0.2564 \sum_{n=1,3,5}^{\infty} \frac{(-1)^{n-1}(\omega_n t - \sin \omega_n t)}{\omega_n^3} \qquad n = 1,3,5\ldots \qquad (9)$$

The moment of inertia of the beam cross section is given by

$$I = \frac{bh^3}{12} = \frac{0.040 \times 0.005^3}{12} = 4.17 \times 10^{-10} \, m^4 \qquad (10)$$

Using the wave number for a simple supported beam $k_n = n\pi/L$, the undamped natural frequencies of the beam are given by

$$\omega_n = \left(\frac{n\pi}{\ell}\right)^2 \sqrt{\frac{EI}{\rho A}}$$

$$= \left(\frac{n\pi}{\ell}\right)^2 \sqrt{\frac{200 \times 10^9 \times 4.17 \times 10^{-10}}{7800 \times 0.04 \times 0.005}} = 288.66 \, n^2 \qquad (11)$$

From Equation 9, the response at the center of the beam at $t = 5$ is given by

$$y(\ell/2,5) = 0.2564 \left[\begin{array}{c} \dfrac{288.66 \times 1^2 \times 5 - \sin 288.66 \times 5}{(288.66 \times 1^2)^3} \\[3mm] + \dfrac{288.66 \times 3^2 \times 5 - \sin 288.66 \times 3^2 \times 5}{(288.66 \times 3^2)^3} \\[3mm] + \dfrac{288.66 \times 5^2 \times 5 - \sin 288.66 \times 5^2 \times 5}{(288.66 \times 5^2)^3} \end{array} \right] = 15.4 \; \mu m \qquad (12)$$

Example 5.2

A long steel rod ($E = 200$ GPa, density = 7800 kg/m³) of 0.5 m length and 25 mm diameter is mounted as a simply supported beam as shown in Figure 5.9.

- a. Determine the first four natural frequencies in Hz.
- b. Determine the generalized mass for the first four resonant frequencies.
- c. If a unit impulse is applied at the center of the beam, determine the generalized force for the first four resonant frequencies.
- d. What is the percentage contribution of the first four modes toward the response at the center of the beam due to the force of part c?
- e. Determine the number of modes of this beam at octave-band frequencies from 31.5 to 1000 Hz.
- f. If a mass m moves along the beam from one end to another at a velocity v, determine the velocities that result in maximum response?

ANSWERS

a. The area of the beam cross section is given by

$$A = \frac{\pi d^2}{4} = \frac{\pi \times (25 \times 10^{-3})^2}{4} = 4.91 \times 10^{-4} \; m^2 \qquad (1)$$

Moment of inertia of the beam cross section is given by

$$I = \frac{\pi d^4}{64} = \frac{\pi \times (0.025)^4}{64} = 1.92 \times 10^{-8} \; m^4 \qquad (2)$$

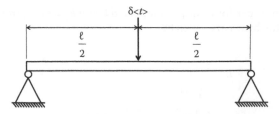

FIGURE 5.9
Simply supported beam of Example 5.2.

The density per unit length of the beam is given by

$$\rho_\ell = \rho A = 7800 \times 4.91 \times 10^{-4} = 3.83 \text{ kg/m} \tag{3}$$

The wave number for a simple supported beam is given by

$$k_{b_n} = \frac{n\pi}{\ell}, n = 1,2,3... \tag{4}$$

Resonance frequencies of the beam are given by

$$\omega_n = \sqrt{\frac{EI}{\rho_\ell}} \times k_{b_n}^2 = n^2 \sqrt{\frac{200 \times 10^9 \times 1.92 \times 10^{-8}}{3.83}} \times \left(\frac{\pi}{0.5}\right)^2 \tag{5}$$

$$= 1250 \, n^2$$

Using Equation 5, the first four resonance frequencies are computed and are as shown in Table 5.2.

b. The mode shape of a simple supported beam is given by

$$\phi_n(x) = \sin\frac{n\pi x}{\ell} \quad n = 1,2,3... \tag{6}$$

From Equation 5.49, the generalized mass is given by

$$m_n = \int_0^\ell \rho A \phi_n^2(x) \, dx \tag{7}$$

$$= \rho A \int_0^\ell \sin^2\left(\frac{n\pi x}{\ell}\right) dx = \frac{\rho_\ell \ell}{2}$$

c. The unit impulse force acting at the center can be defined by

$$f(x,t) = \delta\left\langle x - \frac{\ell}{2}\right\rangle \delta\langle t\rangle \tag{8}$$

TABLE 5.2

Resonant Frequencies of the Simple Supported Beam

Mode Number, n	Resonant Frequency (rad/s)	Resonance Frequency (Hz)
1	1250	198.94
2	5000	795.77
3	11,250	1790.49
4	20,000	3183.09

From Equation 5.50, the generalized force is given by

$$Q_n(t) = \frac{1}{m_n} \int_0^\ell f(x,t)\phi_n(x)dx$$

$$= \frac{2}{\rho_\ell \ell} \int_0^\ell \delta\left\langle x - \frac{\ell}{2} \right\rangle \delta\langle t \rangle \sin\left(\frac{n\pi x}{\ell}\right)dx \qquad (9)$$

$$= \frac{2}{\rho_\ell \ell} \sin\left(\frac{n\pi}{2}\right)\delta\langle t \rangle, n = 1,2,3\ldots$$

$$= (-1)^{\frac{n-1}{2}} \frac{2}{\rho_\ell \ell}\delta\langle t \rangle, n = 1,3,5\ldots$$

The uncoupled equations of motion are given by

$$\ddot{q}_n + \omega_n^2 q_n = (-1)^{\frac{n-1}{2}} \frac{2}{\rho_\ell \ell}\delta\langle t \rangle, n = 1,3,5\ldots \qquad (10)$$

From Equation 1.51, the generalized displacement is given by

$$q_n(t) = \frac{(-1)^{\frac{n-1}{2}} 2\sin\omega_n t}{\rho_\ell \ell \omega_n}, n = 1,3,5\ldots \qquad (11)$$

From Equation 5.38, and Equations 6 and 11 the displacement at any location is given by

$$y(x,t) = \sum_{n=1}^\infty \phi_n(x)q_n(t)$$

$$= \sum_{n=1,3,5}^\infty \sin\left(\frac{n\pi x}{\ell}\right)\frac{(-1)^{\frac{n-1}{2}} 2\sin\omega_n t}{\rho_\ell \ell \omega_n} \qquad (12)$$

From Equation 11, the displacement response at the middle of the beam is given by

$$y\left(\frac{\ell}{2},t\right) = \frac{2}{\rho_\ell \ell} \sum_{n=1,3,5}^\infty \frac{(-1)^{n-1}\sin\omega_n t}{\omega_n} \qquad (13)$$

From Equations 5 and 13 the displacement response at the middle of the beam is given by

$$y\left(\frac{\ell}{2},t\right) = \frac{2}{\rho_\ell \ell} \sum_{n=1,3,5}^\infty \frac{(-1)^{n-1}\sin\omega_n t}{1250\,n^2} \qquad (14)$$

FIGURE 5.10
Simply supported beam with moving load.

d. From Equation 14, the peak amplitude of response at various resonance frequencies is given by

$$\frac{A_3}{A_1} = \frac{1}{9}, \frac{A_5}{A_1} = \frac{1}{25} \tag{15}$$

e. As shown in Figure 5.10, the force excitation for a moving load of mass moving at velocity v is given by

$$f(x,t) = mg\delta\langle x - vt\rangle \tag{16}$$

The generalized force for the above force acting on the beam is given by

$$Q_n(t) = \frac{1}{m_n} \int_0^\ell f(x,t)\phi_n(x)dx$$

$$= \frac{2}{\rho_\ell \ell} \int_0^\ell mg\delta\langle x - vt\rangle \sin\left(\frac{n\pi x}{\ell}\right)dx \tag{17}$$

$$= \frac{2}{\rho_\ell \ell} \sin\left(\frac{nvt}{2}\right)$$

The uncoupled equations of motion are given by

$$\ddot{q}_n + \omega_n^2 q_n = \frac{2mg}{\rho_\ell \ell} \sin\left(\frac{nvt}{\ell}\right) \tag{18}$$

The generalized displacement is given by

$$q_n(t) = \frac{2mg}{\rho_\ell \ell \left[\left(\frac{nv}{\ell}\right)^2 - \omega_n^2\right]} \sin\left(\frac{nvt}{\ell}\right) \tag{19}$$

The equation for displacement due to the moving load is given by

$$y(x,t) = \frac{2mg}{\rho_\ell \ell} \sum_{n=1}^\infty \frac{\sin\left(\frac{nvt}{\ell}\right)}{\left[\left(\frac{nv}{\ell}\right)^2 - \omega_n^2\right]} \tag{20}$$

When $\left(\dfrac{nv}{\ell}\right) = \omega_n$ the denominator of Equation 20 becomes zero resulting in a resonant type of vibration, and the value of velocity that can cause this type of vibration is given by

$$v_n = 1250\ell n \qquad\qquad (21)$$

From Equation 21, typical values of velocity are 625 m/s (2250 kmph) and 1250 m/s (4500 kmph), which are impossible for the present problem.

Example 5.3

A steel cantilever (E = 200 GPa, ρ = 7800 kg/m³) of 500 mm length, 50 mm width, and 5 mm thickness, as shown in Figure 5.11, is subjected to an impulse by dropping a 100 g rigid sphere from a height of 200 mm. Determine the maximum free-end deflection considering *only the first mode*. Assume a deflection shape for the fundamental frequency, $\phi(x) = \left(1 - \dfrac{x}{\ell}\right)^2$ and measure the distance along the beam with origin at the free end.

Assume that the sphere will move away from the beam after giving only one impulse.

The given mode shape satisfies most of the boundary conditions of the cantilever beam.

The velocity at the time of impact is given by

$$v(0) = \sqrt{2gh} = \sqrt{2 \times 9.81 \times 0.2} = 1.98 \text{ m/s} \qquad\qquad (1)$$

The impulse imparted by the rigid sphere onto the beam is given by

$$mv(0) = 0.1 \times 1.98 = 0.198 \text{ kg-m/s} \qquad\qquad (2)$$

The mass per unit length of the beam is given by

$$\rho_\ell = \rho A = 7800 \times 0.05 \times 0.005 = 1.95 \text{ kg/m} \qquad\qquad (3)$$

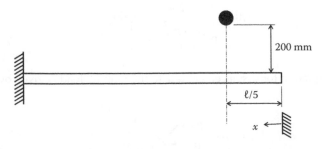

FIGURE 5.11
Cantilever beam with impulsive load.

The generalized mass of the first mode is given by

$$
\begin{aligned}
m_1 &= \int_0^\ell \rho_\ell \phi_1^2 \, dx \\
&= \int_0^\ell \rho_\ell \left(1 - \frac{x}{\ell}\right)^4 dx = 0.195
\end{aligned}
\tag{4}
$$

The force excitation is given by

$$
F(t) = mv(0)\delta\langle t \rangle
\tag{5}
$$

The generalized force is given by

$$
\begin{aligned}
Q_1 &= \frac{1}{m_1} \int_0^\ell mv(0)\delta\langle t \rangle \delta\left\langle x - \frac{\ell}{5} \right\rangle \left(1 - \frac{x}{\ell}\right)^2 dx \\
&= 0.65\delta\langle t \rangle
\end{aligned}
\tag{6}
$$

Since $k_1\ell = 1.875$ for a cantilever beam, the wave number for the first mode is $1.875/0.5 = 3.75$ rad/m.

The moment of inertia of the beam cross section is given by

$$
I_{zz} = \frac{bh^3}{12} = \frac{0.05 \times 0.005^3}{12} = 5.208 \times 10^{-10} \, \text{m}^4
\tag{7}
$$

The first resonant frequency is given by

$$
\omega_1 = k_1^2 \sqrt{\frac{EI_{zz}}{\rho_\ell}} = 3.75^2 \sqrt{\frac{200 \times 10^9 \times 5.208 \times 10^{-10}}{1.95}} = 102.8 \, \text{rad/s} \ (16.36 \, \text{Hz})
\tag{8}
$$

The uncoupled equation of motion for the first mode is given by

$$
\ddot{q}_1 + \omega_1^2 q_1 = Q_1(t)
\tag{9}
$$

Using the resonant frequency and generalized force, Equation 9 becomes

$$
\ddot{q}_1 + 102.8^2 q_1 = 0.65\delta\langle t \rangle
\tag{10}
$$

From Equation 1.51, the impulse response function of Equation 10 is given by

$$
h(t) = \frac{\sin \omega_1 t}{\omega_1}
\tag{11}
$$

From Equations 10 and 11, the response becomes

$$q_1(t) = 0.65h(t) = \frac{0.65\sin 102.8t}{102.8}$$

$$= 6.323\sin 102.8t \, \text{mm} \tag{12}$$

From Equation 5.24, the actual response is given by

$$y(0) = \phi_1(0)q_1(t)$$

$$= 6.323\sin 102.8t \, \text{mm} \tag{13}$$

Therefore the peak amplitude of displacement response is 6.3 mm.

5.4 Wave Approach to Flexural Vibration

The differential equation for the flexural vibration of an Euler-Bernoulli beam is given by

$$EI\frac{\partial^4 y}{\partial x^4} + \rho_\ell \frac{\partial^2 y}{\partial t^2} = f(x,t) \tag{5.52}$$

The displacement response can be expressed in terms of the following functions as

$$y(x,t) = \phi(x)q(t) \tag{5.53}$$

For free vibration, $f(x,t) = 0$, and Equation 5.53 when substituted in Equation 5.38 results in and using $k_B = k_b$

$$\phi^{iv}(x) = k_B^4\phi(x) \tag{5.54}$$

where k_B is the bending wave number given by

$$k_B^4 = \frac{\omega^2 \rho_\ell}{EI} \tag{5.55}$$

Applying boundary conditions to the solution of Equation 5.54 results in natural frequencies and their corresponding mode shapes; the bending wave number and the corresponding natural frequencies will be related by Equation 5.55.

The main focus of this section, however, is to use the wave approach to flexural vibration of beams that yields useful results for vibro-acoustics. Equation 5.54 can be written as a complex differential equation, given by

$$\tilde{\phi}^{iv}(x) = k_B^4\tilde{\phi}(x) \tag{5.56}$$

The solution to Equation 5.56 results in

$$\tilde{\phi}(x) = \bar{A}e^{jk_Bx} + \bar{B}e^{-jk_Bx} + \bar{C}e^{k_Bx} + \bar{D}e^{-k_Bx} \qquad (5.57)$$

The first and third terms are left-moving waves, and the second and fourth terms are right-moving waves. The third and fourth terms exponentially decay with distance and are known as evanescent waves. Evanescent waves do not carry any net energy nor do they relate to airborne sound far away from the beam and vibration away from the source. However, the airborne sound produced by evanescent waves has been used in recent times for obtaining information from near-field measurements such as acoustic holography; however, the measurements will have to be conducted very close to the beam. But we do not consider them in further analysis.

Corresponding to a harmonic frequency ω, the complex solution for displacement becomes

$$\tilde{y}(x,t) = \bar{A}e^{j(\omega t + k_Bx)} + \bar{B}e^{j(\omega t - k_Bx)} \qquad (5.58)$$

Except for the direction of vibration, Equation 5.58 representing flexural vibration is very similar to the equation representing longitudinal vibration except for the frequency-dependent speed of wave propagation. In case of longitudinal vibration, waves of all frequencies travel at the same speed; whereas in the case of flexural waves, the speed of waves is related to frequency. Therefore, the phenomenon of flexural wave propagation is entirely different to that of longitudinal wave propagation. Let us first derive the equations for wave speeds of flexural and longitudinal vibration.

Equation 5.55 can be written as

$$k_B^4 = \frac{\omega^2 \rho_\ell}{EA\kappa^2} \qquad (5.59)$$

where A is the area of cross section and κ is the radius of gyration.

In terms of the longitudinal wave speed, $c_L = \sqrt{\dfrac{E}{\rho}}$, Equation 5.59 becomes

$$k_B^4 = \frac{\omega^2}{c_L^2\kappa^2} \; ; k_B = \sqrt{\frac{\omega}{c_L\kappa}} \qquad (5.60)$$

The bending wave number, bending wave speed, and frequency are related by

$$c_B = \frac{\omega}{k_B} \qquad (5.61)$$

From Equations 5.60 and 5.61, the bending wave speed is given by

$$c_B = \sqrt{\omega c_L \kappa} \qquad (5.62)$$

Therefore, bending wave speeds are frequency dependent, unlike those of longitudinal waves that are constant. This leads to two major classifications of waves: dispersive

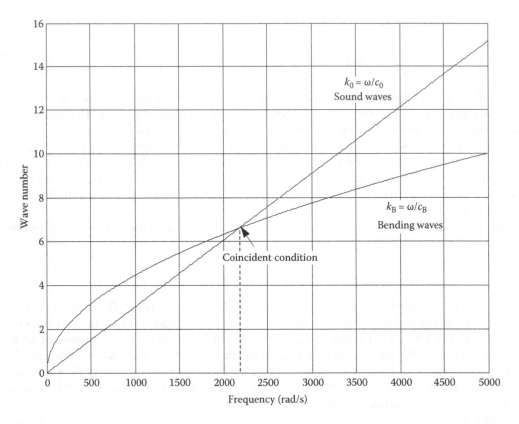

FIGURE 5.12
Variation of wave number versus frequency for various wave types: sound waves and bending waves (sound speed, 330 m/s; longitudinal wave speed, 5000 m/s; and radius of gyration, 10 mm).

waves that are dependent on frequency and nondispersive waves that are independent of frequency.

Figure 5.12 shows the variation of wave number with frequency for various wave types: longitudinal sound waves and flexural beam vibrations. For the longitudinal sound waves, the variation is linear and the speeds of these waves are independent of frequency. However, for bending waves, the variation of wave number with frequency is not linear, because the bending waves are frequency dependent. At a specific frequency, the wave number of both wave types is the same; this is known as coincident condition and the frequency is known as critical frequency. It will be later proved that bending waves radiate maximum sound at this frequency. Since bending waves are frequency dependent, all frequencies contained in a typical pulse travel at different speeds, resulting in its distortion as it travels through the structure.

5.5 Phase Velocity and Group Velocity

Consider the following equation:

$$y(x, t) = \sin(\omega t - kx) \tag{5.63}$$

Equation 5.63 satisfies the flexural vibration Equation 5.52 for a single frequency excitation; it can also satisfy the equation for longitudinal vibration, although the direction of vibration and corresponding wave numbers are different. Equation 5.63 indicates that for a single frequency excitation, the time variation and spatial variation will have to adjust themselves for wave propagation.

Now, let us consider the flexural vibration corresponding to two excitation frequencies, ω_1 and ω_2; let k_1 and k_2 be the corresponding wave numbers. The response becomes

$$y(x,t) = \sin(\omega_1 t - k_1 x) + \sin(\omega_2 t - k_2 x) \tag{5.64}$$

If the frequencies ω_1 and ω_2 are close to each other, Equation 5.64 becomes

$$y(x,t) = 2\sin\left[\left(\frac{\omega_1 + \omega_2}{2}\right)t - \left(\frac{k_1 + k_2}{2}\right)x\right]\cos\left[\left(\frac{\omega_1 - \omega_2}{2}\right)t - \left(\frac{k_1 - k_2}{2}\right)x\right] \tag{5.65}$$

The sinusoidal component is the carrier wave at the average frequency and average wave number. The amplitude of this wave is modulated by a cosine wave at a much lower frequency $\left(\dfrac{\omega_1 - \omega_2}{2}\right)$ and lower wave number $\left(\dfrac{k_1 - k_2}{2}\right)$. The lower frequency cosine wave is the envelope. Equation 5.65 is plotted for two closely spaced frequencies, 10 and 11 Hz, as shown in Figure 5.13.

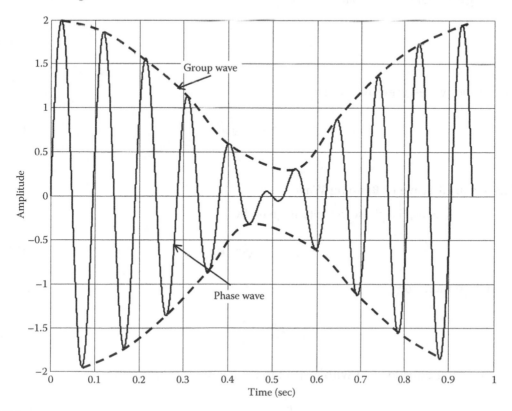

FIGURE 5.13
Combination of two waves of 10 Hz and 11 Hz.

The speed of the carrier wave is given by

$$c_p = \frac{\omega_1 + \omega_2}{k_1 + k_2} \tag{5.66}$$

which is known as the phase velocity, because this is the speed at which one has to travel to keep up with the carrier wave.

The speed of the envelope wave is given by

$$c_g = \frac{\omega_1 - \omega_2}{k_1 - k_2} = \frac{\Delta\omega}{\Delta k} = \frac{d\omega}{dk} \tag{5.67}$$

which is known as group velocity, because this is the velocity at which the average energy is transported. Since resonance frequencies can be very close to each other, group velocity is equal to the variation of frequency with the wave number.

5.5.1 Longitudinal Waves

For longitudinal waves in bars and sound waves, the wave number linearly varies with frequency. Therefore, phase velocity and group velocity are equal.

For longitudinal waves in bars, they are given by

$$c_p = c_g = c_L = \sqrt{\frac{E}{\rho}} \tag{5.68}$$

For sound waves, they are given by

$$c_p = c_g = c_o = \sqrt{\gamma R T} \tag{5.69}$$

5.5.2 Flexural Wave or Bending Waves

For bending waves, the phase velocity is given by

$$c_p = c_B = \sqrt{\omega c_L \kappa} \tag{5.70}$$

From Equation 5.60,

$$k_B^2 = \frac{\omega}{c_L \kappa} \tag{5.71}$$

Differentiating Equation 5.71 with respect to the bending wave number

$$\frac{d\omega}{dk_B} = 2 k_B c_L \kappa \tag{5.72}$$

Since $c_B = \dfrac{\omega}{k_B}$, Equation 5.72 becomes

$$c_g = \frac{2\omega c_L \kappa}{c_B} \tag{5.73}$$

From Equations 5.62 and 5.73

$$c_g = \frac{2\omega c_L \kappa}{\sqrt{\omega c_L \kappa}} = 2c_B \tag{5.74}$$

Therefore, the group velocity of bending waves is twice the bending wave speed. Group velocity has a smaller frequency and hence a longer wavelength and travels twice as fast as the bending waves.

5.6 Modal Density

While analyzing structural elements at higher frequencies, the number of resonances within a frequency band is a reliable parameter than the actual value of resonances. This is expressed as modal density and is illustrated with the example of a simply supported beam under flexural vibration. The wave number for the nth natural frequency is related to the length of the beam as follows:

$$k_{B_n} = \frac{n\pi}{\ell} \tag{5.75}$$

where L is the length of the beam. The product of the wave number and length of the beam, $k_{B_n}\ell$ is a nondimensional quantity, which is only dependent on the boundary conditions and independent of the physical dimensions of the beam.

It would be of interest to know how many resonant modes are present up to a certain wave number. The wave numbers for different resonant frequencies are plotted along a straight line as shown in Figure 5.14. It can be seen that the wave numbers are separated by π/L. The number of modes, present up to a wave number, is given by the number of

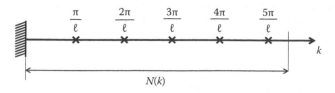

FIGURE 5.14
One-dimensional wave number.

equispaced divisions of π/L. If $N(k)$ are the number of modes present up to a wave number k, they are obtained as integer ratios of k and π/L as follows:

$$N(k) = \frac{k\ell}{\pi} \tag{5.76}$$

The number of modes per unit wave number, known as modal density, is given by the following equation:

$$n(k) = \frac{\ell}{\pi} \tag{5.77}$$

where $n(k)$ is the modal density in wave number. The modal density in frequency $n(\omega)$ is given by

$$n(\omega) = \frac{n(k)}{\dfrac{d\omega}{dk}} = \frac{n(k)}{c_g} \tag{5.78}$$

Equation 5.78 is a general equation and $c_g = \dfrac{d\omega}{dk}$ is the group velocity. The above equations are valid for any type of waves, hence the subscript is dropped.

Therefore, using Equations 5.74 and 5.78, the bending wave modal density in hertz is given by

$$n(f) = n(\omega)\frac{d\omega}{df} = \frac{\ell}{\pi c_g}2\pi = \frac{\ell}{c_B} = \frac{\ell}{\sqrt{\omega c_L \kappa}} \tag{5.79}$$

The average separation of the modes is the reciprocal of modal density. For bending waves, the average separation frequency, $\overline{\delta f_b}$, is given by

$$\overline{\delta f_b} = \frac{\sqrt{\omega c_L \kappa}}{\ell} \tag{5.80}$$

Example 5.4

A long steel beam ($E = 200$ GPa, density $= 7800$ kg/m³) of 0.5 m length and 25 mm diameter is mounted as a simply supported beam. Determine the number of modes of this beam at octave-band frequencies from 31.5 to 1000 Hz.

The radius of gyration of the beam cross section is given by

$$\kappa = \sqrt{\frac{I_{zz}}{A}} = \sqrt{\frac{\dfrac{\pi d^4}{64}}{\dfrac{\pi d^2}{4}}} = \frac{d}{4} = \frac{25}{4} = 6.25 \text{ mm} \tag{1}$$

TABLE 5.3

Results of Example 5.4

No.	Octave Band Center Frequency, f_c (Hz)	Δf (bandwidth) (Hz)	c_B, bending wave speed (m/s)	c_g, group velocity (m/s)	$n(f)$, modal density (modes/Hz)	Number of modes, $\Delta n(f)$
1	31.5	22	79.14	158.29	0.0032	0.0695
2	63	44	111.93	223.85	0.0022	0.0983
3	125	88	157.66	315.32	0.0016	0.1395
4	250	176	222.96	445.93	0.0011	0.1973
5	500	352	315.32	630.64	0.0008	0.2791
6	1000	704	445.93	891.85	0.0006	0.3947

From Equation 2.38 (see Chapter 2), the longitudinal wave speed is given by

$$c_L = \sqrt{\frac{E}{\rho}} = \sqrt{\frac{200 \times 10^9}{7800}} = 5064 \text{ m/s} \tag{2}$$

From Equation 5.70, the bending wave speed is given by

$$c_B = \sqrt{2\pi f_c c_L \kappa} = \sqrt{2\pi \times 5064 \times 6.25 \times 10^{-3}} \sqrt{f_c} = 14.10\sqrt{f_c} \tag{3}$$

From Equation 3 and Equation 5.74, the group velocity for bending waves is given by

$$c_g = 2c_B = 28.2\sqrt{f_c} \tag{4}$$

From Equation 5.79, the modal density for bending waves in the beam is given by

$$n(f) = \frac{\ell}{c_g} = \frac{0.5}{28.02\sqrt{f_c}} = \frac{0.018}{\sqrt{f_c}} \tag{5}$$

The number of modes within an octave band of center frequency, f_c, is given by

$$\Delta n(f) = \Delta f \times n(f) = \frac{0.018\Delta f}{\sqrt{f_c}} \tag{6}$$

The results of the example are shown in Table 5.3.

5.7 Conclusions

A detailed discussion on the flexural vibration of a beam was presented in this chapter. Equations for natural frequencies and mode shapes were derived. In addition, orthogonality conditions were derived that will be useful for determining the response of a beam to any arbitrary excitation. The wave approach was applied to flexural vibration in beams and equations for bending wave speed were derived and compared with that of sound waves to obtain the useful concept of critical frequency. The concept of group and phase velocity was introduced, which is very useful in computing the modal density of any structure. Since modal density is one of the important parameters of any statistical energy analysis (SEA) subsystem, it is useful in later chapters. Now that we have studied the flexural vibration of a beam, it will now be easier to study the vibration of plates and shells in the next chapter.

PROBLEMS

5.1. A beam of length ℓ and rectangular uniform cross section of width b and height h is clamped at one end and free at the other end. Determine the ratio of the first natural frequencies when the beam vibrates transversely in the direction of height and width.

5.2. A steel beam of length $\ell = 1$ m and rectangular uniform cross section of width $b = 50$ mm and height $h = 5$ is clamped at one end and free at the other end. If the free end has a maximum displacement of 5 mm during free vibration, determine the general equation for mode shapes.

5.3. Consider the cantilever beam that has a uniform circular cross section of diameter d shown in the following figure that has a concentrated mass M at its tip. Determine the resonant frequencies due to transverse vibration using both modal and wave approaches.

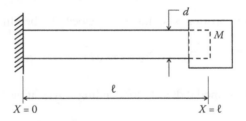

5.4. A steel bar of 2 m length and 40 mm diameter, shown in the following figure, is assumed to be excited with free flexural vibrations. Determine the general equation for its resonant frequencies and compute its first natural frequency if each mass is half the total mass of the shaft. Assume $E = 200$ GPa and the density as 7800 kg/m³ for steel.

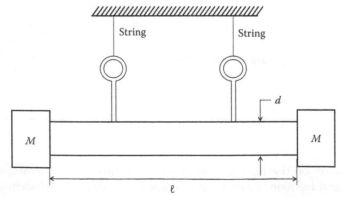

5.5. A 0.5 m long steel beam of rectangular cross section 40 mm width and 5 mm height that is clamped at the left end and supported by a spring of stiffness is shown in the following figure. The stiffness of the spring is half the equivalent stiffness of the beam in the vertical direction. Determine the natural frequencies for transverse vibration of this system using both modal and wave approaches.

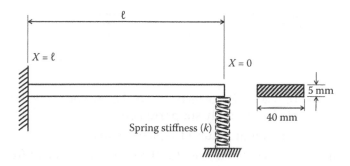

5.6. Derive a general equation for determining the resonance frequencies of a beam in transverse vibration that has a number of concentrated masses at various locations of the beam.

5.7. A simply supported steel beam, shown in the following figure, is of 3 m length and 10 cm diameter and is subjected to two sinusoidal excitation forces of 5 N peak amplitude, and the excitation frequency of each of them is equal to that of its third natural frequency. Determine the location of these two forces so the response is not significant, and determine the midspan deflection using contributions up to the fourth mode. Assume the Young's modulus for steel as 200 GPa and the density as 7800 kg/m³.

5.8. A steel cantilever of 500 mm length, 50 mm width, and 5 mm thickness is subjected to a unit impulse at its free end. Determine the free-end deflection considering only the first mode. Assuming a deflection shape for the fundamental frequency, $\phi(x) = \left(1 - \dfrac{x}{\ell}\right)^2$, measure the distance along the beam with origin at the free end.

5.9. A steel ($E = 200$ GPa, $\rho = 7800$ kg/m³) beam, shown in the following figure, is of 1 m length, and 20 mm diameter and is subjected to an exponential decaying force $100e^{-0.1t}$ at its free end. Determine the free-end deflection considering only the first mode. Assume a deflection shape for the fundamental frequency,

$\phi(x) = \left(1 - \dfrac{x}{\ell}\right)^2$, and measure the distance along the beam with the origin at the

free end.

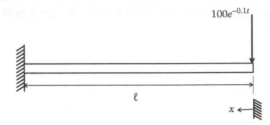

5.10. A long steel rod (E = 200 GPa, density = 7800 kg/m³) of 1 m length and 50 mm diameter is mounted as a simply supported beam.

   a. Determine the first four natural frequencies in hertz.

   b. Determine the generalized mass for the first four resonant frequencies.

   c. If a unit impulse is applied at the center of the beam, determine the generalized force for the first four resonant frequencies.

   d. What is the percentage contribution of the first four modes toward the response at the center of the beam due to the force of part c?

5.11. A tapered steel cantilever, shown in the following figure, is subjected to a sinusoidal excitation of 10 sin 100t N at the tip. Determine the response at the tip considering only one mode by using the approximate mode shape $\phi(x) = \left(1 - \dfrac{x}{\ell}\right)^2$.

Assume D = 100 mm, d = 70 mm, ℓ = 1, E = 200 GPa, and density = 7800 kg/m³

for steel.

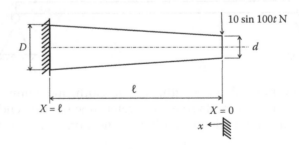

5.12. A 0.5 m long steel beam, shown in the following figure, of rectangular cross section 40 mm width and 5 mm height is clamped at the left end and supported by a spring and damper at the right end. Assume the spring stiffness to be 50% of the bar stiffness in the vertical direction and the damping coefficient is such that the damping factor of the independent single degree of freedom is 0.15 and the mass of the single degree of freedom is 10% of the beam mass. It is subjected to sinusoidal excitation of unit peak amplitude. Determine the equation for power input as a function of frequency.

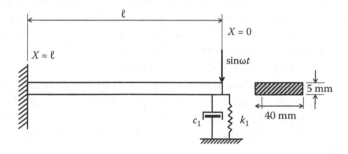

5.13. A long steel rod (E = 200 GPa, density = 7800 kg/m³) of 1 m length and 50 mm diameter is mounted as a simply supported beam. Determine the frequencies beyond which the decay of evanescent waves is beyond 5 dB/m.

5.14. Consider the beam shown in the following figure has a uniform circular cross section of diameter d and is clamped at one end and pinned at the other end. Determine the resonant frequencies due to transverse vibration using both modal and wave approaches.

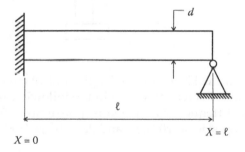

5.15. A 0.5 m long steel beam, shown in the following figure, of rectangular cross section 40 mm width and 5 mm height is supported on both ends by a spring of stiffness that is half the equivalent stiffness of the beam in the vertical direction. Determine the natural frequencies for transverse vibration of this system using both modal and wave approaches.

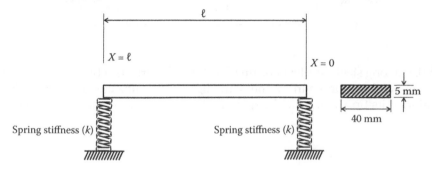

5.16. A steel bar of 2 m length and 40 mm diameter is assumed to be excited with free flexural vibrations for the following end conditions: free-free, fixed-free, and pinned-pinned. Determine the bending wave speeds for resonant frequencies up to 5 kHz for each set of end conditions. If the surrounding air has a characteristic impedance of 400 rayl and speed of sound 340 m/s, determine the number of bending modes above and below the speed of sound in each case. Assume $E = 200$ GPa and density as 7800 kg/m^3 for steel.

5.17. A 0.5 m long steel beam, shown in the following figure, of rectangular cross section 40 mm width and 5 mm height is clamped at the left end and supported by a spring and damper at the right end. Design a tuned damper system to reduce the first resonance response of the system. Assume that the mass, m_1, is 10% of the beam mass.

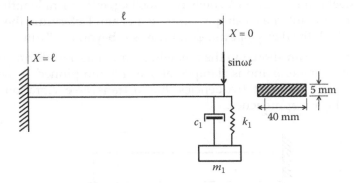

5.18. A simply supported steel beam of 3 m length and 10 cm diameter is subjected to sinusoidal excitation forces as shown in the following figure. Determine the space averaged and frequency averaged response up to 5000 Hz. Assume the Young's modulus for steel as 200 GPa and density as 7800 kg/m^3.

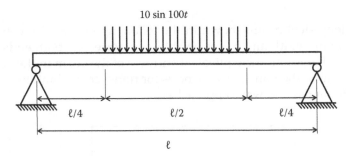

5.19. A 0.5 m long steel cantilever beam of rectangular cross section 40 mm width and 5 mm height is shown in the following figure. If it subjected to a unit pulse excitation at its tip, determine the space and time averaged response of the system.

Assume the Young's modulus for steel as 200 GPa, density as 7800 kg/m³, and a loss factor of 0.01 at all the resonant frequencies of the beam.

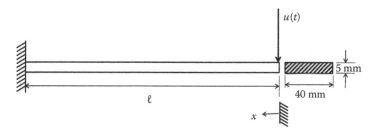

5.20. A simply supported steel beam of 3 m length and 10 cm diameter is the shaft of a turbine that needs flexible rotor balancing. Assume the Young's modulus for steel as 200 GPa and density as 7800 kg/m³. If the turbine runs at 30000 rpm, determine the location of the balancing planes if the steady operating speed is likely to be beyond the third resonant frequency of the beam in flexural vibration.

6

Flexural Vibration of Plates and Shells

Plates are overwhelmingly used in constructing machines and they are, unfortunately, the dominant source of airborne sound that is produced due to their flexural vibration. Therefore, the study of plate vibration is very important from the vibro-acoustic viewpoint. Although there are characteristic differences between plate and beam flexural vibration, the flexural vibration of beams studied in the previous chapter can be easily extended to study flexural vibration of plates.

First, beginning with the general partial differential equation for plates, the flexural response is expressed in terms of natural frequencies and mode shapes. Then, the uncoupled equations of motion are obtained using the orthogonality condition and expressed in terms of generalized coordinates, generalized masses, and generalized forces. Details can be obtained from references, as the procedure is very similar to those of beams. Second, the wave number for flexural vibration is obtained that can be resolved in two dimensions. Then, equations for the speed of bending waves in plates, group velocity, phase velocity, and modal density are obtained.

After plates, shells are also very important elements that are used in constructing aerospace and submarine structures, and their vibration is also briefly discussed.

6.1 Plate Flexural Response Using the Modal Approach

Consider a rectangular plate of length ℓ_1, width ℓ_2, and thickness h with the coordinate system as shown in Figure 6.1. The x and y directions are along the larger plate surface; z is perpendicular to this plate surface. The vibration of the plate in the z-direction, known as flexural vibration, is of interest since it is responsible for airborne sound generation. The plate is assumed to be subjected to a pressure distribution $p(x,y,t)$. Let E be the Young's modulus of elasticity of the plate material and ν its Poisson's ratio.

The equation for flexural vibration can be written as

$$B\left(\frac{\partial^4 z}{\partial x^4} + 2\frac{\partial^4 z}{\partial x^2 \partial y^2} + \frac{\partial^4 z}{\partial y^4}\right) + \rho_a \frac{\partial^2 z}{\partial t^2} = p(x,y,t) \tag{6.1}$$

where B is the flexural rigidity given by

$$B = \left(\frac{E}{1-\nu^2}\right)\frac{h^3}{12} \approx \frac{Eh^3}{12} \tag{6.2}$$

and ρ_a is the density per unit area that is given by the product of the density and thickness of the plate.

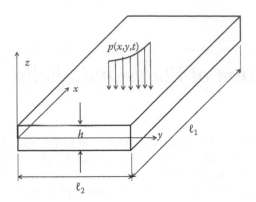

FIGURE 6.1
A rectangular plate subjected to a general pressure excitation.

For free vibration, $p(x,y,t) = 0$. Equation 6.1 now becomes

$$B\left(\frac{\partial^4 z}{\partial x^4} + 2\frac{\partial^4 z}{\partial x^2 \partial y^2} + \frac{\partial^4 z}{\partial y^4}\right) + \rho_a \frac{\partial^2 z}{\partial t^2} = 0 \tag{6.3}$$

A solution of the following form can be assumed to solve Equation 6.3:

$$z(x,y,t) = \phi(x,y)q(t) \tag{6.4}$$

By substituting Equation 6.4 in Equation 6.3, the resulting space-dependent equation can be solved by applying appropriate boundary conditions to compute the resonance frequencies and their corresponding mode shapes. Since the plate is two-dimensional, two indices are used for identifying its resonance frequencies and mode shapes.
Equation 6.4 now becomes

$$z(x,y,t) = \sum_{n=1}^{\infty}\sum_{m=1}^{\infty} \phi_{mn}(x,y)q_{mn}(t) \tag{6.5}$$

where $\phi_{mn}(x, y)$ are mode shapes and q_{mn} are generalized coordinates for the indices m and n.
The mode shapes of the plate also follow similar orthogonality conditions as beams. Based on these conditions, the generalized mass of a plate is given by

$$m_{mn} = \int_0^{\ell_1}\int_0^{\ell_2} \rho_a \phi_{mn}^2(x,y)\,dx\,dy \tag{6.6}$$

and the generalized force is given by

$$
Q_{mn}(t) = \frac{\displaystyle\int_0^{\ell_1}\int_0^{\ell_2} p(x,y,t)\phi_{mn}(x,y)\,dx\,dy}{m_{mn}}
\tag{6.7}
$$

The uncoupled equations of plate vibration in terms of generalized coordinates and forces is given by

$$
\ddot{q}_{mn}(t) + \omega_{mn}^2 q_{mn}(t) = Q_{mn}(t)
\tag{6.8}
$$

where ω_{mn} are the resonant frequencies of the plate.

Equation (6.8) can be solved for various indices of m and n, and then substituted in Equation 6.5 to obtain the total solution.

6.2 Plate Flexural Wave Number

Using Equation 6.4 and its derivatives in Equation 6.3:

$$
B\left(\frac{\partial^4\phi}{\partial x^4} + 2\frac{\partial^4\phi}{\partial x^2\partial y^2} + \frac{\partial^4\phi}{\partial y^4}\right)q(t) + \rho_a\phi\ddot{q}(t) = 0
\tag{6.9}
$$

Let us define the following operator:

$$
\nabla^2(\nabla^2\phi) = \frac{\partial^4\phi}{\partial x^4} + 2\frac{\partial^4\phi}{\partial x^2\partial y^2} + \frac{\partial^4\phi}{\partial y^4}
\tag{6.10}
$$

Using Equation 6.10 in Equation 6.9

$$
-\frac{B\nabla^2(\nabla^2\phi)}{\rho_a\phi} = \frac{\ddot{q}(t)}{q(t)} = -\omega^2
\tag{6.11}
$$

Equation 6.11 results in the following equation:

$$
\left[\nabla^2(\nabla^2) - \frac{\rho_a\omega^2}{B}\right]\phi(x,y) = 0
\tag{6.12}
$$

Equation 6.12 can be written in the complex form as

$$\left[\nabla^2 (\nabla^2) - \frac{\rho_a \omega^2}{B} \right] \tilde{\phi}(x, y) = 0 \tag{6.13}$$

Let us assume the following solution for Equation 6.13 is

$$\tilde{\phi}(x, y) = e^{jk_x x} e^{jk_y y} \tag{6.14}$$

where k_x is the wave number in the x-direction and k_y is the wave number in the y-direction. Various derivatives of Equation 6.14 can be obtained as follows:

$$\frac{\partial^4 \phi}{\partial x^4} = k_x^4 \tilde{\phi}(x, y) \tag{6.15}$$

$$\frac{\partial^4 \phi}{\partial y^4} = k_y^4 \tilde{\phi}(x, y) \tag{6.16}$$

$$\frac{\partial^4 \phi}{\partial x^2 \partial y^2} = k_x^2 k_y^2 \tilde{\phi}(x, y) \tag{6.17}$$

Substituting Equations 6.15 through 6.17 in Equation 6.13, we get

$$\left(k_x^4 + 2k_x^2 k_y^2 + k_y^4 \right) = \frac{\rho_a \omega^2}{B} \tag{6.18}$$

Equation 6.18 can be written as

$$\left(k_x^2 + k_y^2 \right)^2 = \frac{\rho_a \omega^2}{B} \tag{6.19}$$

The bending wave number for plates can be defined as

$$k_B^2 = k_x^2 + k_y^2 \tag{6.20}$$

It can be expressed as a vector as shown in Figure 6.2.

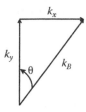

FIGURE 6.2
Plate bending wave number resolved into its components.

6.2.1 Plate Bending Wave Number

From Equations 6.19 and 6.20, the bending wave number of the plate is given by

$$k_B^4 = \frac{\rho_a \omega^2}{B} \tag{6.21}$$

The bending wavelength in terms of the bending wave number is given by

$$\lambda_b = \frac{2\pi}{k_B} \tag{6.22}$$

From Equations 6.21 and 6.22

$$\lambda_b = \frac{2\pi}{k_B}\left(\frac{B}{\rho_a\omega^2}\right)^{\frac{1}{4}} = \frac{2\pi}{\sqrt{2\pi f}}\left(\frac{Eh^3}{12\rho h}\right)^{\frac{1}{4}} = 1.35\sqrt{\frac{hc_L}{f}} \tag{6.23}$$

The critical frequency of a plate is the frequency at which bending wavelengths are same as acoustic wavelengths. The wavelength of airborne sound at a frequency f is given by

$$\lambda_o = \frac{c_o}{f} \tag{6.24}$$

From Equations 6.23 and 6.24, the critical frequency is given by

$$f_{cr} = \frac{c_o^2}{1.82hc_L} \tag{6.25}$$

At frequencies less than the critical frequency, the bending waves are shorter than sound waves; at frequencies higher than the critical frequencies, bending waves are longer than the sound waves. For steel and aluminum having longitudinal wave speeds of 5000 m/s and assuming 344 m/s as the speed of sound in air, the critical frequency can be approximated as $f_{cr} \approx \dfrac{13}{h}$ (h in meters).

Example 6.1

A rectangular steel plate 0.5 × 0.3 × 0.003 m is simply supported. How many resonant frequencies are there below its critical frequency? Given data: $\ell_1 = 0.5$ m, $\ell_2 = 0.5$ m, $h = 3$ mm, $E = 200$ GPa, and $\rho = 7800$ kg/m^3.

From Equation 2.38 (see Chapter 2), the longitudinal speed is given by

$$c_L = \sqrt{\frac{E}{\rho}} = \sqrt{\frac{200 \times 10^9}{7800}} = 5064 \text{ m/s} \tag{1}$$

From Equation 6.25, the critical frequency is given by

$$f_{cr} = \frac{c_o^2}{1.82 h c_L} = \frac{340^2}{1.82 \times 0.003 \times 5064} = 4181 \text{ Hz} (26270 \text{ rad/s}) \tag{2}$$

The mode shape of a simply supported rectangular plate is given by

$$\phi_{mn}(x,y) = \sin\left(\frac{m\pi x}{\ell_1}\right)\sin\left(\frac{n\pi y}{\ell_2}\right) \tag{3}$$

From Equation 3 it can be recognized that the mode shape is a product of two functions, and each of the functions is a product of the wave number and the corresponding coordinate value.

From Equation 3, the resonant wave number along the x-direction of a simply supported plate is given by

$$k_x = \frac{m\pi}{\ell_1} \tag{4}$$

Similarly, from Equation 3, the resonant wave number along the y-direction of a simply supported plate is given by

$$k_y = \frac{n\pi}{\ell_2} \tag{5}$$

From Equations 4 and 5 and Equation (6.20), the resonant bending wave number is given by

$$k_{B_{m,n}}^2 = k_x^2 + k_y^2 = \pi^2 \left(\frac{m^2}{\ell_1^2} + \frac{n^2}{\ell_2^2}\right) \tag{6}$$

From Equations 6.2 and 6.21, the bending wave number and frequencies of a plate are related by

$$k_B^4 = \frac{12\rho\omega^2}{Eh^2} \tag{7}$$

From Equation 7, the resonant frequencies and the corresponding bending wave numbers are given by

$$\omega_{m,n} = \sqrt{\frac{Eh^2 k_{B_{m,n}}^4}{12\rho}} \tag{8}$$

From Equations 6 and 8, the resonant frequencies of the plate in hertz are given by

$$f_{m,n} = \frac{\pi}{2}\left(\frac{m^2}{\ell_1^2} + \frac{n^2}{\ell_2^2}\right)\sqrt{\frac{Eh^2}{12\rho}} \tag{9}$$

By using the given data of plate dimensions and material properties, the resonant frequencies are given by

$$f_{m,n} = 688.84\left(\frac{m^2}{25} + \frac{n^2}{9}\right)\text{Hz} \tag{10}$$

From Equations 2 and 7, the critical bending wave number is given by

$$k_{B_{cr}} = \left(\frac{7800 \times 26270^2 \times 12}{200 \times 10^9 \times 0.003^2}\right) = 77.4 \text{ rad/m} \tag{11}$$

From Equations 6 and 11, the critical bending wave number and the resonant frequency indices can be related by

$$k_{B_{cr}}^2 = 77.4^2 = \frac{\pi^2}{0.0225}\left(\frac{m^2}{0.25} + \frac{n^2}{0.09}\right) \tag{12}$$

Simplifying Equation 12, the following relationship between the indices can be obtained

$$n^2 = \frac{13.66 - 0.09m^2}{0.25} \tag{13}$$

For various values of m, the index value of n can be obtained, and these values are shown in Table 6.1.

For the values of m greater than 12, the value of n becomes an imaginary number and hence not a possible solution. So there are 63 resonant frequencies of the plate below its critical frequency.

TABLE 6.1

Index Values of m and n that
Satisfy Equation 12 of Example 6.1

m	n
1	7
2	7
3	7
4	6
5	6
6	6
7	6
8	5
9	5
10	4
11	3
12	1
Total	63

Example 6.2

A steel plate ($E = 200$ GPa, $\rho = 7800$ kg/m³) of dimensions 300 mm × 300 mm × 5 mm is simply supported. It is subjected to a unit impulse at its center. Determine the center deflection considering responses of modes that are greater than 50% of the response of the mode that gives maximum response.

For a simply supported plate, the mode shapes are given by

$$\phi_{mn}(x,y) = \sin\left(\frac{m\pi x}{\ell_1}\right)\sin\left(\frac{n\pi y}{\ell_2}\right) \tag{1}$$

From Equation 1, the resonant wavenumber along the x-direction of a simply supported plate is given by

$$k_x = \frac{m\pi}{\ell_1} \tag{2}$$

Similarly, from Equation 1, the resonant wavenumber along the y-direction of a simply supported plate is given by

$$k_y = \frac{n\pi}{\ell_2} \tag{3}$$

From Equations 2 and 3, and Equation (6.20), the resonant bending wave number is given by

$$k_{B_{m,n}}^2 = k_x^2 + k_y^2 = \pi^2\left(\frac{m^2}{\ell_1^2} + \frac{n^2}{\ell_2^2}\right) = \frac{\pi^2}{\ell^2}(m^2 + n^2) \tag{4}$$

From Equation 4, the resonant frequencies and the corresponding bending wave numbers are given by

$$\omega_{m,n} = \sqrt{\frac{Eh^2 k_{B_{m,n}}^4}{12\rho}} \tag{5}$$

From Equation 6.6, the generalized mass is given by

$$m_{mn} = \int_0^{\ell_1}\int_0^{\ell_2} \rho_a \phi_{mn}^2(x,y)\,dx\,dy = \rho_a \int_0^{\ell} \sin^2\left(\frac{m\pi x}{\ell_1}\right)dx \int_0^{\ell} \sin^2\left(\frac{n\pi y}{\ell_2}\right)dy$$

$$= \frac{\rho h \ell^2}{4} \tag{6}$$

and the generalized force is given by

$$Q_{mn}(t) = \frac{\displaystyle\int_0^{\ell}\int_0^{\ell} p(x,y,t)\phi_{mn}(x,y)\,dx\,dy}{m_{mn}}$$

$$= \frac{4}{\rho h \ell^2}\int_0^{\ell}\int_0^{\ell} \sin\left(\frac{m\pi x}{\ell}\right)\sin\left(\frac{n\pi y}{\ell}\right)\delta\left\langle x - \frac{\ell}{2}\right\rangle\delta\left\langle y - \frac{\ell}{2}\right\rangle\delta\langle t\rangle\,dx\,dy \tag{7}$$

$$= \frac{4}{\rho h \ell^2}\sin\left(\frac{m\pi}{2}\right)\sin\left(\frac{n\pi}{2}\right)\delta\langle t\rangle$$

Based on Equation 1.51 (see Chapter 1), the impulse response of the generalized response of the plate is given by

$$h_{mn} = \frac{\sin \omega_{mn} t}{\omega_{mn}} \tag{8}$$

Using Equation 8 in the convolution integral Equation 1.55 (see Chapter 1), the generalized response is given by

$$q_{mn} = \frac{1}{\omega_{mn}}\int_0^{t} \sin \omega_{mn}\tau Q_{mn}(t-\tau)\,d\tau$$

$$= \frac{4}{\rho h \ell^2 \omega_{mn}}\sin\left(\frac{m\pi}{2}\right)\sin\left(\frac{n\pi}{2}\right)\sin \omega_{mn} t \tag{9}$$

From Equation 6.5, the response at the center of the plate is given by

$$z\left(\frac{\ell}{2},\frac{\ell}{2},t\right) = \frac{4}{\rho h \ell^2}\sum_{m=1}^{\infty}\sum_{n=1}^{\infty}\frac{\sin^2\left(\dfrac{m\pi}{2}\right)\sin^2\left(\dfrac{n\pi}{2}\right)\sin \omega_{mn} t}{\omega_{mn}} \tag{10}$$

From Equations 6 and 10

$$z\left(\frac{\ell}{2},\frac{\ell}{2},t\right) = \frac{4\ell^2\sqrt{12\rho}}{\rho h\ell^2\pi^2\sqrt{Eh^2}}\sum_{m=1}^{\infty}\sum_{n=1}^{\infty}\frac{\sin^2\left(\frac{m\pi}{2}\right)\sin^2\left(\frac{n\pi}{2}\right)\sin\omega_{mn}t}{m^2+n^2} \tag{11}$$

Since only odd modes contribute toward the response at the plate center, examining the reciprocal of $m^2 + n^2$ shows that only the contribution of the fundamental mode corresponding to $m = 1$ and $n = 1$ is dominant; the contribution from $m = 1$ and $n = 3$ and $m = 3$ and $n = 1$ is only 5% of the dominant response and can thus be neglected.

From Equation 4, the bending wave number of the plate corresponding to the fundamental resonance frequency of the plate is given by.

From Equation 6 the fundamental resonance frequency of the plate is given by

$$k_{B1,1} = \frac{\pi}{\ell}\sqrt{(m^2+n^2)}$$

$$= \frac{\pi}{0.3}\sqrt{(1^2+1^2)} = 14.8096 \text{ rad/m} \tag{12}$$

From Equation 6 the fundamental resonance frequency of the plate is given by

$$\omega_{1,1} = \sqrt{\frac{Eh^2 k_B^4}{12\rho}} = \sqrt{\frac{200\times10^9\times0.005^2\times14.8096^4}{12\times7800}}$$

$$= 1603 \text{ rad/s}(255.125 \text{ Hz}) \tag{13}$$

From Equation 10, the center deflection is given by

$$z\left(\frac{\ell}{2},\frac{\ell}{2},t\right) = \frac{4\times0.3^2\sqrt{12\times7800}}{7800\times0.005\times0.3^2\pi^2\sqrt{200\times10^9\times0.005^2}}$$

$$\frac{\sin^2\left(\frac{\pi}{2}\right)\sin^2\left(\frac{\pi}{2}\right)\sin1603t}{2} = 0.71\sin1603t \text{ mm} \tag{14}$$

6.3 Plate Group Velocity

The bending wave number and bending speed are related by

$$k_B = \frac{\omega}{c_B} \tag{6.26}$$

From Equations 6.21 and 6.26

$$\left(\frac{\omega}{c_B}\right)^4 = \frac{12\rho_a\omega^2}{Eh^3}$$ (6.27)

From Equation 6.27, the bending speed is given by

$$c_B = 1.35\sqrt{fC_L h}$$ (6.28)

Using the bending speed of Equation 6.26 in Equation 6.28

$$k_B^2 = \frac{\omega^2}{1.35^2 \, fc_L h} = \frac{2\pi\omega}{1.35^2 \, c_L h} = \frac{3.45\omega}{c_L h}$$ (6.29)

Differentiating Equation 6.29 with respect to angular frequency:

$$2k_B = \frac{3.45}{C_L h}\frac{d\omega}{dk_b}$$ (6.30)

Since the group velocity $c_g = \dfrac{d\omega}{dk_B}$, Equation 6.30 becomes

$$2k_B = \frac{3.45}{C_L h}c_g$$ (6.31)

From Equation 6.26, Equation 6.31 becomes

$$c_g = \frac{2c_L h\omega}{3.45\,c_B} = \frac{2c_L h\omega}{3.45\times 1.35\sqrt{fc_L h}} = 2\times 1.35\sqrt{fc_L h} = 2c_B$$ (6.32)

Therefore, the group velocity is twice the bending wave speed.

Example 6.3

An infinite steel plate of 5 mm thickness is vibrating at 4000 Hz, with a root mean square (rms) surface velocity of 5 mm/s. Determine the critical frequency, and the bending wave speed and wavelength.

From Equation 2.38, the longitudinal speed is given by

$$c_L = \sqrt{\frac{E}{\rho}} = \sqrt{\frac{200 \times 10^9}{7800}} = 5064 \text{ m/s} \tag{1}$$

From Equation 6.25, the critical frequency is given by

$$f_{cr} = \frac{c_0^2}{1.82 h c_L} = \frac{340^2}{1.82 \times 0.005 \times 5064} = 2509 \text{ Hz} \tag{2}$$

From Equation 6.28, the bending wave speed at 4000 Hz is given by

$$c_B = 1.35\sqrt{f c_L h} = 1.35\sqrt{4000 \times 5064 \times 0.005} = 429.63 \text{ m/s} \tag{3}$$

The bending wave speed, therefore, is much higher than the speed of sound at normal conditions of temperature and pressure.

From Equations 6.23 and 6.28, the bending wavelength of the plate at 4 kHz is given by

$$\lambda_b = \frac{c_B}{f} = \frac{429.63}{4000} = 0.107 \text{ m} \tag{4}$$

6.4 Plate Modal Density

A simply supported plate of lateral dimensions ℓ_1 and ℓ_2 is shown in Figure 6.3. The boundary conditions are that the displacement and bending moment are zero along all the edges of the support. Based on these boundary conditions, the natural frequencies of the plate can be written as

$$\omega_{mn} = k_B^2 \kappa c_L \tag{6.33}$$

where ω_{mn} is the natural frequency corresponding to m and n modes. Both m and n take the values 1, 2, …, ∞; k_B is the bending wave number; κ is the radius of gyration; and c_L is the longitudinal wave speed.

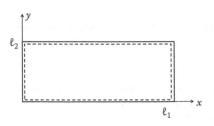

FIGURE 6.3
Simply supported plate.

The mode shapes of the simply supported plate corresponding to the aforementioned natural frequencies are given by

$$\phi_{mn}(x,y) = \sin\left(\frac{m\pi x}{L_1}\right)\sin\left(\frac{n\pi y}{L_2}\right) \tag{6.34}$$

Based on the mode shape represented by Equation (6.34), in order to derive the equation for modal density, a two-dimensional wave number lattice as shown in Figure 6.4 is used. The lattice consists of equispaced divisions in the direction of the lateral dimensions of the plate. The divisions along the ℓ_1 dimension are separated by π/ℓ_1 and the divisions along the ℓ_2 dimension are given by π/ℓ_2. Each grid point of the lattice represents a corresponding natural frequency.

The number of modes inside the quarter circle is given by

$$N(k_b) = \text{total area/area per mode} = \frac{\dfrac{\pi k_B^2}{4}}{\dfrac{\pi}{\ell_1}\dfrac{\pi}{\ell_2}} = \frac{k_B^2 A_p}{4\pi} \tag{6.35}$$

where A_p is the area of the plate.

The modal density in the bending wave number is given by

$$n(k_B) = \frac{dN}{dk_B} = \frac{k_B A_p}{2\pi} \tag{6.36}$$

The modal density in radian frequency is given by

$$n(\omega) = n(k_B)\frac{dk_B}{d\omega} = \frac{k_B A_p}{2\pi}\frac{1}{c_g} = \frac{A_p\sqrt{12}}{4\pi c_L h} \tag{6.37}$$

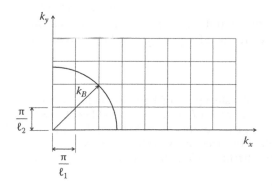

FIGURE 6.4
Two-dimensional wave number lattice.

TABLE 6.2

Number of Modes of a 3 mm Plate in Selected 1/3 Octave Bands

Lower Frequency, f_L (Hz)	Center Frequency, f_C (Hz)	Upper Frequency, f_H (Hz)	Bandwidth, Δ_f	Number of Modes
28	31.5	35	7	1
56	63	71	15	2
111	125	140	29	3
223	250	281	58	6
445	500	561	116	13
891	1000	1122	232	25
1782	2000	2245	463	50
3564	4000	4490	926	100
7127	8000	8980	1853	200

The modal density in hertz is given by

$$n(f) = n(\omega)\frac{d\omega}{df} = \frac{A_p\sqrt{12}}{4\pi c_L h} \times 2\pi$$
$$= \frac{A_p\sqrt{3}}{c_L h}$$

(6.38)

where h is the thickness of the plate.

The average separation frequency (Hz) $\overline{\delta f_p}$ is given by

$$\overline{\delta f_p} = \frac{hc_L}{\sqrt{3}A_p}$$

(6.39)

Table 6.2 shows the number of modes in different 1/3 octave bands for an aluminum plate of area 1 m² and thickness 3 mm. The longitudinal wave speed for aluminum is 5100 m/s and the average separation frequency from Equation 6.39 is 9 Hz. One can see that the number of modes in each frequency band increase with the center frequency.

Example 6.4

Design the thickness of a steel plate of 1 square meter area that will have 500 modes in the 1/3 octave band of center frequency of 1000 Hz (f_L = 891 Hz and f_H = 1122 Hz). What is the critical frequency of this plate? What is the ratio of bending wave speed to acoustic wave speed at this frequency? What is the wave number? What is the modal density in modes per hertz (modes/Hz) and modes per radians per second (modes/rad/s)?

$\Delta N(f)$ = 500 modes, f_L = 891 Hz and f_H =1122 Hz, Δf = 1122 − 891 = 231 Hz
Expected modal density of this plate, $n(f)$ = 500/231 = 2.165 modes/Hz

The longitudinal speed is given by

$$c_L = \sqrt{\frac{E}{\rho}} = \sqrt{\frac{200 \times 10^9}{7800}} = 5064 \text{ m/s} \tag{1}$$

From Equation 6.38, the plate thickness is given by

$$h = \frac{\sqrt{3}A_p}{n(f)c_L} = \frac{\sqrt{3} \times 1}{2.165 \times 5064} = 0.158 \text{ mm} \tag{2}$$

From Equation 6.28, the bending wave speed of the aforementioned plate at 1000 Hz is given by

$$c_B = 1.35\sqrt{fc_L h} = 1.35\sqrt{1000 \times 5064 \times 0.158 \times 10^{-3}} = 38.19 \text{ m/s} \tag{3}$$

By using the speed of sound $c_o = 340$ m/s, the critical frequency of the plate is given by

$$f_c = \frac{c_o^2}{1.82 c_L} = \frac{340^2}{1.82 \times 0.158 \times 10^{-3} \times 5064} = 80 \text{ kHz} \tag{4}$$

The ratio of bending wave speed and sound speed is given by

$$\frac{c_B}{c_o} = \frac{38.19}{340} = 0.1123 \tag{5}$$

From Equation 6.26, the bending wave number of the preceding plate at 1000 Hz is given by

$$k_B = \frac{\omega}{c_B} = \frac{2\pi \times 1000}{38.19} = 164.54 \text{ rad/m} \tag{6}$$

From Equation 6.37, the modal density in radians per second (rad/s) is given by

$$n(\omega) = \frac{n(f)}{\dfrac{d\omega}{df}} = \frac{2.165}{2\pi} = 0.345 \text{ modes/rad/s} \tag{7}$$

From Equation 6.36, the modal density in per unit wave number is given by

$$n(k_B) = \frac{k_b A_p}{2\pi} = \frac{164.54 \times 1}{2\pi} = 26.19 \text{ modes/wavenumber} \tag{8}$$

Example 6.5

Refer to Example 6.1.

a. Determine the modal density and average frequency of separation of this plate. In addition, determine the number of resonant modes in the 1/3 octave band center frequency of 1 kHz.
b. Physically count the number of resonance frequencies in the preceding band and explain why there is a difference.
c. Determine the average frequency of separation of resonance frequencies in the frequency band of part c.
d. Determine the frequency of vibration at which the evanescent waves decay at the rate of 10 dB/m.

A computationally intensive approach to obtain the same answer would be to compute the resonant frequencies of the plate below its critical frequency of 4181 Hz and count them. They are shown in Table 6.3.

ANSWERS

a. From Equation (6.38), the modal density is given by

$$n(f) = \frac{\sqrt{3}A_p}{c_L h} = \frac{\sqrt{3} \times 0.5 \times 0.3}{5064 \times 0.003} = 0.0171 \text{ modes/Hz} \tag{1}$$

From Equation 6.39, the average frequency separation is given by

$$\overline{\delta f}_p = \frac{1}{n(f)} = \frac{1}{0.0171} = 58.47 \text{ Hz} \tag{2}$$

TABLE 6.3

Resonance Frequencies (Hz) below the Critical Frequency Shown in the Shaded Region

m	n							
	1	2	3	4	5	6	7	8
1	105	335	719	1257	1949	2794	3793	4946
2	188	418	802	1340	2032	2877	3876	5029
3	326	556	941	1479	2170	3016	4015	5167
4	520	750	1134	1672	2364	3209	4208	5361
5	769	999	1383	1921	2613	3458	4457	5610
6	1073	1303	1688	2226	2917	3763	4762	5914
7	1432	1663	2047	2585	3277	4122	5121	6274
8	1847	2078	2462	3000	3692	4537	5536	6689
9	2318	2548	2933	3471	4162	5008	6007	7159
10	2843	3074	3458	3996	4688	5533	6532	7685
11	3424	3655	4039	4577	5269	6114	7113	8266
12	4061	4291	4676	5214	5905	6751	7750	8902
13	4752	4983	5367	5905	6597	7442	8441	9594

The bandwidth of the 1 kHz octave band is given by

$$\Delta f_{1000\ Hz} = 1115 - 885 = 230\ Hz \tag{3}$$

The number of modes in the 1 kHz 1/3 octave band is given by

$$\Delta N = \Delta f \times n(f) = 230 \times 0.0171 \approx 4\ modes \tag{4}$$

b. By physical counting of resonance frequencies in Table 6.3, however, there are only two modes in this band due to the averaging effect.
c. Using the resonance frequencies of Table 6.3, the average separation frequency in the 1 kHz frequency band is given by

$$\frac{(941 - 802) + (999 - 941)}{2} = 98.5\ Hz \tag{5}$$

d. The bending wave number corresponding to a 10 dB decay of evanescent waves is given by $20 \log e^{-k_B} = -10\ dB$.

$$k_B = 1.15\ rad/m \tag{6}$$

Frequency corresponding to the preceding wave number is given by

$$\omega = \sqrt{\frac{200 \times 10^9 \times 0.003^2 \times 1.15^4}{12 \times 7800}} = 5.8\ rad/s\ (0.92\ Hz) \tag{7}$$

6.5 Vibration of Shells

A cylindrical shell is commonly used in modeling machinery, aircraft, and spacecraft structures. Therefore, extensive research work is reported in the literature on the vibration of cylindrical shells. The objective of this section, however, is to mainly highlight the differences and similarities between a plate and a shell.

A cylindrical shell has both flexural and circumferential modes of vibration. On similar lines of plate vibration, the natural frequencies of a simply supported cylindrical shell can be derived and is given by the following equation:

$$f_{m,n} = \frac{c_L}{2\pi r} \left[\frac{(1-\upsilon^2)\sigma^4}{(\sigma^2 + n^2)^2} + \frac{h^2}{12r^2} \left\{ (\sigma^2 + n^2)^2 - \frac{n^2(4-\upsilon) - (2+\upsilon)}{2(1-\upsilon)} \right\} \right]^{0.5} \tag{6.40}$$

where $\sigma = m\pi r/L$, r is the radius of the shell, h is the shell thickness, L its length, c_L is the longitudinal wave speed, υ is Poisson's ratio, and m and n are the axial and circumferential mode numbers, respectively. Using this equation one can determine the natural frequencies of machinery, modeled as a thin cylindrical shell.

The modal density of a shell structure can be obtained from the plate model by assuming that the shell is constructed by rolling it in the form of a cylinder with its axis along the y-axis of Figure 6.5a. This shell will have a height of L_2 and a diameter of $d = L_1/\pi$ as shown in Figure 6.5b. The mode shapes in the y direction will be same as before: $\sin n\pi y/L_2$. In the x direction, however, there must be a periodic dependence on the differential angle $(2/d)dx$ subtended at the center of the cylinder. The mode shapes for the cylinder, which contain both sine and cosine terms for circumferential modes, can then be written as (Lyon 1986)

$$\phi_{mn}(x,y) = \sin\left(\frac{n\pi y}{L_2}\right)\left[\cos\left(\frac{2m\pi x}{L_1}\right) + \sin\left(\frac{2m\pi x}{L_1}\right)\right] \tag{6.41}$$

where the wave number k_x is the coefficient of x: $2\,m/d = 2\pi m/L_1$.

As shown in Equation 6.41, the mode numbers in the y direction are separated by π/L_2 similar to a plate, but the mode numbers in the circumferential direction are separated by $2\pi/L_1$, which is twice the separation that was there for a flat plate. Therefore, some modes were lost in comparison to a plate in the process of converting from a plate to shell. However, observe that both sine and cosine mode shapes are present for circumferential modes. Therefore, each lattice point in this new wave-number space corresponds to a pair of resonant modes, which are indicated as shown in Figure 6.6.

Due to the shell curvature, the speed of waves will be faster along the y direction than in the circumferential direction. Hence, the number of lattice points enclosed by a constant-frequency locus will be somewhat smaller than that for the flat plate. Therefore, at the lower frequencies, the modal density and total mode count of cylinders are less

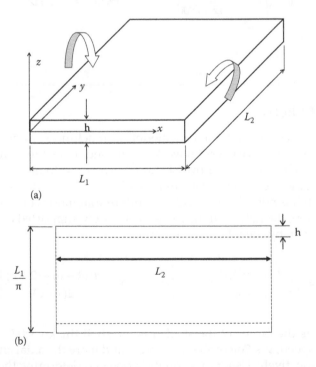

(a)

(b)

FIGURE 6.5
(a) Rectangular plate that will be converted into a shell. (b) Shell obtained from the rectangular plate.

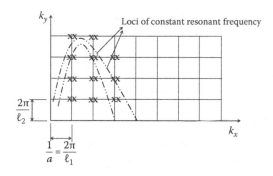

FIGURE 6.6
Wave lattice of a shell. (From Lyon, R.H., *Machinery Noise and Diagnostics*, Butterworth, Boston, 1986. With permission.)

than those of a flat plate of the same area and it increases with frequency up to a certain point (Lyon 1986).

The stiffening effect of curvature can be explained with respect to the in-plane membrane stresses in the cylinder. At low frequencies of wave propagation that will have larger wavelengths, longer than the circumference of the cylinder, the in-plane membrane stresses will be effective. At very high frequencies, since the wavelengths become very small, the membrane stresses become ineffective and hence the behavior of the shell reverts to the behavior of the corresponding flat plate. The transition occurs when the longitudinal wavelength equals the circumference of the cylinder at the ring frequency, given by

$$f_{ring} = \frac{c_L}{\pi d} \tag{6.42}$$

The modal densities of a plate and its equivalent shell are compared in Figure 6.7. Up to the ring frequency given by Equation 6.42, the modal densities of the shell keep increasing, and after reaching a peak at the ring frequency, the modal densities remain constant and

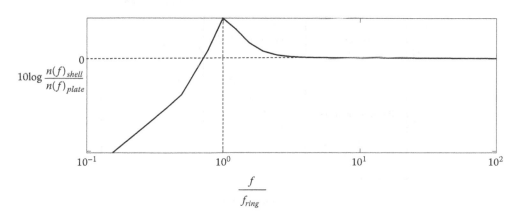

FIGURE 6.7
Comparison of modal densities of a plate and its equivalent shell.

will be same as the flat plate from which the shell was constructed from. See Szechenyi (1971) for details on modal density of unstiffened cylinders.

6.6 Transmission Loss

Barrier materials are used to separate two spaces from any acoustic interaction and they prevent noise from moving from one space to another. For example, noise transmission between two rooms or from outside to inside is generally controlled by the material in between these spaces. Therefore, barrier materials are used as part of doors, windows, walls, roofs, and so on. We can say that sound absorbing materials (discussed in Chapter 8) are defined within a given space, and barrier materials are defined for sound transmission across two spaces. The capability of a barrier material to prevent sound transmission across two spaces can be defined by transmission loss.

A typical plot of frequency characteristics of transmission loss is shown in Figure 6.8. At very low frequencies, the transmission loss is very low and it shows variations due to the panel resonances that are related to the stiffness of the material. However, in the mass-controlled region, the transmission loss increases linearly with frequency at the rate of 6 dB/octave. It is in this region that the transmission loss can be conveniently computed based on the material density. In the critical frequency region, the bending waves of the material attain the speed of sound and hence there is a distinct drop in the transmission loss in this region. Thereafter, the transmission loss increases linearly at the rate of 10 dB/octave in the damping controlled region. Due to the characteristic behavior of materials near the critical frequency region, the transmission loss can be approximated as a constant value in the transition from mass controlled to the damping controlled region. The range of this constant transmission loss region is known as the plateau region that is material dependent. Initially, transmission loss equation for the mass-controlled region is derived. Then the plateau method is discussed that will allow the computation of transmission loss beyond the mass controlled region. The

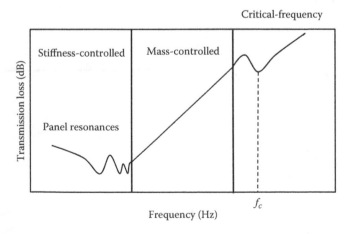

FIGURE 6.8
Frequency characteristics of transmission loss.

transmission loss computed in this section is called the nonresonance transmission loss, which is acceptable in case of barrier materials used in building structures that are very thick. However, with the barrier materials in the case of lightweight structures used in the aerospace industry, resonance transmission loss becomes important; resonance transmission loss is discussed in Chapter 10.

First, we shall discuss a general configuration shown in Figure 6.9. It shows two fluid media separated by a solid interface that is a common situation in many noise control problems. In most situations the fluid will be air and the solid will be a structural material, popularly known as barrier materials. Barrier materials play a very important role in noise control.

A sound wave propagating in the first fluid medium strikes the solid medium; a part of this is reflected, and the rest travels inside the solid medium, and gets reflected at the end of the solid interface, and the rest is transmitted to the second fluid medium. A part of this is transmitted to the second fluid medium. The global coordinate system for all three media is located at the interface of the first fluid and solid interface, and it is in the positive x direction. Therefore, for the first fluid medium, the positive coordinate for wave propagation is toward the left.

Acoustic pressure in the first fluid media is a combination of incident and reflected waves, given by

$$\tilde{p}_1(x,t) = \overline{A}_1 e^{j(\omega t - k_1 x)} + \overline{B}_1 e^{j(\omega t + k_1 x)} \tag{6.43}$$

Equation 6.43 is based on the coordinate system for the first fluid. It should be noted that the first term is the reflected wave and the second term is the incident wave.

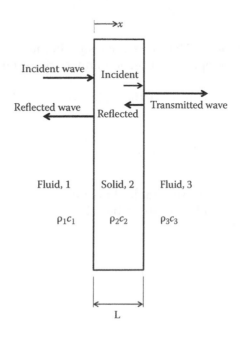

FIGURE 6.9
Interface of three media.

The amplitude of stress in the solid medium is given by

$$\tilde{p}_2(x,t) = \bar{A}_2 e^{j(\omega t - k_2 x)} + \bar{B}_2 e^{j(\omega t + k_2 x)} \tag{6.44}$$

In Equation 6.44, the first term is the right moving wave that is incident at the interface between the solid medium and the second fluid medium and the second term represents the reflection from the same interface.

The acoustic disturbance that enters the second fluid medium is the transmitted wave and its pressure is given by

$$\tilde{p}_3(x,t) = \bar{A}_3 e^{j[\omega t - k_3(x-L)]} \tag{6.45}$$

The particle velocity of the disturbance that travels in the first fluid medium is given by

$$\tilde{u}_1(x,t) = \frac{1}{\bar{Z}_1} \left[\bar{A}_1 e^{j(\omega t - k_1 x)} - \bar{B}_1 e^{j(\omega t + k_1 x)} \right] \tag{6.46}$$

The second term of Equation 6.46 has a negative sign because it represents the incident wave that is traveling in the opposite direction. $\bar{Z}_1$ represents the impedance of the first fluid medium.

The velocity of the solid medium is given by

$$\tilde{u}_2(x,t) = \frac{1}{\bar{Z}_2} \left[\bar{A}_2 e^{j(\omega t - k_2 x)} - \bar{B}_2 e^{j(\omega t + k_2 x)} \right] \tag{6.47}$$

The second term of Equation 6.47 has a negative sign because it represents the reflected wave that is traveling in the opposite direction and $\bar{Z}_2$ represents the impedance of the solid medium.

The particle velocity of the disturbance that enters the second fluid medium is given by

$$\tilde{u}_3(x,t) = \frac{1}{\bar{Z}_3} \bar{A}_3 e^{j[\omega t - k_3(x-L)]} \tag{6.48}$$

where $\bar{Z}_3$ is the impedance of the second fluid medium.

The following boundary conditions are to be satisfied at the interface of the first fluid and solid interface at $x = 0$:

$$\tilde{p}_1 = \tilde{p}_2 \tag{6.49}$$

and

$$\tilde{u}_1 = \tilde{u}_2 \tag{6.50}$$

From Equations 6.43, 6.44, and 6.49

$$\bar{A}_1 + \bar{B}_1 = \bar{A}_2 + \bar{B}_2 \tag{6.51}$$

From Equations 6.46, 6.47, and 6.50

$$\frac{\bar{A}_1 - \bar{B}_1}{\bar{Z}_1} = \frac{\bar{A}_2 - \bar{B}_2}{\bar{Z}_2} \tag{6.52}$$

At $x = L$, the interface of the solid and second fluid media, the following equations are satisfied:

$$\tilde{p}_2 = \tilde{p}_3 \tag{6.53}$$

and

$$\tilde{u}_2 = \tilde{u}_3 \tag{6.54}$$

From Equations 6.44, 6.45, and 6.53

$$\bar{A}_2 e^{-jk_2 L} + \bar{B}_2 e^{jk_2 L} = \bar{A}_3 \tag{6.55}$$

From Equations 6.47, 6.48, and 6.54

$$\frac{1}{\bar{Z}_2}\left[\bar{A}_2 e^{-jk_2 L} - \bar{B}_2 e^{jk_2 L}\right] = \frac{\bar{A}_3}{\bar{Z}_3} \tag{6.56}$$

From Equations 6.51, 6.52, 6.55, and 6.56

$$\begin{bmatrix} 1 & 1 & -1 & -1 \\ 1 & -1 & \dfrac{-\bar{Z}_1}{\bar{Z}_2} & \dfrac{+\bar{Z}_1}{\bar{Z}_2} \\ 0 & 0 & e^{-jk_2 L} & e^{jk_2 L} \\ 0 & 0 & \dfrac{e^{-jk_2 L}}{\bar{Z}_2} & \dfrac{-e^{jk_2 L}}{\bar{Z}_2} \end{bmatrix} \begin{Bmatrix} \bar{A}_1 \\ \bar{B}_1 \\ \bar{A}_2 \\ \bar{B}_2 \end{Bmatrix} = \begin{Bmatrix} 0 \\ 0 \\ \bar{A}_3 \\ \bar{A}_3/\bar{Z}_3 \end{Bmatrix} \tag{6.57}$$

Equation 6.57 can be further simplified as follows:

$$\bar{A}_2 = \frac{e^{jk_2 L}}{2} \bar{A}_3 \left(1 + \frac{\bar{Z}_2}{\bar{Z}_3}\right) \tag{6.58}$$

$$\bar{B}_2 = \frac{e^{-jk_2L}}{2} \bar{A}_3 \left(1 - \frac{\bar{Z}_2}{\bar{Z}_3}\right)$$

(6.59)

$$2\bar{A}_1 = \bar{A}_2 \left(1 + \frac{\bar{Z}_1}{\bar{Z}_2}\right) + \bar{B}_2 \left(1 - \frac{\bar{Z}_1}{\bar{Z}_2}\right)$$

(6.60)

$$\bar{A}_1 = \frac{\bar{A}_3}{4} e^{jk_2L} \left(1 + \frac{\bar{Z}_1}{\bar{Z}_2}\right)\left(1 + \frac{\bar{Z}_2}{\bar{Z}_3}\right)$$

$$+ \frac{\bar{A}_3}{4} e^{-jk_2L} \left(1 - \frac{\bar{Z}_2}{\bar{Z}_3}\right)\left(1 - \frac{\bar{Z}_1}{\bar{Z}_2}\right)$$

(6.61)

From Equation 6.61

$$\frac{\bar{A}_1}{\bar{A}_3} = \frac{1}{4}\left\{ e^{jk_2L} \left(1 + \frac{\bar{Z}_1}{\bar{Z}_2}\right)\left(1 + \frac{\bar{Z}_2}{\bar{Z}_3}\right) + e^{-jk_2L}\left(1 - \frac{\bar{Z}_2}{\bar{Z}_3}\right)\left(1 - \frac{\bar{Z}_1}{\bar{Z}_2}\right)\right\}$$

(6.62)

Equation 6.62 represents the amplitude of the acoustic disturbances in the fluid media. Generally the same fluid medium will be on both sides of the solid interface. Therefore, $\bar{Z}_1 = \bar{Z}_3$, for most of the practical cases. In addition, the impedance of the fluid medium is much less than that of the solid medium. Therefore, $\dfrac{\bar{Z}_1}{\bar{Z}_2} \ll 1$.

If k_2L is small (<0.25), $\sin k_2L = k_2L \cos k_2L = 1$.

For the preceding simplifications, Equation 6.62 becomes

$$\frac{\bar{A}_1}{\bar{A}_3} = \frac{1}{4}\left\{ (1 + jk_2L)\left(1 + \frac{\bar{Z}_2}{\bar{Z}_1}\right)(1 - jk_2L)\left(1 - \frac{\bar{Z}_2}{\bar{Z}_1}\right)\right\}$$

$$= \frac{1}{4}\left(2 + 2jk_2L\frac{\bar{Z}_2}{\bar{Z}_1}\right) = 1 + \frac{jk_2L}{2}\frac{\bar{Z}_2}{\bar{Z}_1}$$

(6.63)

The transmission coefficient is defined as the ratio of transmitted sound power to the incident sound power. Since the same surface area is considered on both sides, it can also be expressed as the transmitted sound intensity to incident sound intensity:

$$\tau_t = \frac{I_{tr}}{I_{in}}$$

(6.64)

Since the fluid media on both sides of the solid interface is the same, the transmission coefficient is the ratio of the square of the transmitted acoustic pressure to the square of the incident acoustic pressure. In terms of the amplitude of the acoustic pressures of Equation 6.63, the transmission coefficient can be expressed as

$$\tau_t = \left| \frac{A_3}{A_1} \right|^2 = \frac{1}{1 + \left(\dfrac{k_2 L\, Z_2}{2 Z_1} \right)^2}$$

(6.65)

The impedance of the fluid medium is given by

$$\bar{Z}_1 = \rho_1 c_1$$

(6.66)

and the impedance of the solid medium is given by

$$\bar{Z}_2 = \rho_2 c_2$$

(6.67)

The wave number of the solid medium is given by

$$k_2 = \frac{2\pi f}{c_2}$$

(6.68)

From Equations 6.65 through 6.68, the transmission coefficient is given by

$$\tau_t = \frac{1}{1 + \left(\dfrac{\pi \rho_2 f L}{\rho_1 c_1} \right)^2}$$

(6.69)

By defining $M_S = \rho_2 L$ as area density, and assuming air is the medium on both sides of the partition, the reciprocal of the transmission coefficient is given by

$$\frac{1}{\tau_t} = 1 + \left(\frac{\pi M_S f}{\rho_o c_o} \right)^2$$

(6.70)

For frequencies greater than 60 Hz, $1 \ll \left(\dfrac{\pi M_S f}{\rho_o c_o} \right)^2$, and Equation 6.70 thus becomes

$$\frac{1}{\tau_t} \approx \left(\frac{\pi M_S f}{\rho_o c_o} \right)^2$$

(6.71)

Transmission loss is defined as the reciprocal of the transmission coefficient

$$TL = 10 \log \frac{1}{\tau_t} \ \text{dB}$$

(6.72)

Equation 6.71 is substituted in Equation 6.72 to obtain

$$TL = 20 \log \frac{\pi M_s f}{\rho_o c_o} \tag{6.73}$$

Assuming the characteristic impedance of air $\rho_o c_o = 400$ rayl, Equation 6.73 can be expanded as

$$TL = 20 \log f + 20 \log M_s - 42 \, dB \tag{6.74}$$

Equation 6.74 represents approximate transmission loss in the mass controlled region for normal incidence sound waves. For random incidence of sound, which is more common in practice, the actual transmission loss can be 5 dB less than predicted by Equation 6.74 for normal incidence.

Therefore, for random incidence, the transmission loss becomes

$$TL = 20 \log f + 20 \log M_s - 47 \, dB \tag{6.75}$$

From Equation 6.75, it can be computed that doubling the frequency results in a change of 6 dB in transmission loss. Similarly, doubling of thickness increases the transmission loss by 6 dB across the frequencies.

In the critical frequency region when the bending wave speeds become equal to the speed of sound, there is a significant reduction in transmission loss predicted by Equation 6.75. Hence the region around the critical frequency is the region of bandwidth $\Delta f_p = f_2 - f_1$ and the ratio of these frequencies are frequency dependent and can be obtained from Watters (1959).

At frequencies above the critical frequency, known as the damping controlled region, the transmission loss increases by 10 dB/octave.

Based on the preceding discussion, an approximate transmission loss from the plateau method is shown in Figure 6.10. See Barron (2003) for additional information.

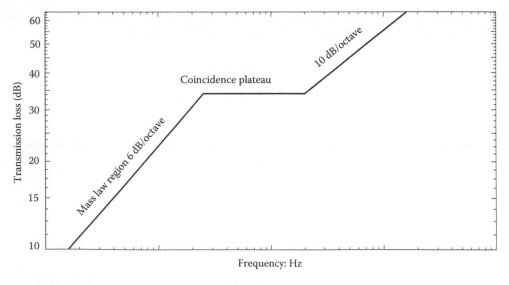

FIGURE 6.10
Approximate transmission loss using the plateau method.

Example 6.6

Compute the transmission loss of a 4 mm thick steel sheet (4 m × 4 m) from 63 Hz to 16000 Hz. Assume Young's modulus, E = 200 GPa; and density, ρ = 7800 kg/m³ for steel.

The approximate critical frequency of the steel sheet is given by $f_c \approx \dfrac{13}{h} = \dfrac{13}{0.004} = 3250$ Hz.

From Watters (1959), TL_p = 40 dB for steel and the corresponding Δf_p = 3.5 and $\dfrac{f_2}{f_1}$ = 11 for steel.

The area density is M_s = 7800 × 0.004 = 31.2 kg/m².

From Equation 6.75, the transmission loss at 63 Hz is given by $TL = 20\log 63 + 20\log 31.2 - 47$ dB = 19 dB.

We can add 6 dB for the transmission loss next to the consecutive octave bands as shown in Table 6.4.

From Table 6.4, it is clear that the transmission loss enters the plateau region between 500 and 1000 Hz. Since TL_p = 40 dB for steel, the transmission loss has to increase by 3 dB before it reaches 40 dB, and the lower frequency of the plateau region is given by

$$f_1 = 500 + \frac{3}{6} \times 500 = 750 \text{ Hz} \tag{1}$$

Since $\dfrac{f_2}{f_1}$ = 11 for steel, the transmission loss will be 40 dB from 750 Hz to 8250 Hz. After 8250 Hz, the transmission loss increases by 10 dB per octave. Since the transmission loss at 16500 Hz will be 50 dB, the transmission loss at 16000 Hz is given by

$$TL(16000\,\text{Hz}) = 50 \times \frac{16000 - 8250}{8250} = 47 \text{ dB} \tag{2}$$

The transmission loss of steel in the frequency range from 63 Hz to 16000 Hz is shown in Table 6.5.

For most noise control applications, the acoustical engineer does not need to calculate the transmission loss of barrier materials. For barrier materials, transmission loss has been measured in the laboratory. The results will be available with the manufacturers. ASTM Standard E90 is used for conducting such experiments. The basic procedure is to mount the test specimen as a partition between two reverberation rooms. Sound is

TABLE 6.4

Transmission Loss in the Mass Controlled Region (Example 6.6) (at octave band center frequencies)

Frequency (Hz)	63	125	250	500	1000
Transmission loss (dB)	19	25	31	37	42

TABLE 6.5

Transmission Loss of a 4 mm Steel Sheet from 63 to 16000 Hz (at octave band center frequencies except for 750 and 8250 Hz)

Frequency (Hz)	63	125	250	500	750	1000	2000	4000	8000	8250	16000
Transmission loss (dB)	19	25	31	37	40	40	40	40	40	40	47

introduced in one of the rooms and sound pressure measurements are made to ascertain the noise reduction between the source room and the receiving room. The rooms are so constructed that significant transmission is through the specimen only; all other paths must have a much larger transmission loss than the specimen that is being tested.

The transmission loss is given by

$$TL = NR + 10\log\frac{S}{A} \tag{6.76}$$

where S is the total area of sound transmitting surface; A is the total absorption in the receiving room, in Sabins; and NR is the noise reduction between the two rooms (sound pressure level difference in decibels).

A single number, known as the sound transmission class (STC), can also represent transmission loss. STC is obtained by fitting a specified transmission loss curve to the test data and its value at 500 Hz is the STC value. These STC values give good measures of comparison in architectural acoustics.

6.7 Conclusions

The modal approach that was applied to study the flexural vibration of beams was extended to study the flexural vibration of plates in this chapter. The bending wave number of plates is now expressed in the form of a vector that has components in the coordinate directions. This concept will be useful in studying the vibro-acoustics of plates. The modal density and group velocity were presented and they will play a major role in statistical energy analysis. The shell structures were also studied and it was shown that the shells behave similar to plates after the ring frequency. The transmission loss (limp mass) of barrier materials was studied, which is useful in many noise control applications; resonance transmission loss that is relevant for lightweight structures will be studied in Chapter 10.

PROBLEMS

6.1. A steel plate of dimensions 300 × 300 × 10 mm is simply supported. It is uniformly subjected to a sinusoidal pressure of amplitude 1 Mpa and 100 Hz frequency. Determine the center deflection.

6.2. Consider the steel plate of Problem 6.1. If a concentrated mass that is 30% of the plate mass is centrally placed, determine the first ten resonance frequencies of the plate.

6.3. Use the impedance method to solve for the natural frequencies of Problem 6.2.

6.4. Consider the steel plate of Problem 6.1. Determine the value of a spring that is centrally placed below the plate that would increase the first resonance frequency of the plate by 20%.

6.5. Consider the steel plate of exercise Problem 6.1. If the parallel longer edges are clamped and the other parallel edges are free, determine the first resonance frequency of the plate.

6.6. Design a centrally located tuned vibration absorber to the steel plate of Problem 6.1 to attenuate its response at the first resonance frequency.

6.7. Determine the natural frequencies and mode shapes of a thin circular plate of radius r and thickness h that is clamped along its rim.

6.8. Consider a thin circular steel plate of 100 mm diameter and 3 mm thick. Determine its fundamental frequency if it is supported only at the rim and plot the corresponding mode shape.

6.9. Refer to the Problem 6.1.

   a. Determine the bending wave speed and wavelength, them compare them with the acoustic wave speed and acoustic wavelength of surrounding air at 20°C.

   b. Determine the modal density in wave number, radians per second, and hertz, and the average separation frequency in hertz.

   c. Determine the plate critical frequency.

6.10. A steel plate ($E = 200$ GPa, $\rho = 7800$ kg/m³) of dimensions 300 mm × 300 mm × 5 mm is simply supported. It is subjected to a unit impulse at its center.

   a. Determine the center deflection considering responses of modes that are greater than 50% of the response of the mode that gives maximum response. (Already worked in Example 6.2.)

   b. Determine the bending wave speed and wavelength corresponding to significant responses of part a, then compare them with the acoustic wave speed and acoustic wavelength of surrounding air.

   c. Determine the modal density in wave number, radians per second, and hertz, and the average separation frequency in hertz corresponding to significant frequencies.

   d. If the plate is kept in a room having 2000 acoustic modes, at what octave band excitation frequency does the plate have a similar number of modes?

   e. Determine the plate critical frequency.

6.11. A rectangular steel plate 0.5 × 0.3 × 0.003 m is simply supported ($E = 200$ GPa, $\rho = 7800$ kg/m³). If a unit impulse is applied at the center of the plate, determine its deflection at $x = \ell_1/3$ and $y = \ell_2/3$.

6.12. Refer to Problem 6.11.

   a. Compute the list of resonance frequencies below 1 kHz and list them according to the two-dimensional indices and mark them on the wave lattice diagram.

   b. Determine the group velocity and phase velocity of this plate at 1 kHz and determine its modal density using group velocity.

6.13. Design the thickness of a steel plate of 1 square meter area that will have 500 modes in the 1/3 octave band of center frequency of 1000 Hz ($f_L = 891$ Hz and $f_H = 1122$ Hz).

   a. What is the critical frequency of this plate?

   b. What is the ratio of bending wave speed to acoustic wave speed at this frequency?

   c. What is the wave number?

   d. What is the modal density in modes per hertz (modes/Hz) and modes per radians per second (modes/rad/s)?

6.14. Evaluate the bending wave speed at 1000 Hz and the critical frequencies of 5 mm flat plates made of various metallic materials. How do you take advantage of the differing bending wave speeds for a practical application?

6.15. Consider Example 6.5. Plot a graph of k_x and k_y of all natural frequencies below critical frequency.

6.16. Consider Example 6.5. Plot a graph of the bending wave speeds of all natural frequencies below critical frequency.

6.17. Consider a cylindrical shell of 3 m internal diameter, 5 mm thick and 10 m height.

    a. Compute its resonant frequencies up to 5 kHz.

    b. Determine its ring frequency and critical frequency.

6.18. Consider Problem 6.17. Determine the modal density of the shell up to 5 kHz and compare it with that of a 5 mm thick flat plate.

6.19. Consider a shell of 1 m inside diameter and thickness 10 mm and a length of 5 m. Determine the radial and circumferential natural frequencies of this shell up to 5000 Hz.

6.20. A composite wall consists of areas A_1, A_2, A_3, ..., A_n having transmission loss values TL_1, TL_2, ..., TL_n. Determine the average transmission loss of this wall.

7

Sound Sources

Determining the source strength of airborne sound is the most difficult part of any noise reduction program of a machine. However, it is possible to model certain sources so that their characteristic features will help identify the source of sound in a machine so we can take appropriate measures to reduce sound. Among these sources, monopole, dipole, and quadrupole are very important. Therefore, we shall discuss these sources in detail in this chapter. In addition, a baffled piston, which serves as an important acoustic source in many applications, is also discussed. This discussion will help define radiation impedance that is an important characteristic of sound–structure interaction in the later chapters.

For modeling the source of sound, the concept of velocity potential can be advantageously used. Defining a velocity potential function helps us to extend the equations obtained in Chapter 3 to model and analyze sound sources. We shall, however, first extend the one-dimensional wave equation to three dimensions and then define the velocity potential. This will eventually lead to the definition of volume velocity that is useful in quantifying the strength of an acoustic source. The velocity potential then will be used to quantify three important sound sources: monopoles, dipoles, and quadrupoles. Then the discussion is further extended to acoustic radiation by a baffled piston.

7.1 Velocity Potential

We have to first define the concept of velocity potential and then express dynamic pressure and particle velocity in terms of velocity potential. Velocity potential can be easily defined by using the curl function in three dimensions. Let us first derive the wave equation in three dimensions by using the one-dimensional wave equation.

From Equation 3.29 (see Chapter 3) the one-dimensional wave equation in terms of dynamic pressure and density is given by

$$\frac{\partial^2 p_d}{\partial x^2} = \frac{\partial^2 \rho_d}{\partial t^2} \tag{7.1}$$

Since the dynamic pressure is varying as the second derivative of the space variable in Equation 7.1, by using the Laplacian operator, the three-dimensional equation can be obtained from it as follows:

$$\nabla^2 p_d = \frac{\partial^2 \rho_d}{\partial t^2} \tag{7.2}$$

where the Laplacian operator is given by

$$\nabla^2 = \frac{\partial^2}{\partial x^2} + \frac{\partial^2}{\partial y^2} + \frac{\partial^2}{\partial z^2} \tag{7.3}$$

From Equation 3.26 (Chapter 3), the one-dimensional equation in terms of dynamic pressure and particle velocity, the following equation can be written:

$$\frac{\partial p_d}{\partial x} + \rho_o \frac{\partial u}{\partial t} = 0 \tag{7.4}$$

Equation 7.4 can also be extended to three dimensions using the gradient operator as

$$\vec{\nabla} p_d + \rho_o \frac{\partial \mathbf{u}}{\partial t} = 0 \tag{7.5}$$

The gradient operator is given by

$$\vec{\nabla} = \frac{\partial}{\partial x} \hat{\mathbf{i}} + \frac{\partial}{\partial y} \hat{\mathbf{j}} + \frac{\partial}{\partial z} \hat{\mathbf{k}} \tag{7.6}$$

where **i**, **j**, and **k** are unit vectors representing the coordinate directions representing the Cartesian coordinates.

The particle velocity vector in Equation 7.5 is given by

$$\mathbf{u} = u_x \hat{\mathbf{i}} + u_y \hat{\mathbf{j}} + u_z \hat{\mathbf{k}} \tag{7.7}$$

From Equation 3.30 (Chapter 3), the one-dimensional wave equation in terms of dynamic pressure is given by

$$\frac{\partial^2 p_d}{\partial x^2} = \frac{1}{c_o^2} \frac{\partial^2 p_d}{\partial t^2} \tag{7.8}$$

where c_o is the velocity of airborne sound in air.

Equation 7.8 can be expressed in three dimensions using the Laplacian operator as

$$\nabla^2 p_d = \frac{1}{c_o^2} \frac{\partial^2 p_d}{\partial t^2} \tag{7.9}$$

For a scalar function, the curl of its gradient is equal to zero and it can be expressed as

$$\vec{\nabla} \times \vec{\nabla} \phi = 0 \tag{7.10}$$

By using Equation 7.10, taking the curl of Equation 7.5, the following equation can be written:

$$\vec{\nabla} \times \left(\vec{\nabla} p_d + \rho_o \frac{\partial \mathbf{u}}{\partial t} \right) = 0 \tag{7.11}$$

By using Equation 7.10, the first term of Equation 7.11 is zero. Thus Equation 7.11 becomes

$$\rho_o \vec{\nabla} \times \frac{\partial \mathbf{u}}{\partial t} = 0 \tag{7.12}$$

Equation 7.12 can be rewritten as

$$\frac{\partial}{\partial t}(\vec{\nabla} \times \mathbf{u}) = 0 \tag{7.13}$$

Since the particle velocity is a dynamic variable

$$\vec{\nabla} \times \mathbf{u} = 0 \tag{7.14}$$

Therefore, the particle velocity vector is irrotational and can thus be expressed as the gradient of a scalar function ϕ, given by

$$\mathbf{u} = \vec{\nabla}\phi \tag{7.15}$$

From Equations 7.5 and 7.15

$$\vec{\nabla} p_d + \rho_o \frac{\partial \vec{\nabla}\phi}{\partial t} = 0 \tag{7.16}$$

Equation 7.16 can be written as

$$\vec{\nabla} \left(p_d + \rho_o \frac{\partial \phi}{\partial t} \right) = 0 \tag{7.17}$$

Since the dynamic pressure and velocity potentials are dynamic variables, without wave propagation, Equation 7.17 becomes

$$p_d = -\rho_o \frac{\partial \phi}{\partial t} \tag{7.18}$$

Equation 7.18 relates dynamic pressure with time rate of change of velocity potential. Therefore, the velocity potential also must satisfy the wave equation as follows:

$$\nabla^2 \phi = \frac{1}{c_o^2} \frac{\partial^2 \phi}{\partial t^2} \tag{7.19}$$

From Equation 7.5, the particle velocity and the gradient of dynamic pressure are related by

$$\mathbf{u} = -\frac{1}{\rho_o} \int \vec{\nabla} p_d \, dt \tag{7.20}$$

The constants of integration are zero due to dynamic variables that exist only due to wave propagation.

7.2 Monopoles

Consider a pulsating sphere of radius a that has a velocity amplitude U_a (Figure 7.1).
The complex velocity is given by

$$\tilde{u}_a = U_a e^{j\omega t} \tag{7.21}$$

The wave equation in terms of velocity potential in spherical coordinates is given by

$$\nabla^2 [r\phi(r,t)] = \frac{1}{c_o^2} \frac{\partial^2 [r\phi(r,t)]}{\partial t^2} \tag{7.22}$$

The solution is given by

$$\phi(r,t) = \frac{G_1(c_o t - r)}{r} + \frac{G_2(c_o t + r)}{r} \tag{7.23}$$

The first term on the right represents an outward moving wave and the second term represents an inward moving wave. However, a pulsating sphere can only send outward moving waves in a non-reflective environment.

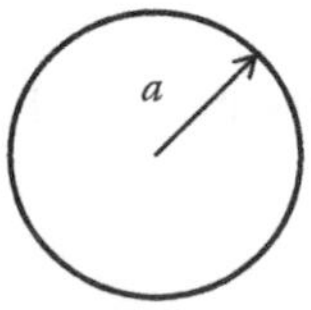

FIGURE 7.1
Monopole source.

The complex exponential of the velocity potential of an outward moving is given by

$$\tilde{\phi}(r,t) = \frac{\bar{A}}{r} e^{j(\omega t - k_o r)} \tag{7.24}$$

From Equations 7.15 and 7.24, the particle velocity at any radius is given by

$$\tilde{u}_r = \frac{\partial \tilde{\phi}}{\partial r} = \frac{\bar{A}}{r} e^{j(\omega t - k_o r)} \left(-\frac{1}{r} - jk_o \right) \tag{7.25}$$

Equating Equations 7.21 and 7.25

$$\bar{A} = -\left(\frac{U_a a^2}{1 + jk_o a} \right) e^{jk_o a} \tag{7.26}$$

From Equations 7.24 and 7.26, the velocity potential of a pulsating sphere is given by

$$\tilde{\phi}(r,t) = -\frac{1}{r} \left(\frac{U_a a^2}{1 + jk_o a} \right) e^{j[\omega t - k_o (r-a)]} \tag{7.27}$$

The source strength, also known as volume velocity, is a product of surface area and its velocity and is given by

$$\tilde{Q}(t) = 4\pi a^2 U_a e^{j\omega t} = Q_p e^{j\omega t} \tag{7.28}$$

Q_p is the peak value of the source strength given by

$$Q_p = 4\pi a^2 U_a \tag{7.29}$$

From Equations 7.28 and 7.29, the velocity potential in terms of source strength is given by

$$\tilde{\phi}(r,t) = -\frac{\tilde{Q}(t)}{4\pi r} \left(\frac{1}{1 + jk_o a} \right) e^{-j[k_o (r-a)]} \tag{7.30}$$

From Equations 7.18 and 7.30, the dynamic pressure distribution due to a pulsating sphere in terms of source strength is given by

$$\tilde{p}_d = \frac{\tilde{Q}(t)}{4\pi r} \left(\frac{j\omega \rho_o}{1 + jk_o a} \right) e^{-j[k_o (r-a)]} \tag{7.31}$$

Using the relation $c_o = \omega/k_o$, Equation 7.31 becomes

$$\tilde{p}_d = \frac{\tilde{Q}(t)}{4\pi r}\left(\frac{jk_o\rho_o c_o}{1+jk_o a}\right)e^{-j[k_o(r-a)]} \tag{7.32}$$

From Equations 7.15 and 7.30, the particle velocity distribution in terms of source strength is given by

$$\tilde{u}(r,t) = \frac{\tilde{Q}(t)}{4\pi r^2}\left(\frac{1+jk_o r}{1+jk_o a}\right)e^{-j[k_o(r-a)]} \tag{7.33}$$

The specific acoustic impedance of the monopole is given by the ratio of dynamic pressure and particle velocity. Therefore, from Equations 7.32 and 7.33, the specific acoustic impedance is given by

$$\bar{Z}_S = \frac{\tilde{p}_d}{\tilde{u}} = \rho_o c_o \frac{jk_o r}{1+jk_o r} \tag{7.34}$$

which is same as Equation 3.105 derived earlier in Chapter 3.

From Equations 7.32 and 7.33, the sound intensity of a monopole source is given by

$$\mathbf{I}(r) = \frac{1}{2}[\tilde{p}_d\tilde{u}^*(r,t)] = \frac{Q_{rms}^2 k_o^2 \rho_o c_o}{16\pi^2 r^2\left(1+k_o^2 a^2\right)} \tag{7.35}$$

Since the sound intensity due to a monopole is uniform in all directions, its sound power can be obtained by multiplying by the spherical surface area at the corresponding radius and is given by

$$\pi_M = \frac{Q_{rms}^2 k_o^2 \rho_o c_o}{4\pi\left(1+k_o^2 a^2\right)} \tag{7.36}$$

See Blackstock (2000), Kinsler and Frey (1962), and Norton and Karczub (2003) for further details.

Example 7.1

The surface of a monopole source is sinusoidally pulsating at 500 Hz that gives a sound pressure level of 100 dB at 5 m from the source and the circumference of the source is same as the wavelength at this frequency. Assume nominal density of air as 1.2 kg/m³ and speed of sound as 340 m/s. What is the size of the monopole source and its volume velocity and sound power, root mean square (rms) velocity of the sphere's surface and the particle velocity at a distance of 10 m from the source?

From Equation 3.42 (Chapter 3), the wave number is given by

$$k_o = \frac{2\pi f}{c_o} = \frac{2\pi 500}{340} = 9.24 \text{ rad/m} \tag{1}$$

And extending Equation 2.55 (Chapter 2) to airborne sound waves, the wavelength is given by

$$\lambda_o = \frac{2\pi}{k_o} = \frac{2\pi}{9.24} = 0.68 \text{ m} \tag{2}$$

From the given condition that the monopole circumference should be equal to the wavelength, the monopole radius is given by

$$a = \frac{\lambda}{2\pi} = \frac{0.68}{2\pi} = 0.1082 \text{ m} = 10.82 \text{ cm} \tag{3}$$

From Equation 3.57 (Chapter 3), the dynamic pressure corresponding to 100 dB sound pressure level is given by

$$p_{rms} = p_{ref}\sqrt{10^{\frac{L_p}{10}}} = 20 \times 10^{-6}\sqrt{10^{\frac{100}{10}}} = 2 \text{ Pa} \tag{4}$$

From Equation 7.32, the volume velocity of the monopole is given by

$$Q_{rms} = \frac{4\pi p_{rms}r}{k_o\rho_o c_o}\sqrt{1+(k_o a)^2} = \frac{4 \times \pi \times 2 \times 5}{9.24 \times 1.2 \times 340}\sqrt{1+(9.24 \times 0.1082)^2} = 0.047 \text{ m}^3/\text{s} \tag{5}$$

From Equation 7.36, the sound power of this monopole is given by

$$\Pi = \frac{Q_{rms}^2 k_o^2 \rho_o c_o}{4\pi\left[1+(k_o a)^2\right]} = \frac{0.047^2 9.24^2 \times 1.2 \times 340}{4\pi\left[1+(9.24 \times 0.1082)^2\right]} = 3.06 \text{ W} \tag{6}$$

From Equation 7.29, the plate surface velocity is given by

$$U_{a_{rms}} = \frac{Q_{rms}}{4\pi a^2} = \frac{0.047}{4\pi \times 0.1082^2} = 0.32 \text{ m/s} \tag{7}$$

From Equation 3.105, the impedance magnitude of the spherical waves at $r = 10$ m is given by

$$|Z| = \frac{\rho_o c_o k_o r_2}{\sqrt{1+(k_o r_2)^2}} = \frac{1.2 \times 340 \times 9.24 \times 10}{\sqrt{1+(9.24 \times 10)^2}} = 407.98 \text{ rayl} \tag{8}$$

Since the dynamic pressure of spherical waves is inversely proportional to the square of distance from the source, the dynamic pressure at $r = 10$ m is given by

$$p_{rms}^2(r_2 = 10 \text{ m}) = \frac{p_{rms}^2(r_1 = 5 \text{ m})}{\left(\dfrac{r_2}{r_1}\right)^2} = \frac{4}{2^2} = 1 \text{ Pa}^2 \tag{9}$$

From Equation 3.57, the sound pressure level at $r_2 = 10$ m is given by

$$L_p = 10 \log \frac{p_{rms}^2}{p_{ref}^2} = 10 \log \frac{1}{(20 \times 10^{-6})^2} = 94 \text{ dB} \tag{10}$$

Using the value of impedance magnitude from Equation 8 and Equation 3.104, the particle velocity at $r_2 = 10$ is given by

$$u_{rms} = \frac{p_{rms}}{|Z|} = \frac{1}{407.98} = 2.45 \text{ mm/s} \tag{11}$$

7.3 Dipoles

A dipole is a sound source model that is composed of two monopoles in close proximity to each other and oscillating out of phase with each other as shown in Figure 7.2. Consider the two monopoles separated by a distance $2d$. A line is drawn perpendicular to the midpoint of the line joining these monopoles. A point that is at a distance r from the midpoint and at a height H is considered for analysis.

The velocity potential of the dipole can be expressed as the sum of the velocity potentials of each monopole and is given by

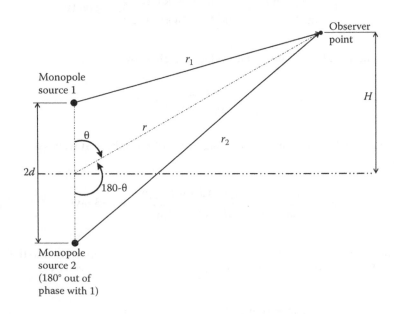

FIGURE 7.2
Dipole source.

$$\phi(r,t) = \phi_1(r,t) + \phi_2(r,t) \tag{7.37}$$

The distance r_1 is given by

$$r_1^2 = r^2 + d^2 - 2rd\cos\theta \tag{7.38}$$

The distance r_2 is given by

$$r_2^2 = r^2 + d^2 + 2rd\cos\theta \tag{7.39}$$

At distances much greater than the distance between the monopoles ($r \gg d$), the preceding distances can be approximated as

$$r_1 = r - d\cos\theta \tag{7.40}$$

and

$$r_2 = r + d\cos\theta \tag{7.41}$$

From Equation 7.30 neglecting the radii of monopoles, the complex velocity potential of the dipole, in terms of the complex source strengths of the monopoles it is comprised of, is given by

$$\tilde{\phi}(r,t) = -\frac{\tilde{Q}_1(t)}{4\pi r_1}e^{-jk_o r_1} - \frac{\tilde{Q}_2(t)}{4\pi r_2}e^{-jk_o r_2} \tag{7.42}$$

Using Equations 7.40 and 7.41 in Equation 7.42, we have

$$\tilde{\phi}(r,t) = -\frac{e^{-jk_o r}}{4\pi r}\left[\tilde{Q}_1(t)e^{jk_o d\cos\theta} + \tilde{Q}_2(t)e^{-jk_o d\cos\theta}\right] \tag{7.43}$$

The complex source strengths of each of the monopoles can be expressed as

$$\tilde{Q}_1(t) = Q_p e^{j(\omega t + \pi/2)} \tag{7.44}$$

and

$$\tilde{Q}_2(t) = Q_p e^{j(\omega t - \pi/2)} \tag{7.45}$$

Using Equations 7.44 and 7.45, Equation 7.43 becomes

$$\tilde{\phi}(r,t) = -\frac{Q_p e^{j(\omega t - k_o r)}}{4\pi r}\left[e^{j(k_o d\cos\theta + \pi/2)} + e^{-j(k_o d\cos\theta + \pi/2)}\right] \tag{7.46}$$

Using the complex exponential relationship for a sine function, Equation 7.46 becomes

$$\tilde{\phi}(r,t) = -\frac{Q_p e^{j(\omega t - k_o r)}}{2\pi r} \sin(k_o d \cos\theta) \tag{7.47}$$

For $k_o d \ll 1$, Equation 7.47 becomes

$$\tilde{\phi}(r,t) = -\frac{Q_p e^{j(\omega t - k_o r)}}{2\pi r} k_o d \cos\theta \tag{7.48}$$

From Equations 7.18 and 7.48, the dynamic pressure due to a dipole is given by

$$\tilde{p}_d = -\frac{j\rho_o \omega Q_p e^{j(\omega t - k_o r)}}{2\pi r} k_o d \cos\theta \tag{7.49}$$

From Equations 7.15 and 7.48, the particle velocity due to a dipole is given by

$$\tilde{u}(r,t) = -\frac{Q_p k_o d \cos\theta e^{j(\omega t - k_o r)}}{2\pi}\left(\frac{1 + jk_o r}{r^2}\right) \tag{7.50}$$

The sound intensity due to a dipole source can be obtained using Equations 7.49 and 7.50 and is given by

$$\mathbf{I}(r) = \frac{1}{2}\left[\tilde{p}_d \tilde{u}^*(r,t)\right] = \frac{\rho_o c_o Q_p^2 k_o^4 (d\cos\theta)^2}{8\pi^2 r^2} \tag{7.51}$$

The far-field sound radiation at a distance r can be obtained by integrating Equation 7.51 with respect to θ and then multiplying with the corresponding spherical area. The sound power radiated by the dipole is given by

$$\pi_D = \frac{Q_{rms}^2 k_o^4 d^2 \rho_o c_o}{3\pi} \tag{7.52}$$

From Equations 7.36 and 7.52, for small radii of the monopole sources, the ratio of sound power radiated by a dipole and a monopole is given by

$$\frac{\pi_D}{\pi_M} = \frac{4k_o^2 d^2}{3} = \left(\frac{d}{\lambda}\right)^2 \frac{16\pi^2}{3} \tag{7.53}$$

See Blackstock (2000), Kinsler and Frey (1962), and Norton and Karczub (2003) for further details.

Example 7.2

A monopole sound source of source strength 0.08 m³/s (rms) and radius 30 mm (or $a = 0.03$ m) radiates at 1000 Hz in air. A dipole source was created by placing a similar monopole at a distance of 10 mm from it, and the two sources were tuned to radiate 180° out of phase with each other. Estimate the sound power radiated by the resulting dipole and that radiated by each of the individual monopoles. Assume impedance of air as 400 rayl and velocity of sound in air as 340 m/s.

The center distance between the monopoles is calculated as $2d = 0.010 + 2 \times 0.03 = 0.07$ m.

From Equation 3.42, the wavenumber of airborne sound at 1000 Hz and the given sound speed is given by

$$k_o = \frac{2\pi f}{c_o} = \frac{2\pi \times 1000}{340} = 18.48 \text{ rad/m} \tag{1}$$

From Equation 7.36, the sound power radiated by each monopole is given by

$$\begin{aligned} \Pi_M &= \frac{Q_{rms}^2 k_o^2 \rho_o c_o}{4\pi\left[1+(k_o a)^2\right]} \\ &= \frac{0.08^2 \times 18.48^2 \times 400}{4\pi\left[1+(18.48 \times 0.03)^2\right]} = 53.22 \text{ W} \end{aligned} \tag{2}$$

From Equation 7.52, the sound power radiated by the dipole is given by

$$\Pi_D = \frac{Q_{rms}^2 k_o^4 d^2 \rho_o c_o}{3\pi} = \frac{0.08^2 \times 18.48^4 \times 0.035^2 \times 400}{3\pi} = 38.81 \text{ W} \tag{3}$$

7.4 Monopoles Near a Rigid, Reflecting Surface

In many practical situations, sound sources are placed very close to a hard reflecting surface on the ground, and in the far field they can be treated as single point sources. However, properties of the ground surface will have to be accounted for as they are related to the extent of reflection. These ground effects are more pronounced when the distance of these sources from the ground is less than one wavelength of the corresponding frequency (see Problem 7.10). For such a condition, if the ground surface is a hard reflecting surface, this results in the monopole and its image of equal strengths vibrating in phase with each other. The resulting velocity potential can be obtained from Equation 7.44 as

$$\tilde{\phi}(r,t) = -\frac{Q_p e^{j(\omega t - k_o r)}}{2\pi r}\cos(k_o d \cos\theta) \tag{7.54}$$

For $d \ll \lambda$ and $kd \ll 1$, Equation 7.54 becomes

$$\tilde{\phi}(r,t) = -\frac{Q_p e^{j(\omega t - k_o r)}}{2\pi r} \tag{7.55}$$

Therefore, from Equation 7.55 it is clear that a sound source very close to a hard reflecting surface results in doubling of the velocity potential of a single monopole that is far away from reflecting surfaces. Similar effects are discussed for multiple reflection surfaces in the next chapter on room acoustics.

From Equations 7.18 and 7.55, the dynamic pressure due to a monopole placed very close to a hard reflecting surface is given by

$$\tilde{p}_d = \frac{j\omega Q_p \rho_o e^{j(\omega t - k_o r)}}{2\pi r} \tag{7.56}$$

and from Equations 7.15 and 7.55 the particle velocity is given by

$$\tilde{u}_r(r,t) = \frac{Q_p e^{j(\omega t - k_o r)}}{2\pi r^2}(jk_o r + 1) \tag{7.57}$$

From Equations 7.49, 7.56, and 7.57, the sound intensity of the aforementioned monopole is given by

$$I(r) = \frac{1}{2}\text{Re}\left(\tilde{p}_d \tilde{u}_r^*\right) = \frac{1}{2}\left(\frac{j\omega Q_p \rho_o e^{j(\omega t - k_o r)}}{2\pi r}\frac{Q_p e^{-j(\omega t - k_o r)}}{2\pi r^2}(-jk_o r + 1)\right)$$
$$= \frac{Q_p^2 k_o^2 \rho_o c_o}{8\pi^2 r^2} \tag{7.58}$$

The sound power of the aforementioned monopole source can be obtained by integrating the sound intensity of Equation 7.58 over only the hemispheric space and it is given by

$$\Pi = \frac{Q_p^2 k_o^2 \rho_o c_o}{8\pi^2 r^2}2\pi r^2 = \frac{Q_p^2 k_o^2 \rho_o c_o}{4\pi r} = \frac{Q_{rms}^2 k_o^2 \rho_o c_o}{2\pi r} \tag{7.59}$$

Comparing Equation 7.52, for a very small monopole source, and Equation 7.59 it is clear that a monopole placed very close to a hard reflecting surface doubles its sound power, but the surface vibration velocity has remained the same and therefore, can be referred to as constant volume sources.

See Blackstock (2000), Kinsler and Frey (1962), and Norton and Karczub (2003) for further details.

7.5 Sound Radiated from a Vibrating Piston in a Rigid Baffle

This is a classical problem discussed in several reference books and papers. In this problem it is assumed that a rigid piston is vibrating or a layer of air is vibrating, and there is a rigid baffle around it. This model of sound radiation is especially useful in studying sound radiation of loud speakers and from vibrating sources surrounded by a rigid structure.

7.5.1 Far-Field Effects

Consider a rigid flat piston mounted on a rigid baffle so its surface is flush with the baffle (Figure 7.3). The axis of the piston is perpendicular to the baffle and the observer point is at a distance r and angle θ with respect to the piston axis. The sound radiated by the piston can be modeled in terms of a large number of point monopoles that are radiating together. However, each of the monopoles is radiating from very close to the hard surface of the baffle. If Q_p is the strength of each elemental monopole source, from Equation 7.32, the dynamic pressure is given by

$$\tilde{p}_d = \frac{jk_oc_o\rho_oe^{j(\omega t-k_or)}}{2\pi r}Q_p \tag{7.60}$$

In order to integrate all the monopole sources on the rigid piston, the coordinates are defined as shown in Figure 7.4. The z coordinate is perpendicular to the piston surface and coincides with the piston axis. The x and y coordinates are perpendicular to the z-axis and are in the plane of the piston as shown. Consider an elemental area of the piston that

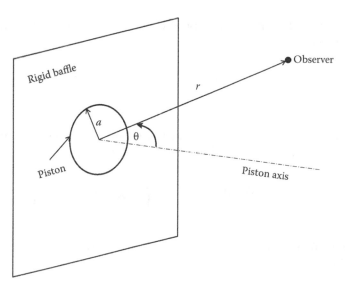

FIGURE 7.3
A piston mounted on a rigid baffle.

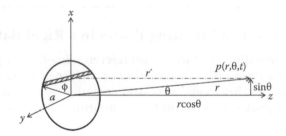

FIGURE 7.4
Coordinates of the piston.

makes an angle of ϕ and the observer point is at a distance of r from the piston center and makes an angle θ.

If U_p is the peak surface velocity of the piston and dS is the elemental surface area of the strip shown in the hatched area of Figure 7.4, the dynamic pressure due to monopoles along this strip is given by

$$d\tilde{p}_d = \frac{jk_oc_o\rho_o a \sin\phi e^{j(\omega t - k_o r')}}{\pi r'}U_p dx \tag{7.61}$$

r' is the perpendicular distance from the observation point and the elemental area and it is assumed that they are along the same straight line and parallel to the piston axis. It is expressed as

$$r' \approx r - a\sin\theta\cos\phi \tag{7.62}$$

r' of Equation 7.62 is used in the numerator of Equation 7.61, but $r' \approx r$ in the denominator. Equation 7.61 now becomes

$$d\tilde{p}_d = \frac{jk_oc_o\rho_o aU_p e^{j(\omega t - k_o r)}e^{j(k_o a\sin\theta\cos\phi)}\sin\phi}{\pi r}dx \tag{7.63}$$

In order to obtain the dynamic pressure due to the entire piston in the far field, Equation 7.63 must be integrated for values of x between $-a$ and a. It is given by

$$\tilde{p}_d(r,\theta,t) = \frac{jk_oc_o\rho_o aU_p e^{j(\omega t - k_o r)}}{\pi r}\int_{-a}^{a} e^{j(k_o a\sin\theta\cos\phi)}\sin\phi\, dx \tag{7.64}$$

By using the transformation $x = a\cos\phi$ in Equation 7.64, the dynamic pressure distribution can be obtained as

$$\tilde{p}_d(r,\theta,t) = \frac{jk_oc_o\rho_o a^2 U_p e^{j(\omega t - k_o r)}}{\pi r}\int_{0}^{\pi} e^{j(k_o a\sin\theta\cos\phi)}\sin^2\phi\, d\phi \tag{7.65}$$

By symmetry, $\displaystyle\int_0^\pi \sin(ka\sin\theta\cos\phi)\sin^2\phi\,d\phi = 0$ and $\displaystyle\int_0^\pi \cos(k_o a\sin\theta\cos\phi)\sin^2\phi\,d\phi =$

$\dfrac{\pi J_1(k_o a\sin\theta)}{k_o a\sin\theta} = \dfrac{\pi J_1(v)}{v}$, where $v = ka\sin\theta$. Equation 7.65 now becomes

$$\tilde{p}_d(r,\theta,t) = \frac{jk_o c_o \rho_o a^2 U_p e^{j(\omega t - k_o r)}}{2\pi r}\frac{2\pi J_1(v)}{v} \tag{7.66}$$

Therefore, $\dfrac{2J_1(v)}{v}$ is the directivity factor that depends on the wave number, area of the piston and the angle made by the observer point with the piston center. As shown in Figure 7.5, it has a value of 1 along the center line of the piston axis and then varies for various wave numbers, piston radii , and angle locations of the observer point.

The directivity factor for a sound-radiating piston in a rigid baffle can thus be expressed in decibels as

$$H(\theta) = 20\log\left|\frac{2J_1(v)}{v}\right| dB \tag{7.67}$$

The directivity factor of Equation 7.67 is plotted for various values of $k_o a$ in Figure 7.6. For the large value of $ka = 18$ there are several lobes, and the directivity factor falls very sharply

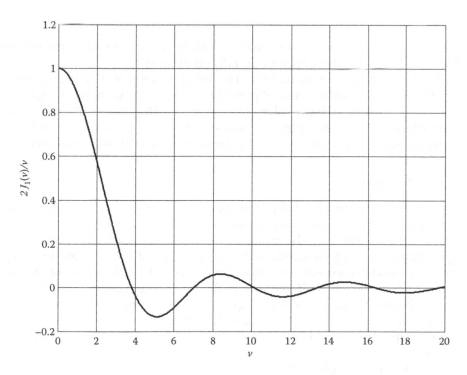

FIGURE 7.5
Directivity factor of a piston mounted on a rigid baffle.

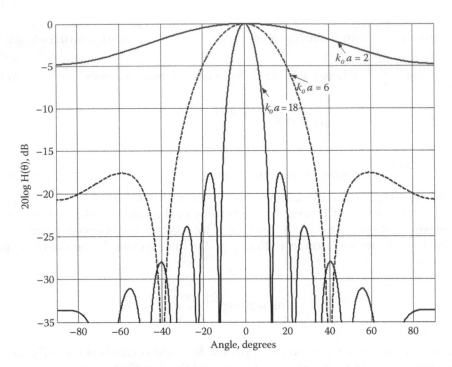

FIGURE 7.6
Directivity factor for various values of $k_o a$.

within ±10° around the center axis of the piston, and the several lobes beyond ±10° are less than −15 dB. For $ka = 6$, the directivity factor falls off more slowly and they drop to −10 dB between ±25°. The directivity factor for $ka = 2$ falls off the slowest of all the other cases and it reaches −5 dB between the entire range of possible angles (±90°). As shown in Figure 7.6, the number of lobes decreases with the value of ka. Because of the aforementioned behavior of the directivity factor, low frequency loud speakers can be very big and still be unidirectional, whereas high frequency loud speakers need to be small to be relatively omnidirectional.

7.5.2 Near-Field Effects

The near-field effects due to a baffled piston are equally important. This would allow for the consideration of the radiation impedance, which exists in addition to the mechanical impedance of the structure. The radiation impedance is due to the proximity of the vibrating surface to the fluid surrounding the piston, which affects the vibration response. Hence the total sound pressure at every point on the piston will be a sum of the pressure due to the vibrating element itself and due to the radiated pressures of all the elements on the piston. Figure 7.7 shows a single-degree-of-freedom system with mechanical and radiation impedance connected in parallel. Radiation impedance of flexible structures is discussed in Chapter 9 on sound–structures interaction, which is more involved than that of a rigid piston.

The piston velocity, $\tilde{U}_p$, is given by

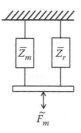

FIGURE 7.7
Mechanical and radiation impedance.

$$\tilde{U}_p = \frac{\tilde{F}_m}{\bar{Z}_m + \bar{Z}_r} \tag{7.68}$$

where $\tilde{F}_m$ is the mechanical force applied on the piston to drive it; $\tilde{Z}_m$ is the mechanical impedance of the piston that includes its mass and stiffness and damping of the structural support; and $\tilde{Z}_r$ is the radiation impedance. Depending on the density of the surrounding fluid, radiation impedance can result in fluid loading.

The radiation impedance is given by

$$\bar{Z}_r = \frac{\tilde{F}_p}{\tilde{U}_p} \tag{7.69}$$

where $\tilde{F}_p$ force on the piston due to acoustic pressure fluctuations. The radiation impedance can thus be obtained by integrating the elemental pressure distribution over the surface area of the piston to obtain the sound pressure at a point and then integrating over the area to get the force due to acoustic pressure fluctuations.

The radiation impedance can be obtained as (see Kinsler and Frey, 1962, for details)

$$\bar{Z}_r = \rho_o c_o \pi z^2 [R_1(2kz) + jX_1(2kz)] \tag{7.70}$$

where the resistive function $R_1(2kz)$ is given by

$$R_1(2kz) = \frac{(2kz)^2}{2.4} - \frac{(2kz)^4}{2.4^2.6} + \frac{(2kz)^6}{2.4^2.6^2.8} - \cdots \tag{7.71}$$

and the reactive function $X_1(2kz)$ is given by

$$X_1(2kz) = \frac{4}{\pi} \left\{ \frac{2kz}{3} - \frac{(2kz)^3}{3^2.5} + \frac{(2kz)^5}{3^2.5^2.7} - \right\} \tag{7.72}$$

Equations 7.71 and 7.72 are shown in Figure 7.8.

By accounting for the radiation impedance, the piston velocity equation can be written as

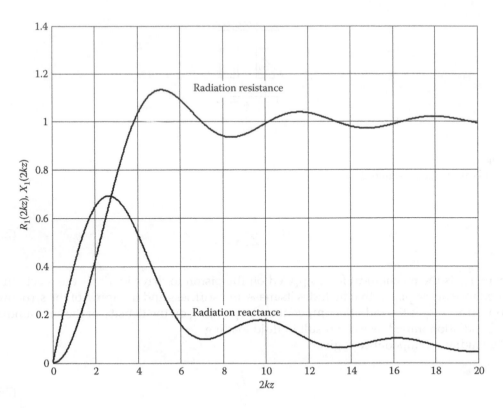

FIGURE 7.8
Resistive and reactive functions for the radiation impedance of a circular piston.

$$\tilde{U}_p = \frac{\tilde{F}_m}{R_m + j\left(\omega m - \dfrac{k}{\omega}\right) + \rho_o c_o \pi z^2 [R_1(2kz) + jX_1(2kz)]}$$ (7.73)

The sound power radiated by the piston can be estimated using Equation 1.96 (see Chapter 1):

$$\langle W \rangle = \frac{1}{2}\left|\tilde{U}_p\right|^2 \mathrm{Re}\left(\bar{Z}_m + \bar{Z}_r\right)$$ (7.74)

From Equations 7.73 and 7.74

$$\langle W \rangle = \frac{1}{2}\left|\tilde{U}_p\right|^2 \left(R_m + \rho_o c_o \pi z^2 R_1(2kz)\right)$$ (7.75)

The first term relates to the mechanical power that is dissipated and the second term represents the sound power that is radiated into the surrounding medium.

See Blackstock (2000), Kinsler and Frey (1962), and Norton and Karczub (2003) for further details.

Example 7.3

A flat piston of 0.4 m diameter is mounted on an infinite baffle so it radiates on one side of an infinite baffle into air at a frequency of 500 Hz. What must be the velocity amplitude to radiate 1 watt of acoustic power, and what is the force amplitude to produce this velocity amplitude? Assume the piston has a mass of 0.020 kg, stiffness of 1750 N/m, the speed of sound in air (c_o) as 340 m/s, and damping factor of 0.5.

Wavenumber, $k = k_o = \dfrac{\omega}{c_o} = \dfrac{2\pi.500}{340} = 9.24$ rad/s

Piston mass, M = 0.020 kg

Stiffness, K = 1750 N/m

The product of the wavenumber and the piston diameter is given by

$$2kz = 2 \times 9.24 \times 0.2 = 3.69$$

The normalized resistance is given by Equation 7.71:

$$
\begin{aligned}
R_1(2kz) &= \frac{(2kz)^2}{2.4} - \frac{(2kz)^4}{2.4^2.6} + \frac{(2kz)^6}{2.4^2.6^2.8} - \cdots \\
&= \frac{(3.69)^2}{2.4} - \frac{(3.69)^4}{2.4^2.6} + \frac{(3.69)^6}{2.4^2.6^2\,8} - \cdots = 1.012
\end{aligned}
\tag{1}
$$

The normalized reactance is given by Equation 7.72:

$$
\begin{aligned}
X_1(2kz) &= \frac{4}{\pi}\left\{ \frac{2kz}{3} - \frac{(2kz)^3}{3^2.5} + \frac{(2kz)^5}{3^2.5^2.7} - \right\} \\
&= \frac{4}{\pi}\left\{ \frac{3.69}{3} - \frac{(3.69)^3}{3^2.5} + \frac{(3.69)^5}{3^2.5^2.7} - \right\} = 0.5768
\end{aligned}
\tag{2}
$$

From Equation 7.75

$$
\begin{aligned}
U_p &= \sqrt{\frac{2\Pi}{\left[R_m + \rho_o c_o \pi z^2 R_1(2kz) \right]}} \\
&= \sqrt{\frac{2 \times 1}{\left[5.9161 + 400\pi \times 0.2^2 \times 1.0122 \right]}} = 0.1877 \text{ m/s}
\end{aligned}
\tag{3}
$$

From Equation 7.73

$$F_m = U_p \sqrt{\left[R_m + \rho_o c_o \pi z^2 R_1(2kz)\right]^2 + \left[\omega m - \frac{k}{\omega} + \rho_o c_o \pi z^2 X_1(2kz)\right]^2}$$

$$= 0.1877 \sqrt{[5.9161 + 400\pi \times 0.2^2 \times 1.0122]^2 + \left[3141.6 \times 0.02 - \frac{1750}{3141.6} + 400\pi \times 0.2^2 \times 0.5768\right]^2}$$

$$= 20.17 \text{ N}$$

(4)

7.6 Quadrupoles

Just like dipoles were obtained combining two monopoles vibrating out of phase, it is possible to obtain quadrupoles by combining two dipoles vibrating out of phase. Since a dipole has one axis, the quadrupole has two axes. Since there are two dipoles vibrating out of phase, there is no mass or momentum variation across a quadrupole. However, a quadrupole applies stress on the medium that results in the production of sound. Quadrupole sound sources are generally found in gas flow due to viscous stresses within gas. Quadrupoles can be formed in two ways: lateral quadrupoles as shown in Figure 7.9a and longitudinal quadrupoles as shown in Figure 7.10a. The respective directivity patterns of the quadrupoles are shown in Figure 7.9b and Figure 7.10b.

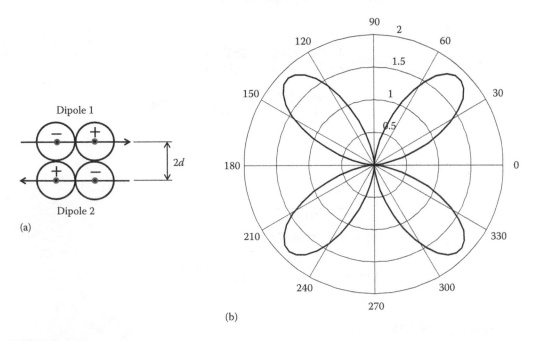

FIGURE 7.9
(a) Lateral quadrupole. (b) Directivity of a lateral quadrupole.

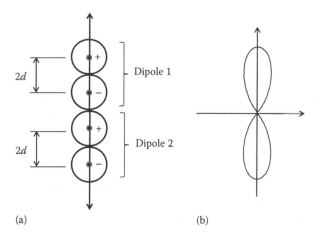

FIGURE 7.10
(a) Longitudinal quadrupole. (b) Directivity of a longitudinal quadrupole.

The sound power of quadrupoles can be obtained by considering the action of two dipoles vibrating out of phase with each other. The sound power of a lateral quadrupole can be obtained as (see Problem 7.13)

$$\pi_{quad\_lat} = \frac{4Q_{rms}^2 \rho_o c_o d^4 k_o^6}{15\pi} \tag{7.76}$$

and the sound power of longitudinal quadrupole can be obtained as

$$\Pi_{quad\_long} = \frac{4Q_{rms}^2 \rho_o c_o d^4 k_o^6}{5\pi} \tag{7.77}$$

Q_{rms} is the source strength of each of the identical monopoles of the quadrupole and d is the distance between them.

By using dimensional analysis it can be proven that the sound power of both quadrupoles is proportional to the eighth power of flow velocity in comparison to the dipole sound power being proportional to the sixth power of flow velocity.

$$\frac{\Pi_{quad\_long}}{\pi_M} = \frac{4Q_{rms}^2 \rho_o c_o d^4 k_o^6}{5\pi} \frac{4\pi\left(1 + k_o^2 a^2\right)}{Q_{rms}^2 k_o^2 \rho_o c_o} \approx d^4 k_o^4 \approx \left\{\frac{d}{\lambda}\right\}^4 \tag{7.78}$$

The same ratio would result by using the sound power of a lateral quadrupole. It shows that quadrupoles are less efficient than dipoles. Therefore, monopoles are the most efficient and quadrupoles are the least efficient radiators of sound.

See Blackstock (2000), Kinsler and Frey (1962), and Norton and Karczub (2003) for further details.

7.7 Conclusions

A detailed discussion on important sound sources—monopole, dipole, and quadrupole—were presented in this chapter by using the concept of velocity potential. In addition, a baffled piston was discussed that will help in taking the discussion further on sound–structure interaction due to vibration of flexible structures in Chapter 9. These sources will be useful in simulating many practical sound environments.

PROBLEMS

7.1. A spherical monopole source of unknown radius generates 10 mW of sound power at a frequency of 1000 Hz. Assume that the rms velocity of the pulsating spherical surface is 5 mm/s, the characteristic impedance of air as 400 rayl, and the velocity of sound in air as 340 m/s.

a. Determine the radius of the sphere.

b. Determine the sound power level of this source.

c. Determine the sound pressure level at 10 m from the center of the sphere, neglecting the sphere's dimensions.

7.2. Dynamic pressure and particle velocity measurements were conducted at a certain distance from a monopole source vibrating at 100 Hz and they are as follows:

$$\tilde{p} = (2.6579 + j0.9674)e^{j200\pi t} \text{ Pa}$$

$$\tilde{u} = (7.3 + j0.6)e^{j200\pi t} \text{ mm/s}$$

$$\rho_o = 1.16 \text{ kg/m}^3; \ c_o = 344 \text{ m/s}$$

(Assume that the dynamic pressure and particle velocity are measured using a multichannel analyzer whose channels are synchronized with each other.)

a. The monopole source has a circumference that is the same as the wavelength of sound at 100 Hz. Determine the rms surface velocity of the monopole.

b. For the same surface velocity of the monopole, determine the sound pressure level at a distance of 4 m, when the frequency of vibration of the monopole changes to 500 Hz. What is your conclusion regarding the dimensions of the monopole to its frequency of vibration and the resulting sound pressure level?

7.3. A monopole of 30 mm radius is pulsating at a frequency of 1000 Hz, and the peak amplitude of its surface velocity is 10 mm/s. Derive an equation for complex potential using these parameters to obtain equations for particle velocity and dynamic pressure at a certain distance r. Also determine the particle velocity at a distance of 10 m and the corresponding sound pressure level. What is the volume velocity of this source? What is the sound power of this source? Assume 400 rayl as the sound wave impedance and the velocity of sound as $c_o = 340$ m/s.

7.4. A point monopole generates a sound pressure level of 60 dB at a distance of 1 m at a frequency of 200 Hz. Calculate its volume velocity amplitude and sound power. In addition, calculate the rms radial particle velocity at that distance. Check your estimated sound power by integrating the radial intensity over a spherical surface of 1 m radius.

7.5. Dynamic pressure and particle velocity measurements were conducted at a certain distance from a monopole source vibrating at 1000 Hz and they are as follows:

$$\tilde{p} = (1.33 + j0.48)e^{j2000\pi t} \text{ Pa}$$

$$\tilde{u} = (3.386 + j1.173)e^{j2000\pi t} \text{ mm/s}$$

$$\rho_o = 1.16 \text{ kg/m}^3; \ c_o = 344 \text{ m/s}$$

a. Determine the distance from the source at which these measurements were made.

b. Determine the sound power of this monopole source.

c. What is the radius of this monopole to produce a volume velocity of 0.001 m³/s?

d. Based on part c, what is the surface velocity of this monopole? (Assume that the dynamic pressure and particle velocity are measured using a multi-channel analyzer whose channels are synchronized with each other.)

7.6. Determine the sound intensity based on the dynamic pressure and particle velocity of Problem 7.5.

7.7. Consider a point spherical source inside a sphere of radius r that is filled with a fluid of density ρ_1 and speed of sound c_1. It is surrounded by a fluid of density ρ_2 and speed of sound c_2 (see following figure). If the first fluid is hydrogen and the second fluid is air, and if both of them are at 20°C, determine the transmission and reflection coefficients at 1 kHz. Compare them with the transmission and reflection coefficients if the interface is a plane surface.

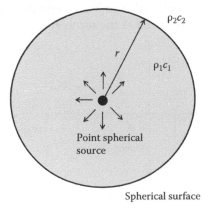

Spherical surface

7.8. A sinusoidally pulsating sphere of 200 mm radius is required to produce a sound pressure level of 65 dB at a distance of 200 m at the following frequencies:

30 Hz, 300 Hz, and 3 kHz. Determine the velocity amplitude required for each frequency.

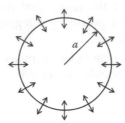

7.9. The spherical enclosure shown in the above figure is filled with a fluid and its outer surface is sinusoidally pulsating. Determine the pressure field and particle velocity inside the enclosure for $r < a$.

7.10. Consider a monopole placed at a height h from the air–water interface as shown in the following figure.

    a. Determine the condition with respect to its wavelength for it to be an efficient radiator in the far field inside water.

    b. What is the effect on power radiated if the interface becomes a solid surface?

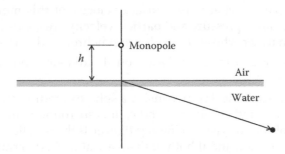

7.11. Consider the monopole below the air–water interface shown in the following figure at a height h. If an impulse action is used as a monopole source, which can be modeled as a shock pulse of T seconds duration, determine the time waveform of the signal received at the receiver.

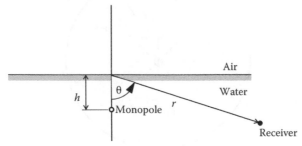

7.12. Determine the decay of sound pressure level from a dipole as a function of kr.

7.13. Consider the dipole placed at a height h above a solid surface as shown in the following figures. Determine under what conditions these dipoles converge to a quadrupole.

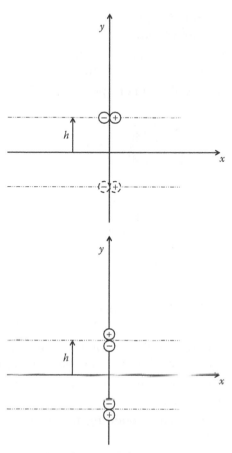

7.14. A plane circular piston placed in an infinite baffle sends out pressure pulses in water. The piston diameter is 500 mm and generates a sound pressure level of 100 dB (ref 1 μPa) at a distance of 750 m along the axis of the piston.

    a. Determine the range of angles at which the pressure in the far field is zero.

    b. Determine the rms velocity of the piston.

    c. Determine the sound pressure level at the same location for doubling the piston velocity.

7.15. A plane circular piston of 300 mm diameter is placed in an infinite baffle that sends pressure pulses in water. If it radiates 50 W of acoustic power at 15 kHz, determine the piston velocity and mass loading. In addition, determine the sound pressure level (ref 1 μPa) at a distance of 500 m along the axis.

7.16. A plane circular piston of 300 mm diameter is placed in an infinite baffle that radiates airborne sound. If the maximum displacement amplitude of the piston is 0.15 mm, determine the acoustic power radiated, radiation mass, and sound pressure level at a distance of 2 m along the axis.

7.17. A plane circular piston of 500 mm diameter is placed in an infinite baffle that operates in water. The piston radiates 100 W of sound power at 20 kHz. Determine the velocity amplitude of the piston and the radiation mass loading. If the piston has a mass of 0.15 kg, which is backed up by a spring of stiffness

1000 N/m, and a dashpot of damping factor of 0.15, estimate the force required to produce the velocity amplitude.

7.18. A baffled circular piston is excited by white noise excitation. Derive equations for sound pressure in the far field as a function of frequency and distance.

7.19. Obtain the mechanical equivalent of a pulsating sphere shown in the following figure.

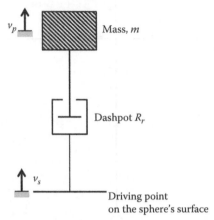

7.20. The sonic boom, caused by objects moving faster than sound, are acoustical transients called N-waves, as shown in the following figure.

a. Obtain the amplitude spectrum of this wave using the Fourier transform of Equation 4.17 (see Chapter 4).

b. Obtain the frequency corresponding to the prominent frequency of the spectrum for $T = 0.1$ s.

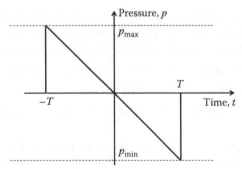

8

Room Acoustics

The interaction of sound with the surfaces of an enclosed space comes under the purview of room acoustics. Room acoustics in earlier times was also called architectural acoustics and building acoustics, when the science and engineering of pleasant sound was the main objective. That was the time in which interest in machine-generated noise was not very important. Reducing sound from noisy machinery inside an enclosed space is much different from modifying an enclosed space to obtain pleasant sound. I would prefer to use room acoustics with respect to machinery noise and architectural acoustics for producing the right note in an auditorium. The most important approach to changing room acoustics is the use of sound-absorbing materials, which is the subject of this chapter.

As discussed earlier, determining the sound power of a sound source is important for its objective characterization. Once the sound power is known, the resulting sound pressure levels in any environment can be easily determined. Therefore, by introducing the concept of room constant, which is related to the average absorption coefficient, necessary equations are derived that can help relate sound power levels to sound pressure levels. In the process of deriving these equations it becomes inevitable to derive the equation for energy density that will be useful in Chapter 10 for computing the total energy of an acoustic system. The same equations can also be used to determine sound power by conducting sound pressure level measurements in various types of environments.

Although much of the chapter appears to be only relevant to a machine that has already been built, it is important to note that the sound power level in various bands can be examined to see whether any of them can be reduced by using the principles of vibro-acoustics or by using acoustic absorption materials or barrier materials.

First, we discuss acoustic absorption materials and experimental determination of the acoustic absorption coefficient. Then we define the concept of the average acoustic absorption coefficient that is representative of the sum total of all acoustic absorption materials within an environment. Then we derive equations for energy density that will help define the concept of Room constant. Subsequently, equations are derived for relating sound power levels to sound pressure levels in various types of environment and determination of sound power from sound pressure level measurements. Finally, the modal density of an acoustic volume is presented.

8.1 Sound Absorption Using Acoustic Materials

In order to improve the acoustics of the Boston Symphony Hall in 1904, W. Sabine conducted several experiments to obtain characteristics of the sound absorbing materials in an enclosed space. The obtain characteristic properties of these acoustic materials now play a very important role in reducing machinery noise. Acoustic materials absorb sound energy within an enclosed space as shown in Figure 8.1. Generally, porous and fibrous materials

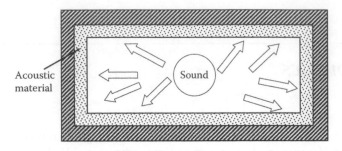

FIGURE 8.1
Absorption of sound by acoustic material within an enclosed space.

are excellent absorbers of sound, and the absorption occurs as a transfer between aerodynamic and thermodynamic energies. In the case of porous materials, there is adiabatic and isothermal heat transfer due to gaseous expansions and rarefactions along the pore edges. For fibrous materials, the fibers are forced to move and a temperature rise occurs due to fiber bending and friction. Acoustic energy is converted into heat in both cases. However, the rise in temperature due to heating is negligible.

Sound absorption is characterized by an absorption coefficient. Denoted by α, it quantitatively describes acoustic absorption and it varies between 0 and 1. $\alpha = 0$ represents complete reflection of acoustic energy, as in the case of hard materials like brick wall and concrete. $\alpha = 1$ represents complete absorption of incident energy. Most acoustic materials used in noise control will have a value between 0 and 1 for various frequencies. For every material, the acoustic absorption coefficient is frequency dependent and it is generally measured in octave or 1/3 octave bands, depending on the frequency analysis of the sound level meters used in sound measurement.

Before we discuss the details of measuring the acoustic absorption coefficient of a material, let us look at various possibilities of noise reduction in a typical machine, as shown in Figure 8.2. Sound is generated in a machine due to either vibration of an external surface that is converted to sound (vibro-acoustics) or due to flow, and there are various possibilities of reducing this sound. Sound produced by flow can be reduced by using a muffler or silencer; low frequency sound can be reduced by a reactive muffler that reduces sound by designed area changes of the flow path and high frequency sound is reduced by using acoustic absorption material.

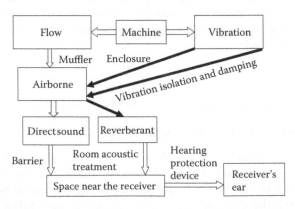

FIGURE 8.2
Various possibilities of noise reduction in a machine.

Whereas, sound generated due to vibration can be reduced by an enclosure that is lined with acoustic absorption material. Sometimes the sound generated due to vibro-acoustic phenomena can be reduced by providing damping to the vibrating surfaces of the machine. Vibration from a machine can travel to other areas around the machine and can result in secondary radiation of sound, which can be reduced by proper vibration isolation. Once the sound leaves a machine, it is called direct airborne sound. If the machine is kept inside a room, additional sound is generated due to multiple reflections, depending on the extent of acoustic absorption along the room surfaces. This additional sound is known as reverberant sound and it can be nearly 10 dB more than the direct airborne sound produced by the machine. Therefore, treating the room surfaces of the room in which the machine is kept can reduce sound pressure levels to the extent of 10 dB. The airborne sound, both direct and reverberant, can be further reduced only by the use of hearing protection devices before they reach the human ear. From the earlier discussion it is clear that absorption of sound using acoustic absorption materials plays a very important role in many noise control applications.

There are various methods of measuring the acoustic absorption coefficient. The simplest is the impedance tube method using a small sample of the acoustic absorption material. Since acoustic materials of large surface areas are generally used in most applications, the room method of measuring the acoustic absorption coefficient, which is discussed later, yields a better estimate. However, the impedance tube method is a very quick method of obtaining acoustic absorption coefficient.

8.2 Acoustic Absorption Coefficient (Tube Method)

Figure 8.3 shows an experimental setup for measuring acoustic absorption coefficient using the Tube method (Sabine 1942). The sample material is placed at the end of a tube. Sound waves generated by the variable frequency source establish a standing waves in the tube. They are reflected when they strike the sample and the reflection depends on

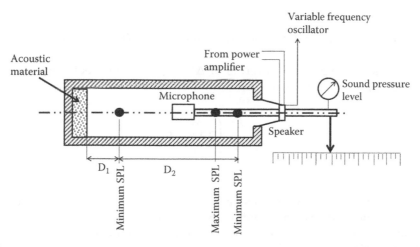

FIGURE 8.3
Experimental setup for measuring the normal acoustic absorption coefficient.

the absorption coefficient of the sample. The standing wave pattern has some character-istic features that are useful in measuring the acoustic absorption coefficient. Therefore, at a certain section of the tube closer to the sample, minimum sound pressure level is obtained and at a further distance away, maximum sound pressure level is obtained. These sound pressure levels can be measured by a movable microphone whose location can be accurately measured from outside. A sound level meter indicates the sound pres-sure levels. When the sound source at a certain frequency is switched on, the microphone is moved to obtain SPL_{low} at a distance L_1 and SPL_{high} at a distance L_2. These readings are obtained at various frequencies from which the corresponding acoustic absorption coefficient can be obtained.

In summary, the following parameters are measured with the movable microphone, at the desired frequencies of the audio oscillator:

ΔL = Difference in decibels between the maximum and minimum sound pressure levels

D_1 = Distance from the face of specimen to the nearest minimum of the standing wave pattern

D_2 = Distance from the first to the second minimum of the standing wave pattern

Since the sound source of Figure 8.3 results in the normal incidence on the material, the acoustic absorption coefficient obtained is known as the normal incidence acoustic absorp-tion coefficient. It can be shown that it is related to the properties of the material by

$$\alpha_n = 1 - \left(\frac{z/\rho_o c_o - 1}{z/\rho_o c_o + 1} \right)^2 \tag{8.1}$$

where z is the specific normal acoustic impedance and $\rho_o c_o$ is the characteristic impedance of the medium (air) that is present inside the tube.

The specific normal acoustic impedance has two components, expressed as

$$z = r + jx \tag{8.2}$$

where r is the specific normal acoustic resistance and x is the specific normal acoustic reactance.

Equation 8.1 can be expressed in terms of ΔL, the difference between maximum and minimum sound pressure levels, as

$$\alpha_n = 1 - \left(\frac{10^{\frac{\Delta L}{20}} - 1}{10^{\frac{L}{20}} + 1} \right)^2 \tag{8.3}$$

The normal acoustic absorption coefficient α_n of Equation 8.3 is plotted in Figure 8.4.

Normal incidence absorption coefficient is not an accurate measure of acous-tic absorption. Because, in practice, sound can be incident on an acoustic absorbent material from various angles; depending on the angle of incidence, it can have dif-ferent levels of sound absorption. Therefore, an acoustic absorption coefficient that

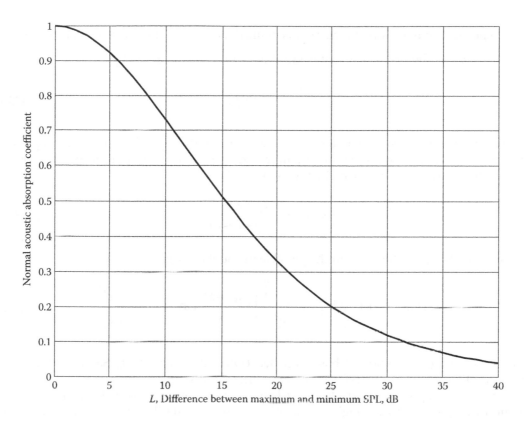

FIGURE 8.4

Normal acoustic absorption coefficient.

is determined based on random incidence is more relevant for practical applications. In any case, based on the distances D_1 and D_2, random incidence acoustic absorption coefficient can be calculated.

Dividing specific normal acoustic impedance of Equation 8.2 by characteristic impedance of air

$$\frac{z}{\rho_o c_o} = \frac{r}{\rho_o c_o} + \frac{jx}{\rho_o c_o} = \cot h(A + jB) \tag{8.4}$$

where

$$A = \cot h^{-1}[10^{L/20}] \tag{8.5}$$

and

$$B = \pi \left(\frac{1}{2} - \frac{D_1}{D_2} \right) \tag{8.6}$$

The specific acoustic impedance ratio is defined as $\xi = \dfrac{Z}{\rho c}$. The reciprocal of this quantity is known as the specific acoustic admittance ratio and is given by

$$\frac{1}{\xi} = \eta = \mu + j\kappa \tag{8.7}$$

where μ is specific acoustic conductance ratio and κ is the specific acoustic susceptance ratio. In terms of the aforementioned parameters, it is possible to obtain the absorption coefficient in any direction, given by (Beranek 1960)

$$\alpha(\theta) = \frac{4\mu\cos\theta}{\kappa^2 + (\mu + \cos\theta)^2} \tag{8.8}$$

Averaging Equation 8.8 over all the possible incident angles,

$$\alpha_{stat} = 8\mu\left[1 - \mu\ln\left(1 + \frac{2\mu + 1}{|\eta|^2}\right) + \frac{\mu^2 - k^2}{\kappa}\tan^{-1}\left(\frac{\kappa}{|\eta|^2 + \mu}\right)\right] \tag{8.9}$$

The preceding discussion is based on Bell (1982), Beranek (1960), Sabine (1942), and Watson (1927), which can be referred to for further details.

For fibrous and porous materials, the resistive component can also be determined experimentally by passing air through the material and measuring the pressure loss across the material. It is given by

$$r = \frac{SP}{U} \tag{8.10}$$

where P is the air pressure difference across the test specimen Pa, U is volume velocity of the airflow m^3/s, and S is the area of the specimen.

For homogeneous materials of thickness T, the resistivity per unit thickness is the preferred parameter:

$$r_o = \frac{SP}{TU} \tag{8.11}$$

Example 8.1

The following results were obtained during the acoustic absorption coefficient measurement using an impedance tube at 500 Hz. Given $\Delta L = 7$ dB, $D_1 = 120$ mm, and $D_2 = 280$ mm, and determine α_n, $\dfrac{r}{\rho c}$, and $\dfrac{x}{\rho c}$.

From Equation 8.3, the normal acoustic absorption coefficient is given by

$$\alpha_n = 1 - \left(\frac{10^{\frac{7}{20}} - 1}{10^{\frac{7}{20}} + 1}\right)^2 = 0.85 \tag{1}$$

$$\mathrm{Cot}\,h\,A = 10^{\frac{7}{20}} = 2.51 \tag{2}$$

$$A = 0.4805 \tag{3}$$

$$B = \pi\left(\frac{1}{2} - \frac{120}{280}\right) = 0.2244 \tag{4}$$

$$\cot h(A + jB) = 1.87 - 0.73\hat{j} \tag{5}$$

$$\frac{r}{\rho c} = 1.87,\ \frac{x}{\rho c} = -0.73 \tag{6}$$

8.3 Room Method of Measuring Absorption Coefficient

The tube method of determining the acoustic absorption coefficient is based on a very small sample of the material. Since this small space does not allow sound waves to fall on the specimen from all directions, which is a practical situation, the coefficient so measured will not be of much practical value. Hence, a better method would be to measure acoustic absorption of large area samples. This is known as the room method, which is explained as follows.

In the room method, the decay rate of sound is measured in a large sample of the material under study (ANSI/ASTM C423-77). By comparing the decay rate to that of the bare room, statistical absorption coefficients α_{stat}, at the desired frequency bands, can be determined using the following Sabine's formula. The absorption coefficients so determined are more representative of practical application than those obtained from the tube method.

$$T_{60}^{j} = \frac{0.161\,V}{S_T\bar{\alpha}_j + 4\times V\,\alpha_{air}^{j}} \tag{8.12}$$

T_{60}^{j} = Reverberation time for jth band
V = Room volume (m³)
S_T = Total surface area (m²)
$\bar{\alpha}_j$ = Average absorption coefficient
α_{air}^{j} = Absorption due to air can be obtained from references

Although the tube method cannot give a very accurate value of the acoustic absorption coefficient, it can give a quick measure of order of magnitude comparison between various materials.

Before we proceed further and relate sound power and the room constant, we need to define the energy density of an enclosed space.

8.4 Room Constant

In any volume consisting of different acoustic absorption materials of various areas, the total absorption of airborne sound is the sum total of the acoustic properties of all the materials in that volume. Therefore, an average acoustic absorption coefficient for the entire volume needs to be defined and it has to be computed in various band frequencies of interest.

Consider an enclosed volume of various surface areas S_1, S_2, ..., S_n with respective absorption coefficients α_1, α_2, ..., α_n corresponding to the jth band.

The average absorption coefficient $\bar{\alpha}_j$ is given by

$$\bar{\alpha}_j = \frac{1}{S_T} \sum_{i=1}^{n} \alpha_{ij} S_i \qquad (8.13)$$

where α_{ij} is the absorption coefficient of the ith surface in the jth frequency band and S_T is the total surface area equal to $S_1 + S_2 +, ..., S_n$.

The acoustic absorption coefficient is a number between 0 and 1, and thus cannot give a direct indication of the extent of noise reduction due to absorption. But by combining the average acoustic absorption coefficient with the total surface area of all the acoustic materials in the volume, a better description of the total noise that can be absorbed in the volume can be obtained. It is known as Room constant and it has the dimensions of surface area.

The Room constant R_j, for the jth band is given by

$$R_j = S_T \frac{\bar{\alpha}_j}{1 - \bar{\alpha}_j} \quad \bar{\alpha}_j \neq 1 \qquad (8.14)$$

Equation 8.14 will be proven later in Equation 8.49 in the context of deriving the equation for energy density.

When the individual acoustic absorbing materials are irregular in shape, instead of being flat surfaces lined with acoustic absorption material, a better descriptor is in Sabins. Therefore, total absorption, expressed as Sabins, corresponding to the jth band

$$= \sum_{i=1}^{N} S_i \bar{\alpha}_j \qquad (8.15)$$

Both Equations 8.13 and 8.14 will be useful in practice.

The Room constant and Sabin are very commonly used parameters related to acoustic absorption within an enclosure. In the preceding equations it is important to consider all the surface areas within the enclosure, such as extended surfaces and partitions. If the acoustic absorption coefficients are very small, Sabins Equation 8.15 can be taken as the room constant.

Example 8.2

An auditorium has dimensions of 15 m × 12 m × 4 m high. The walls are having plywood paneling, the floor is covered with a carpet, and the ceiling is having perforated tiles; acoustic absorption coefficients are shown in Table 8.1. There are 15 windows of ordinary glass, each 2 m × 1.00 m and the windows are permanently closed. When the auditorium is in use, there are 150 people seated on wooden chairs. Determine the room constant for the following conditions: (A) for the empty room when there are only chairs and no people, and (B) when the room is full with audience seated on chairs.

Length $L_1 = 15$ m, width $L_2 = 12$ m, height $L_3 = 4$ m

Let α_{S_empty} be the total Sabins of the room when the room is not occupied and α_{S_full} be the total Sabins of the room when the room is full.

$$\alpha_{S\_empty}(500\,\text{Hz}) = (L_1 \times L_2 \times \alpha_2) + (L_1 \times L_2 \times \alpha_3) + \alpha_1[2 \times L3(L_1 + L_2) - 15 \times 2 \times 1] + (150 \times \alpha_{S2}) + (15 \times \alpha_4 \times 2 \times 1)$$

$$= (15 \times 12 \times 0.75) + (15 \times 12 \times 0.2) + 0.15[2 \times 4 \times (15 + 12) - 15 \times 2 \times 1] + (150 \times 0.015) + (15 \times 0.15 \times 2 \times 1)$$

$$= 205.65 \text{ Sabins} \tag{1}$$

$$\alpha_{S\_full}(500\,\text{Hz}) = \alpha_{S\_empty}(500\,\text{Hz}) - 150 \times \alpha_{S2} + 150 \times \alpha_{S1}$$

$$= 205.65 - 150 \times 0.015 + 150 \times 0.35 = 255.9 \text{ Sabins} \tag{2}$$

The average acoustic absorption coefficients for the empty and occupied rooms can be obtained as follows:

$$\bar{\alpha}_{empty} = \frac{\alpha_{S\_empty}}{Area} = \frac{205.65}{2(15 \times 12 + 12 \times 4 + 15 \times 4)} = 0.3570 \tag{3}$$

$$\bar{\alpha}_{full} = \frac{\alpha_{S\_full}}{Area} = \frac{255.9}{2(15 \times 12 + 12 \times 4 + 15 \times 4)} = 0.447 \tag{4}$$

The room constant for the empty room can be obtained as

$$R_1 = S_T\left(\frac{\bar{\alpha}_{empty}}{1 - \bar{\alpha}_{empty}}\right) = 576\left(\frac{0.3570}{1 - 0.3570}\right) = 320 \tag{5}$$

TABLE 8.1
Acoustic Properties of Materials of Example 8.2

Octave Band (Hz)	500
Plywood panel, α_1	0.15
Perforated ceiling, α_2	0.75
Carpet, α_3	0.20
People on chairs, Sabins, α_{S1}	0.35
Glass, α_4	0.15
Wooden chairs, Sabins, α_{S2}	0.015

and the room constant for the occupied room can be obtained as

$$R_2 = S_T \left(\frac{\overline{\alpha}_{full}}{1 - \overline{\alpha}_{full}} \right) = 576 \left(\frac{0.4470}{1 - 0.4470} \right) = 460 \qquad (6)$$

8.5 Energy Density

In the mathematical treatment of sound in an enclosed space, the energy per unit volume from a source transported to different parts of the room is required. It can be derived as follows.

Consider an undisturbed volume of fluid, V_0, that changes to a different volume, V_1, due to the passage of an acoustic disturbance. The change in potential energy due to this change of volume is given by

$$U = -\int_{V_0}^{V_1} p \, dV \qquad (8.16)$$

The negative sign of Equation 8.16 indicates that the positive acoustic fluctuating pressure produces a decrease in fluid volume, thus increasing its potential energy.

Since density is the ratio of mass to volume, its derivative with respect to volume is given by

$$\frac{d\rho}{dV} = -\frac{m}{V^2} = -\frac{\rho}{V} \qquad (8.17)$$

From Equation 3.7 (see Chapter 3), the following equation is written for the variation of dynamic pressure with respect to density of air:

$$\frac{dp_d}{d\rho} = \frac{\gamma p_o}{\rho_o} \qquad (8.18)$$

Multiplying the corresponding sides of Equations 8.17 and 8.18:

$$\frac{dp_d}{dV} = -\frac{\rho \gamma p_o}{V \rho_o} \qquad (8.19)$$

For small changes in volume, Equation 8.16 becomes

$$U = -\int_{V_0}^{V_1} p_d \, dV \qquad (8.20)$$

For small changes in pressure and density, Equation 8.19 becomes

$$\frac{dp_d}{dV} = -\frac{\gamma p_o}{V_o} \qquad (8.21)$$

From Equations 8.20 and 8.21, the change in potential energy when the dynamic pressure changes from 0 to p_d is given by

$$U = \frac{V_o}{\gamma p_o} \int_0^{p_d} p_d \, dp_d = \frac{V_o p_d^2}{2\gamma p_o} \qquad (8.22)$$

From the earlier equation $c_o^2 = \frac{\gamma p_o}{\rho_o}$, Equation 8.22 can be simplified as

$$U = \frac{V_o p_d^2}{2\rho_o c_o^2} \qquad (8.23)$$

The kinetic energy change due to the passage of an acoustic wave is given by

$$T = \frac{V_n \rho_o u^2}{2} \qquad (8.24)$$

For plane waves, the ratio between dynamic pressure and particle velocity is related to the characteristic impedance, $\rho_o c_o$. Using this simplification in Equation 8.24, the kinetic energy becomes

$$T = \frac{V_o p_d^2}{2\rho_o c_o^2} \qquad (8.25)$$

The total energy of the volume due to the passage of acoustic waves is a sum of the potential and kinetic energies, respectively represented by Equations 8.23 and 8.25. The energy per unit volume is therefore given by

$$D = \frac{p_d^2}{\rho_o c_o^2} \qquad (8.26)$$

Equation 8.26 is useful in computing the total energy of an acoustic subsystem of a certain volume and average dynamic pressure.

8.6 Directivity Index

Sound sources whose dimensions are small compared with the wavelengths of sound are omnidirectional; otherwise they are directional. These directional characteristics of a sound source will have to be accounted for when determining its sound power. This will help in predicting sound pressure levels in any environment due to a known sound power source.

The directivity factor, $Q_D(\alpha)$, is the ratio of sound intensity I_α at some distance r from the source and at an angle α to a specific axis of a directional noise source of sound power W kept in a nonreflective environment to the intensity I_s produced at some distance r from a uniformly radiating sound source of equal sound power:

$$Q_D(\alpha) = \frac{I_\alpha}{I_s} = \frac{\langle p_\alpha^2 \rangle}{\langle p_s^2 \rangle} \tag{8.27}$$

From Equation 8.27, the directivity index, which is actually the directivity factor expressed in decibels, DI_α is defined as

$$DI_\alpha = 10 \log Q_D(\alpha) = L_{p\alpha} - L_{ps} \tag{8.28}$$

The sound intensity due to a uniform source of sound power W is given by

$$I_s = \frac{W}{4\pi r^2} = \frac{p_s^2}{\rho_o c_o} \tag{8.29}$$

From Equations 8.27 and 8.29

$$\frac{p_s^2}{\rho_o c_o} Q_D(\alpha) = \frac{p_\alpha^2}{\rho_o c_o} \tag{8.30}$$

$$\frac{p_\alpha^2}{\rho_o c_o} = \frac{W Q_D(\alpha)}{4\pi r^2} \tag{8.31}$$

Dividing by p_{ref}^2 on both sides and taking the logarithm

$$10 \log \frac{p_\alpha^2}{p_{ref}^2} = 10 \log \frac{W}{\dfrac{p_{ref}^2}{\rho_o c_o}} + 10 \log Q_D(\alpha) + 10 \log \frac{1}{4\pi r^2} \tag{8.32}$$

Since $P_{ref} = 20\mu Pa$ and $\rho_o c_o = 400$, by defining $W_{ref} = \dfrac{p_{ref}^2}{\rho_o c_o} = 10^{-12}$ W

$$L_{p\alpha} = L_w + 10\log Q_D(\alpha) - 10\log 4\pi r^2$$
$$= L_w + DI_\alpha - 10\log 4\pi r^2$$

(8.33)

Equation 8.33 will be used later to determine the sound power and directivity index of a sound source in an anechoic room by measuring sound pressure levels around it at a specified distance.

Even for sound sources that radiate uniformly in all directions, if they are placed close to reflecting surfaces, they are restricted from radiating noise in that direction. It can be accounted for by changing the area of the spherical surface surrounding the sound source. Directivity resulting out of constraining a uniform sound source by placing it around reflecting surfaces can be defined as the reciprocal of the fraction of the spherical surface that is available for a uniform source to radiate sound. The directivity factors for various possibilities of reflective surfaces around a sound source are shown in Table 8.2, and they can be expressed as

$$Q_P = \frac{1}{\text{fraction of the spherical surface}}$$

(8.34)

If the sound source itself is radiating nonuniformly, then the directivity index can be obtained by measurements in an anechoic room and then added to Equation 8.34. The directivity factor of Equation 8.27 can be written as a product of two directivity factors—one due to placement close to reflective surfaces and the other due to nonuniform sound radiation:

$$Q_\alpha = Q_P Q_D(\alpha)$$

(8.35)

where Q_P is dependent on how the sound source is placed, that is, hanging in air, placed on the ground away from reflecting surfaces, or near a corner. They are illustrated in Figure 8.5. $Q_D(\alpha)$ depends on the distribution of sound pressure levels around a source, which can be obtained from experiments for a nonuniform source. The sound pressure level in any direction α is now given by

$$L_{p\alpha} = L_w + 10\log Q_P Q_\alpha(D) - 10\log 4\pi r^2$$

(8.36)

TABLE 8.2

Directivity Factor for Various Mounting Positions

Position	Directivity Factor, Q_P	Directivity Index DI_p (10 log Q_P dB)
Free space	1	0
Center of a large flat surface	2	3
Intersection of two large flat surfaces	4	6
Intersection of three flat surfaces	8	9

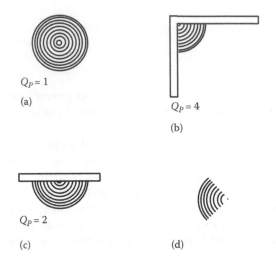

FIGURE 8.5
Directivity factor Q_p due to reflecting surfaces around a source. (a) Uniform spherical radiation with no reflecting surfaces, (b) uniform radiation over ¼ of a sphere, (c) uniform hemispherical radiation with one reflecting surface, and (d) nonuniform radiation and Q_p depends on the machine.

8.7 Sound Power Measurement in Anechoic Rooms

In anechoic rooms, the machine is mounted on a hard reflecting floor and the sound radiated from the machine is not reflected from the walls or roof that are lined with acoustic absorbing material. All the radiated sound energy is absorbed by the sound-absorbing material placed on all the boundaries of the room. It is very expensive to build such an environment, especially for testing bigger size machines. Since there are no reflections, it is possible to define a hypothetical surface around the machine, as shown in Figure 8.6, to conduct sound pressure measurements at different points of the surface and at each octave band center frequency.

The sound power is given by ($Q_p = 2$)

$$L_W = 10 \log \frac{p_m^2}{p_{ref}^2} + 10 \log 2\pi r^2 , \text{dB} \tag{8.37}$$

where p_m^2 is the average mean square value of the dynamic pressure, $p_{ref} = 20 \times 10^{-6} \, N/m^2$, and r is the radius of the hemisphere. If there are n points on the hemisphere at which measurements were made, then the average mean square pressure is arrived at from the following equation:

$$p_m^2 = \frac{1}{n} \sum_{i=1}^{n} p_i^2 \tag{8.38}$$

When the mean sound pressure is determined as described earlier, it is also possible to determine a directivity index (DI). This may be of interest in situations where noise sources do not radiate equally in all directions. From Equation 8.36, since $DI_p = 3$ dB, the directivity index in the ith direction can be calculated from the following formula:

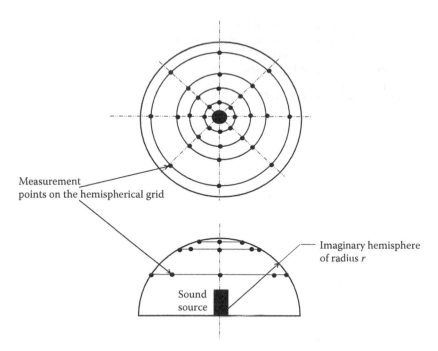

FIGURE 8.6
Sound pressure level measurement in an anechoic room.

$$DI = 10 \log \frac{p_i^2}{p_{ref}^2} - 10 \log \frac{p_m^2}{p_{ref}^2} + 3, dB \tag{8.39}$$

All the measurements must be made with the machine operating under normal conditions and it should be preferably mounted on vibration isolators to prevent secondary radiation of sound. It should be noted that the sound power and directivity index are computed for various octave or 1/3 octave bands.

Example 8.3

The sound pressure level (dB) measurements shown in Table 8.3 are from a machine placed on a concrete floor in an anechoic room, at eight equispaced points at a radius of 2 m from the machine. The dominant frequency is at 500 Hz.

a. Determine the directivity index.
b. Determine the average sound power and sound power level.
c. Determine the sound pressure levels along the 270° direction at a distance of 5 m from the source.

TABLE 8.3

Sound Pressure Levels in Various Directions (Example 8.3)

No.	1	2	3	4	5	6	7	8
Deg	0	45	90	135	180	225	270	315
L_p (dB)	97	87	98	97	90	85	93	85

Figure 8.7 shows the arrangement for measuring sound pressure levels inside an anechoic room. All the inside surfaces of the room except the floor are covered with acoustic absorption material. Measurements are conducted at a certain height above the ground level and around the machine at a radius of 2 m along the angular positions given in the problem.

The average sound pressure level is given by

$$L_{p_m} = 10\log\left(\frac{1}{n}\sum_{i=1}^{n}10^{\frac{L_p^i}{10}}\right) \tag{1}$$

Using Equation 1, the average sound pressure level of the exhaust stack is given by

$$L_{p_m} = 10\log\left[\frac{1}{8}\left(10^{\frac{97}{10}} + 10^{\frac{87}{10}} + 10^{\frac{98}{10}} + 10^{\frac{97}{10}} + 10^{\frac{90}{10}} + 10^{\frac{85}{10}} + 10^{\frac{93}{10}} + 10^{\frac{85}{10}}\right)\right] \tag{2}$$

$$= 94\,\text{dB}$$

a. The directivity index is given by

$$DI_\alpha = L_p^i - L_{p_m} \tag{3}$$

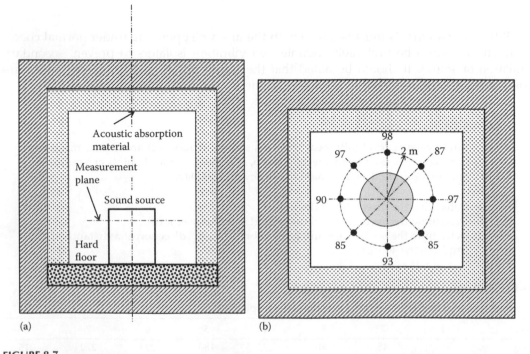

(a)   (b)

FIGURE 8.7

(a) Elevation and (b) plan of the measurement setup inside an anechoic room.

TABLE 8.4

Directive Index of the Sound Source in Different Directions

Angle, α (deg)	0	45	90	135	180	225	270	315
DI_α (dB)	3	−7	4	3	−4	−9	−1	−9
DI_p (dB)	3	3	3	3	3	3	3	3

Using Equation 3, the directivity index in various directions is calculated and tabulated as shown in Table 8.4.

b. Based on the average sound pressure level, the average sound power of the sound source placed on a hard surface is given by

$$L_w = L_{p_m} + 20\log r + 8\,\text{dB}$$
$$= 94 + 20\log 2 + 8 = 108\,\text{dB} \tag{4}$$

Equation 4 is a modification of Equation 8.36. Alternatively, the following equation can also be used to compute the sound power:

$$L_w = L_{Pm} + 10\log 4\pi r^2 - DI_p$$
$$= 94 + 10\log 4\pi 2^2 - 3 \tag{5}$$
$$= 108\,\text{dB}$$

where $DI_p = 3$ dB is the directivity factor arising out of placing the machine on a hard floor.

The sound pressure levels at various angles at a distance r is given by

$$L_p^i = L_W - 10\log(4\pi r^2) + DI_\alpha + DI_p \tag{6}$$

c. The sound pressure levels at a radius of 5 m along the 270° direction is given by

$$L_p^i(270°) = 108 - 10\log(4\pi 5^2) - 1 + 3$$
$$= 85\,\text{dB} \tag{7}$$

8.8 Sound Power Relations in Free Field

A free field, very similar to an anechoic room, provides a nonreflective environment that is commonly encountered in many noise control situations.

There are mainly two types of noise sources that are of interest in machine noise control: point and line. Depending upon whether these sources are placed on the ground or above the ground, the relationship between the sound power level and sound pressure level changes. Using Equation 8.36, depending upon whether the source is placed on the ground or above the ground, the following equations can be obtained for the point and line sources.

8.8.1 Point Source above the Ground

For a point source in free field radiating equally in all directions ($Q_\alpha = 1$) and located above the ground ($Q_P = 1$), the following relation, obtained from Equation 8.36, holds between the sound power of the source and the resulting sound pressure level at a distance r from the source:

$$L_p = L_W - 20 \log r - 11 \, \text{dB} \tag{8.40}$$

From Equation 8.40, for every doubling distance there is a decrease of sound pressure level given by 20 log 2 ≈ 6 dB.

Example 8.4

The sound power of the inlet noise from a compressor is 0.01 watt at 63 Hz and 0.005 watt at 125 Hz (all octave band center frequencies). Determine the sound pressure level at 3 m from the muffler. Assume the exit of muffler to be a point source in a free field without any reflections.

The equation relating sound pressure level and sound power of a point source located in a free field is given by

$$L_p = L_w - 20 \log r - 11 \, \text{dB} \tag{1}$$

The sound pressure level for the 63 Hz octave frequency band is given by

$$L_p(63 \, \text{Hz}) = 10 \log \frac{0.01}{10^{-12}} - 20 \log 3 - 11 \, \text{dB}$$
$$= 79 \, \text{dB} \tag{2}$$

The sound pressure level for the 125 Hz octave frequency band is given by

$$L_p(125 \, \text{Hz}) = 10 \log \frac{0.005}{10^{-12}} - 20 \log 3 - 11 \, \text{dB}$$
$$= 76 \, \text{dB} \tag{3}$$

8.8.2 Point Source on the Ground

For a point source radiating uniformly in all directions (Q_α) located on the ground ($Q_P = 2$), the following relation holds between the sound power of the source and the resulting sound pressure level at a distance r from the source. From Equation 8.36

$$L_p = L_W - 20 \log r - 8 \, \text{dB} \tag{8.41}$$

Figure 8.8 shows the variation of sound pressure levels at several locations away from a point source placed on the ground in a free field. It can be seen that for every doubling of distance, there is a reduction of 6 dB.

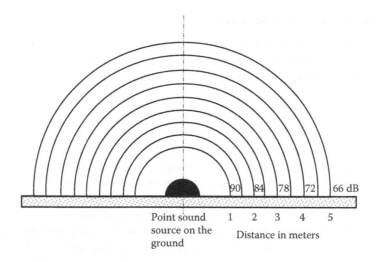

FIGURE 8.8
Point source on the ground.

8.8.3 Line Source Located above the Ground

For a line source uniformly radiating in all directions ($Q_\alpha = 1$) located above the ground ($Q_P = 1$), the area of the sound radiating surface can be taken as $2\pi rL$ and Equation 8.36 can be used by considering the unit length of the source and the sound power per unit length. Then the following relation holds between the sound power of the source and the resulting sound pressure level at a distance r from the source:

$$L_p = L_W - 10 \log r - 8\,\text{dB} \tag{8.42}$$

Figure 8.9 shows a line source located above the ground. A noisy pipeline is a good example of line source and the sound pressure level reduces by 10 log 2 ≈ 3 dB for every doubling distance.

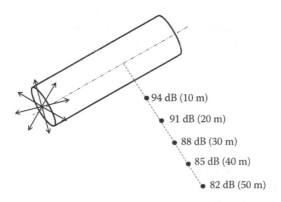

FIGURE 8.9
Line source.

8.8.4 Line Source Located on the Ground

For a line source uniformly radiating in all directions ($Q_\alpha = 1$) and located on the ground ($Q_P = 2$), the following relation holds between the sound power of the source and the resulting sound pressure level at a distance r from the source:

$$L_p = L_W - 10 \log r - 5 \, \text{dB} \qquad (8.43)$$

Therefore, for every doubling of the distance, there is a reduction of 3 dB, for a train modeled as a line source on the ground. The same equation holds for traffic on road. Whereas the train sound is going to last for less than a minute, the traffic sound remains almost the same for several hours.

The preceding equations are useful in estimating sound pressure levels at a distance from sources of known sound power. The reduction in sound pressure levels is mainly due to the decrease of sound intensity away from the sound source. The effect of sound absorption from atmospheric air is also possible and is dependent on the frequency of the sound source and relative humidity of air.

In the more general case, the source is not a point or line source; instead, it has finite values of length, width, and height. In this case, sound pressure and sound power levels are interrelated by the equation

$$L_W = L_p + 10 \log S \qquad (8.44)$$

where S is the surface area (square meters) of the imaginary surface at which sound pressure level measurements were carried out.

8.9 Reverberant Rooms

In most practical situations, the room environment will reflect sound. Therefore, it is necessary to derive relevant equations that will allow computation of sound pressure levels due to known sound power sources and directivity that are kept in enclosed rooms of known room constants or average absorption coefficients $\bar{\alpha}$. Consider a source of sound power level L_w (sound power W) that is placed in an enclosed space of room constant R. Let the sound source be placed in a room of total volume V and total surface area S_T and an average absorption coefficient $\bar{\alpha}$.

8.9.1 Sound Generated due to Reverberation

Consider a sound source of sound power W radiating airborne sound as shown in Figure 8.10. This sound power contributes toward the direct field that is the sound received directly from the source. If the room has an average acoustic absorption coefficient $\bar{\alpha}$, as the sound source continuously radiates sound, the sound power absorbed due to acoustic absorption is $W\bar{\alpha}$. The balance sound power $W(1 - \bar{\alpha})$ undergoes multiple reflections and generates reverberant sound. Therefore, the sound power responsible for generating reverberant energy is given by

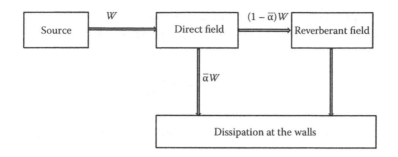

FIGURE 8.10
Power balance for reverberant rooms.

$$W_{rev} = W(1 - \bar{\alpha}) \tag{8.45}$$

From Equation 8.26, the total reverberant sound energy of the room is given by

$$E_{rev} = \frac{\langle p^2 \rangle_{rev} V}{\rho_o c_o^2} \tag{8.46}$$

Every time a sound wave strikes a surface, $\bar{\alpha} E_{rev}$ is lost from the reverberant field. For the given volume and surface area of the room that has air as a medium, statistically there are $\frac{c_o S}{4V}$ reflections per second and the rate of energy loss is a product of reverberant energy lost due to each striking of a surface and the number of reflections per second. It is given by

$$\dot{E}_{rev} = \frac{\bar{\alpha} c_o S_T \langle p^2 \rangle_{rev}}{4 \rho_o c_o^2} = \frac{\bar{\alpha} S_T \langle p^2 \rangle_{rev}}{4 \rho_o c_o} \tag{8.47}$$

The rate of energy loss must equal the input sound power to the reverberant field. Therefore, Equation 8.45 must be equal to Equation 8.47:

$$W(1 - \bar{\alpha}) = \frac{\bar{\alpha} S_T \langle p^2 \rangle_{rev}}{4 \rho_o c_o} \tag{8.48}$$

Simplifying Equation 8.31 and using Equation 8.13 relating the average acoustic absorption coefficient, room constant, and surface area:

$$\left(\frac{\langle p^2 \rangle_{rev}}{\rho_o c_o} \right) = \frac{4(1 - \bar{\alpha})W}{S_T \bar{\alpha}} = \frac{4W}{R} \tag{8.49}$$

Earlier, we had derived Equation 8.31 relating sound power and mean square pressure in a free field when there are no reflections. This mean square pressure can also now be called direct mean square pressure. It can be written as

$$\left(\frac{\langle p \rangle_{direct}^2}{\rho_o c_o} \right) = \frac{W Q_\alpha}{4 \pi r^2} \tag{8.50}$$

The total sound pressure resulting out of direct and reverberant and fields can be written as the summation of Equations 8.49 and 8.50, and is given by

$$\left(\frac{\langle p \rangle^2_{direct}}{\rho_o c_o}\right) + \left(\frac{\langle p \rangle^2_{rev}}{\rho_o c_o}\right) = W\left\{\frac{Q_\alpha}{4\pi r^2} + \frac{4}{R}\right\} \tag{8.51}$$

Dividing both sides of Equation 8.29 by p^2_{ref}, taking the logarithm and multiplying by 10, Equation 8.33 becomes

$$L_p = L_w + 10 \log_{10}\left\{\frac{Q_\alpha}{4\pi r^2} + \frac{4}{R}\right\} \tag{8.52}$$

In Equation 8.52, the term $4/R$ accounts for the reverberant nature of an acoustic environment, and $W_{ref} = \dfrac{p^2_{ref}}{\rho_o c_o} = \dfrac{(20\mu Pa)^2}{400}$ is the reference power equal to 10^{-12} watt (1 pico watt). If R is very large, like in open space or in a room packed with highly absorbent acoustic material that does not allow any reflections, it results in

$$L_p = L_w + 10 \log_{10}\left\{\frac{Q_\alpha}{4\pi r^2}\right\} \tag{8.53}$$

Equation 8.53 has been derived earlier for relating sound power and sound pressure levels in the free field.

8.9.2 Sound Field Variation in Reverberation Rooms

Figure 8.11 shows a machine placed in a typical reverberant environment encountered in practice. The machine is generally placed at one end of the room. Due to reflection from the wall, the sound pressure levels may be very high close to the wall facing the machine. Very close to the machine, there will be many noise sources and all of them will not eventually contribute to the noise away from the machine. This is known as the near field, and we are generally not interested in measuring noise in this field, as it does not progress beyond a certain distance. It is generally taken as 1 m from the machine, and this is also the level that affects the operator of the machine. One meter away from the machine is the free field in which the sound pressure levels decrease as a function of distance. In fact, measurements of the sound pressure level must be made in this region. After moving a certain distance in the free field we get the reverberant region that is close to the other end of the wall. In this region, the sound pressure levels are almost the same. The extent of the free field and reverberant field in a typical room depends on the environment. If there are too many absorbing surfaces like those in anechoic room, most of the room will be a free field. On the other hand, if there are too many hard, reflective surfaces, there will be multiple reflections of sound resulting in a reverberant environment for most of the room. The best way to judge a reverberant room is when the sound pressure levels remain the same at every point.

Figure 8.12 shows the sound level distribution for acoustic spaces of various room constants. For reverberant rooms the Room constant is very small, and for anechoic rooms

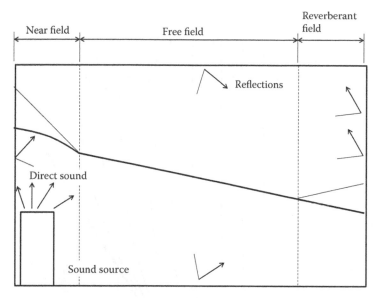

FIGURE 8.11
Near field, free field, and reverberant field.

the room constant is very large, and for an open space the Room constant is extremely large (infinity in mathematics). As seen in Figure 8.12, for very low values of Room constant, the free-field region in which the sound pressure level decreases is very small and then it becomes constant even after moving sufficient distance away from the source. Whereas for large Room constants, the sound pressure level linearly decreases with distance.

Example 8.5

Sound pressure level (dB) measurements of a machine placed on a concrete floor in an anechoic room, at eight equispaced points on a radius of 2 m from the machine, are shown in Table 8.5. The dominant frequency is at 500 Hz.

a. Determine the directivity index.
b. Determine the average sound power and sound power level.
c. If the above machine is placed in the corner of a room of dimensions 4 m × 8 m × 3 m and if the Room constant is 150 m², determine the least sound pressure level (rounded off to the nearest integer) at 500 Hz and its locations within the room. (Use average sound power and draw a sketch showing regions of least sound pressure.)
d. Determine the acoustic absorption coefficient of the room at 500 Hz.

ANSWERS

a. The average sound pressure level is given by

$$L_{p_m} = 10\log\left(\frac{10^{\frac{97}{10}} + 10^{\frac{87}{10}} + 10^{\frac{98}{10}} + 10^{\frac{97}{10}} + 10^{\frac{90}{10}} + 10^{\frac{85}{10}} + 10^{\frac{93}{10}} + 10^{\frac{85}{10}}}{8}\right) = 94\text{ dB} \qquad (1)$$

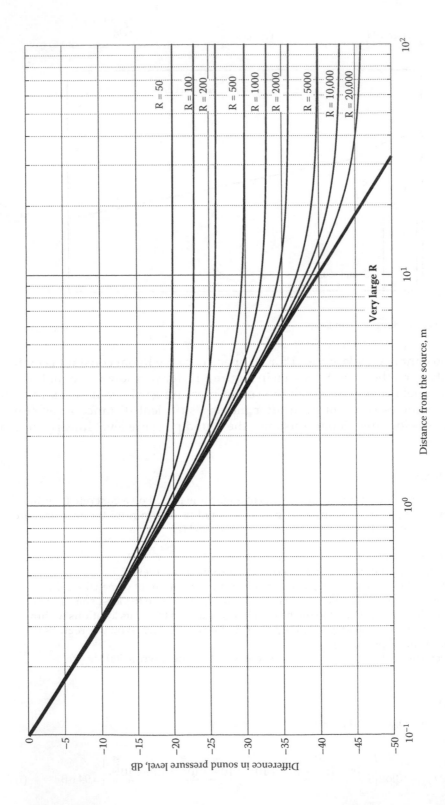

FIGURE 8.12
Sound level distribution in acoustic spaces.

TABLE 8.5

Sound Pressure Levels (Example 8.5)

Location	1	2	3	4	5	6	7	8
L_p (dB)	97	87	98	97	90	85	93	85

TABLE 8.6

Sound Pressure Levels, Directivity Factor, and Directivity Index at Various Locations

Location	1	2	3	4	5	6	7	8
L_p (dB)	97	87	98	97	90	85	93	85
DI_α	2.92	−7.08	3.92	2.92	−4.08	−9.08	−1.08	−9.08

The directivity index due to nonuniform radiation of sound, DI_α, is given by

$$DI_\alpha = L_p - L_{p\_avg} \tag{2}$$

The directivity index from Equation 2 for various locations are shown in Table 8.6.

b. The average sound power level is given by

$$L_w(average) = L_{p_m} - 10\log\left(\frac{Q_P}{4\pi r^2}\right)$$
$$= 94 - 10\log\left(\frac{2}{4\pi 2^2}\right) = 108 \text{ dB} \tag{3}$$

c. When the machine is kept in the corner of a room of room constant 150, the nonuniform sound radiation pattern of the machine is not relevant. The sound pressure levels at various locations from the machine is given by

$$L_p = L_w(average) + 10\log\left(\frac{Q_P}{4\pi r^2} + \frac{4}{R}\right) \tag{4}$$

TABLE 8.7

Sound Pressure Levels at Various Locations of the Reverberant Room

r (m)	1	2	3	4	5	6	7	8	9	10
L_p (dB)	106	101	98	96	95	95	94	94	93	93
r (m)	11	12	13	14	15	16	17	18	19	20
L_p (dB)	93	93	93	93	93	93	93	93	93	93

Note: The sound pressure levels become constant after 8 m.

From Table 8.7 it is clear that the sound pressure levels remain constant at 93 dB after $r = 9$ m.

Area of the room, $S_T = 2[(4 \times 8) + (8 \times 3) + (4 \times 3)] = 136$ m²

d. Average acoustic absorption coefficient = $\bar{\alpha} = \dfrac{R}{S_T + R} = \dfrac{150}{136 + 150} = 0.524$

8.10 Sound Power Measurement in Diffuse Field (Reverberant Room)

In a reverberant room, sound energy is uniformly distributed in the entire room. The uniform distribution of sound energy is possible by providing reflecting, nonparallel surfaces for the entire room. A reverberant room is shown in Figure 8.13.

The machine may be operated under its normal load in a corner of the room, and the sound pressure levels can be measured at any point away from the machine. In this method, the directivity of the source cannot be determined. On the other hand, the diffuseness of the field allows a relatively easy estimation of the sound energy density in the room, because the sound energy density in the room is directly related to the difference between the sound energy emitted by the machine and that absorbed by the room boundaries. A measure of absorption is obtained by determining the room reverberation time T. Reverberation time is the time taken by sound to decay by 60 dB and it depends on the acoustic absorbing environment. When the volume V of the room and its reverberation time are known, the sound power level can be derived as follows.

From Equation 8.12, neglecting the sound absorbed by air, the reverberation time is given by

$$T_{60} = \frac{0.161\,V}{S_T \bar{\alpha}} \tag{8.54}$$

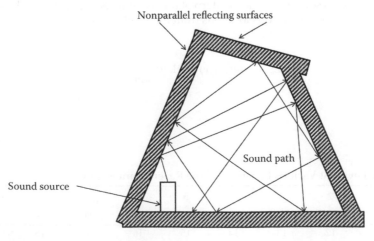

FIGURE 8.13
Sound pressure level measurement in a reverberant room.

From Equation 8.14, for small values of the average absorption coefficient typical of a reverberant room, Equation 8.54 becomes

$$R = \frac{0.161\,V}{T_{60}} \qquad (8.55)$$

In Equation 8.52, for reverberant rooms $\frac{4}{R} \gg \frac{Q_\alpha}{4\pi r^2}$. Then by using the room constant of Equation 8.55, the relationship between sound pressure level and sound power level becomes

$$L_p = L_w + 10\log_{10}\left\{\frac{4T_{60}}{0.161\,V}\right\} \qquad (8.56)$$

Expanding Equation 8.56, the sound power level of a machine whose sound pressure levels are measured inside a reverberant room is given by

$$L_W = L_p - 10\log T_{60} + 10\log V - 14\,\text{dB} \qquad (8.57)$$

Although a single measurement is sufficient, it is wise to take more number of measurements and obtain its mean. The sound pressure measurements must be conducted at various 1/3 octave or octave band center frequencies, as required by the specification. The reverberation time of the room must be determined from experiments for each of the frequency bands, for which sound power has to be estimated.

Example 8.6

A machine was kept in a reverberant room of 100 m³ volume that has a reverberation time of 1.2 s at 500 Hz octave band frequency. The sound pressure level was measured corresponding to 500 Hz at several locations of the room and they are as follows: 95 dB, 96 dB, 94 dB, 95 dB, 96 dB, and 94 dB. Determine the sound power level and sound power of this machine.

The average sound pressure level of this machine in the reverberant room is 95 dB.
From Equation 8.57, the sound power level of the machine can be computed as

$$L_W = 95 - 10\log 1.2 + 10\log 100 - 14 = 100\,\text{dB}$$

8.11 Sound Power Measurement in Semireverberant Rooms

In many practical situations, both the above environments (anechoic and reverberant) may not be possible for determining sound power levels. In such a case, the shop floor itself can be used for determining sound power levels. Although a semireverberant environment is slightly inferior of the two aforementioned environments, it gives reasonably good results. It consists of conducting two sets of sound pressure measurements. One set of measurements is conducted using a noise source whose sound power is already known by conducting experiments in either anechoic or reverberant rooms. The second set of measurements is conducted using the actual machine itself. The sound power levels are obtained from the following equation.

The sound pressure levels due to a machine of unknown sound power level L_w placed inside a semireverberant room of room constant R is given by

$$L_p = L_w + 10 \, \log_{10} \left\{ \frac{Q_\alpha}{4\pi r^2} + \frac{4}{R} \right\} \tag{8.58}$$

The sound pressure levels due to a machine of known sound power level L_{ws} placed inside the same semireverberant room of room constant R is given by

$$L_{ps} = L_{ws} + 10 \, \log_{10} \left\{ \frac{Q_\alpha}{4\pi r^2} + \frac{4}{R} \right\} \tag{8.59}$$

Subtracting Equations 8.58 and 8.59, the sound power level of the unknown machine is given by

$$L_w = L_{ws} + L_p - L_{ps} \tag{8.60}$$

Equation 8.60 can also be used in reverberant rooms even if the reverberation time and volume of the room are not known. But an average of sound pressure levels at various locations of the room will have to be used for reverberant rooms.

Example 8.7

A machine with known sound power level of 100 dB is tested in a semireverberant room that results in a sound pressure level of 80 dB at a certain point away from the source. Then a machine of unknown sound power was located at the same location and then its sound pressure level was measured as 90 dB. Determine the sound power level of the unknown machine. Does it matter to know whether the test was conducted in semireverberant room?

From Equation 8.60, the unknown sound power level is given by

$$L_W = L_{Ws} + L_p - L_{ps}$$

$$= 100 + 90 - 80 = 110 \text{ dB}$$

It does not matter to know any information about the nature of room constant since we are conducting two sets of measurements at the same location.

8.12 Modal Density of an Enclosed Volume

The modal density of a three-dimensional volume enclosure with rigid walls is given by

$$n(f) = \frac{4\pi f^2 V}{c_o^3} + \frac{\pi f A}{2c_o^2} + \frac{P}{8c_o} \tag{8.61}$$

where V is the volume, A is the area of inside surface, and P is the perimeter. For large acoustic volumes, only the first term of Equation 8.61 is used.

8.13 Conclusions

The discussion presented in this chapter is very useful for describing the basic characteristics of an acoustic subsystem. Since the extent of sound absorption is an important aspect of this acoustic subsystem, the absorption properties of various materials were discussed. In addition, the concept of a Room constant is useful in predicting sound pressure levels in a room of known sound power. Furthermore, for many of the vibro-acoustic applications, the energy of the acoustic system is important, which was discussed in this chapter. Although the main objective of this chapter was to provide the basics of room acoustics for studying sound–structure interaction, much of the information can be used to solve many machinery noise problems encountered in practice.

PROBLEMS

8.1. The sound pressure levels shown in the following table were measured at a radius of 5 m from a point source in air at the 500 Hz band: 98, 87, 97, 97, 94, 93, 95, and 89 dB. The angular locations (measured counterclockwise) are shown in the following figure.
   a. Calculate the directivity index and the directivity factor along each of the microphone locations.
   b. Calculate the sound power level.
   c. Determine the sound pressure levels in the 135° direction at 8 m.

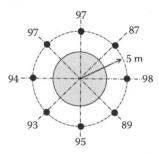

Angle, α (deg)	0	45	90	135	180	225	270	315
SPL (dB)	98	87	97	97	94	93	95	89

8.2. A machine that has a sound power level of 120 dB in the 1 kHz octave band is placed on a hard concrete floor. The dimensions of the room are 12 × 8 × 4 m.

The concrete floor has an acoustic absorption coefficient of 0.02. The walls and the ceiling can be treated with acoustic material. It is expected that the sound pressure levels become constant after a distance of 3 m from the machine after acoustic treatment.

a. Determine the acoustic absorption coefficient of the lining for walls and ceiling.

b. Determine the reverberant sound pressure levels in the room before and after the treatment.

8.3. The sound pressure level (dB) measurements of a machine placed on a concrete floor in an anechoic room, at eight equispaced points on a radius of 1 m from the machine, are shown in the following table. The dominant frequency of the machine sound is at 500 Hz.

a. Determine the directivity index along each direction.

b. Determine the average sound power and sound power level.

c. If the machine is placed in the corner of a room of dimensions 10 m (width) × 25 m (length) × 3 m (height), and if the room constant is 175 m², beyond what distance does the sound pressure level become a constant?

d. Determine the acoustic absorption coefficient of the room at 500 Hz.

Location	1	2	3	4	5	6	7	8
L_p (dB)	97	92	100	97	90	89	93	80

8.4. An auditorium of dimensions 20 m × 50 m × 10 m is required to have an optimum reverberation time for piano concerts. Determine the average absorption coefficients in octave bands from 63 to 4000 Hz for this condition.

8.5. The sound source of Problem 8.1 is placed on the concrete floor of a room of room constant 2000. Determine the sound pressure levels at 4 m along equispaced angular directions.

8.6. A machine that is uniformly radiating sound is used to measure the Room constant of a room. Sound pressure level measurements are conducted at a distance of 4 m and 8m respectively from the machine and they are measured at various octave bands as shown in the following table.

a. Determine the average absorption coefficient in each octave band.

b. Determine the reverberation time in each octave band.

c. Determine the room constant in each octave band.

d. Can this method always work? The room has dimensions of 12 × 8 × 4 m.

Octave Band Center Frequency (Hz)	63	125	250	500	1000	2000	4000	8000
L_p (4 m) (dB)	75	80	85	90	95	90	85	80
L_p (8 m) (dB)	73	77	82	86	90	85	81	76

8.7. A room of dimensions 10 × 8 × 4 m has the following reverberation times (T_{60}) and the sound pressure levels of a machine placed in a corner are as shown in the following table.

a. Determine the room constant of this room in the respective octave bands.

b. Determine the sound power levels of this machine assuming the room to be reverberant.

c. Determine the breakup of direct and reverberant sound pressure levels.

Octave Band Center Frequency (Hz)	125	250	500	1000	2000	4000
			Reverberation Time, T_{60}			
	0.03	0.04	0.06	0.10	0.11	0.12
			Sound Pressure Levels (dB)			
	90	95	100	110	112	109

8.8. A room has the dimensions of 15 (length) × 8 (width) × 4 (height) m. The floor is concrete, the walls are constructed using bricks, and the ceiling consists of acoustic panels. The absorption coefficients of various materials and the sound power of the machine in various octave bands is shown in the following table. It also has six open windows of 1 × 1 m on the walls. Determine the location along the length of the room at which the sound pressure levels become reverberant and their A-weighted levels.

Octave Band Center Frequency (Hz)	125	250	500	1000	2000	4000
			Acoustic Absorption Coefficient			
Bricks, α_b	0.02	0.03	0.035	0.04	0.04	0.06
Concrete, α_f	0.01	0.01	0.02	0.02	0.02	0.02
Ceiling, α_c	0.4	0.4	0.6	0.8	0.7	0.6
Air, α_w	1.0	1.0	1.0	1.0	1.0	1.0
Sound power level of source L_W (dB)	95	100	105	120	115	110

8.9. A machine has a dominant sound power level of 125 dB in the 1 kHz octave band. It has dimensions of 1 × 1 × 1 m and is enclosed within an enclosure of 2 × 2 × 2 m. The average acoustic absorption coefficient inside the enclosure is 0.7 and the enclosure has a transmission loss of 30 dB; both these values are in the 1 kHz octave band. This machine with an enclosure is kept on the concrete floor of a room of dimensions 10 × 12 × 4 m and an average acoustic absorption coefficient of 0.25 in the 1 kHz octave band. Determine the reverberant sound pressure levels in this room and suggest a possible design change of the enclosure (both the lining and transmission loss) to reduce this level by 15 dB.

8.10. The following table shows the sound power levels of a machine in various octave bands. It is required to design an enclosure for this machine so the NC 50 sound pressure levels of the room in which the machine with an enclosure will be placed on a concrete floor. The room has the dimensions of 15 (length) × 8 (width) × 4 (height) m. The floor is concrete, the walls are constructed using bricks, and the ceiling consists of acoustic panels whose acoustic absorption coefficients are shown in the table. The mineral wool that will be used in the enclosure has acoustic absorption coefficients as shown in the table. The machine has dimensions of 1 × 1 × 1 m and is enclosed within an enclosure of 2 × 2 × 2 m. Assume that the machine and the enclosure are placed on the concrete floor and the mineral wool lining for the enclosure is only for the side-walls and roof of the enclosure. Determine the transmission loss of the enclosure wall to ensure that the reverberant sound pressure levels in the room do not exceed the NC 50 levels.

Octave Band Center Frequency (Hz)	125	250	500	1000	2000	4000
	Acoustic Absorption Coefficient					
Bricks, α_b	0.02	0.03	0.035	0.04	0.04	0.06
Concrete α_f	0.01	0.01	0.02	0.02	0.02	0.02
Ceiling α_c	0.4	0.4	0.6	0.8	0.7	0.6
Mineral wool	0.1	0.3	0.7	0.95	1.0	1.0
Sound power level of source, L_W (dB)	95	100	105	120	115	110
NC 50 SPL	64	58	54	51	49	48

8.11. The following figure shows an office attached to a workshop. The table shows the NC 40 sound pressure levels that are expected in the office, reverberant sound pressure levels in the workshop, and average absorption coefficients inside the office and workshop. There is a wooden door of dimensions 1 × 2.5 m and a glass window of dimensions 3 × 1 m. Determine the average transmission loss between the workshop and the office and the thickness of the door and glass, assuming that the wall is a 200 mm brick wall. Both the office and the workshop have a height of 4 m.

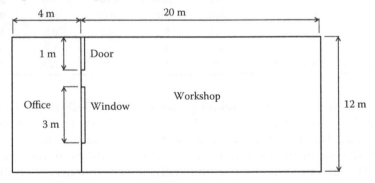

Octave Band Center Frequency (Hz)	125	250	500	1000	2000	4000
	Acoustic Absorption Coefficient					
Workshop	0.02	0.03	0.035	0.04	0.04	0.06
Office	0.2	0.25	0.35	0.6	0.65	0.7
Sound pressure levels in the workshop (dB)	90	95	100	110	112	109
NC 40 SPL	56	50	45	41	39	38

8.12. The following figure shows a plan view of the classroom of a school, one side of which is facing the highway and the noise levels in this room are dominated by the highway noise coming through the walls and windows (see table). The walls are made of 150 mm thick brick wall and two glass windows of 2 m × 1 m with 4 mm thick glass; the room height is 4 m. Determine the overall noise levels in A-weighted decibels (dBA) of the classroom. If the area of both windows is reduced by 50%, what is the new level?

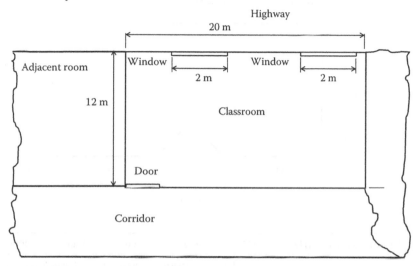

Octave Band Center Frequency (Hz)	125	250	500	1000	2000	4000
	Sound Pressure Levels of the Highway Noise (dB)					
	59	63	68	70	72	65
	Acoustic Absorption Coefficient					
	0.2	0.25	0.35	0.6	0.65	0.7

8.13. A water cooler is located in the corridor outside a room. Closing the only door that is constructed from a 30 mm thick plywood has not provided the required attenuation. There could be acoustic leaks between the corridor and rooms through ducts, but there is no direct evidence to support it. In the following table are the sound pressure level measurements inside the room when the door was closed and open. Why is there no significant reduction even after closing the door? Could you have reached the same conclusion with the help of overall weighted and unweighted sound pressure levels when the door was open and closed?

Octave Band Center Frequency (Hz)	125	250	500	1000	2000	4000
	Sound Pressure Levels (dB)					
Door open	70	63	53	47	52	40
Door closed	70	62	50	45	50	38

8.14. The following figure shows the plan view of a hotel lobby facing highway noise. The height of the ceiling is 6 m. Due to movement of people moving in and out of the lobby, the doors on average opened 75 times per hour; whenever the doors are opened, there is a 5 s exposure to the highway noise. All the walls are made of 150 mm thick brick wall and the glass doors are of 2 m height and 20 mm thick. Assume that the noise enters only through the front wall. What is the L_{eq} dBA for an 18-hour day?

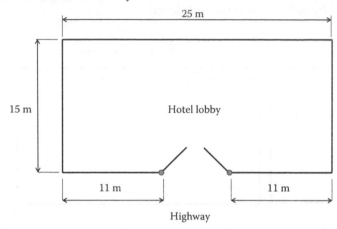

Octave Band Center Frequency (Hz)	125	250	500	1000	2000	4000
	Sound Pressure Levels of the Highway Noise (dB)					
	59	63	68	70	72	65
	Acoustic Absorption Coefficient of the Lobby					
	0.2	0.35	0.5	0.85	0.9	0.9

8.15. Refer to Problem 8.14. What is the L_{eq} dBA if the hotel lobby is turned around to show its back wall to the highway, assuming noise to enter only through the back walls.

8.16. Refer to Problem 8.11. Determine the sound power of the source in the workshop, and sound pressure levels for direct and reverberant fields in the workshop if the reverberation times for the workshop are as shown in the following table.

Octave Band Center Frequency (Hz)	125	250	500	1000	2000	4000
	Reverberation Time, T_{60}					
	0.03	0.04	0.06	0.10	0.11	0.12

8.17. Refer to Problems 8.11 and 8.16. If the sound source in the workshop is kept in the office, what will be the sound pressure levels in the workshop?

8.18. Refer to Problem 8.12. Determine the overall sound pressure levels if each of the glass windows are replaced by double glazed glass windows, each 4 mm thick and 100 mm apart.

8.19. Determine the total energy of the room of Problem 8.5 in the respective octave bands.

8.20. Determine the modal density of a room of dimensions 10 m (width) × 25 m (length) × 3 m (height) in octave bands from center frequencies 63 to 4000 Hz.

9

Sound–Structure Interaction

Airborne sound is produced either by vibrating surfaces or due to flow. However, we will be concerned about the latter in this chapter. But not all vibrating surfaces will result in sound and not all the frequencies of vibration will be faithfully converted to the corresponding frequencies of airborne sound. There are some conditions under which a vibrating body can produce sound. Therefore, it is important to know the conditions under which a vibrating body can produce sound, which is the main objective of this chapter. Initially, the sound produced by a rigid vibrating piston is considered to define the basic concept of radiation resistance. Later, it is extended to a pulsating sphere. But in practice, most vibrating structures support wave propagation, which makes the prediction of sound from such surfaces difficult. An infinite plate in flexural (bending vibration) is chosen to derive the basic conditions of efficient sound radiation. Finite plates, however, have different sound-radiating characteristics than infinite plates. Only physical aspects of sound radiation from vibrating structures that can sustain wave motion are discussed in this chapter, without getting into too many details that can be obtained from other reference texts.

9.1 Sound Radiated by a Rigid Piston

Consider the plane waves generated by a rigid piston moving inside a rigid tube of cross-section area A, as shown in Figure 9.1. Let ρ_o be the nominal density, p_o the ambient pressure, and c_o the speed of sound. The dynamic pressure, p_d, and particle velocity, u, are related by

$$\frac{p_d}{u} = \rho_o c_o \tag{9.1}$$

The time-averaged force exerted by the piston and its product with piston velocity is the sound power radiated by the piston and is given by

$$\pi_{rad} = <p_d A u>_t = <u^2>_t \rho_o c_o A \tag{9.2}$$

$<u^2>_t$ in Equation 9.2 is the mean square velocity.

Before we discuss Equation 9.2 any further, let us note the power dissipated by a pure resistance electric circuit, which is given by

$$\pi_{diss} = I^2 R \tag{9.3}$$

where I is the current and R the resistance.

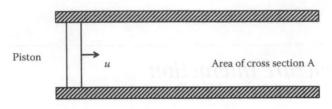

FIGURE 9.1
Sound radiated by a rigid piston.

Applying the force-voltage analogy to Equations 9.2 and 9.3, the coefficient of the mean square velocity in Equation 9.2, $\rho_o c_o A$, can be given a new name—radiation resistance, R_{rad}:

$$R_{rad} = \rho_o c_o A \tag{9.4}$$

Using Equation 9.4, Equation 9.2 can be written as

$$\pi_{rad} = <u^2>_t R_{rad} \tag{9.5}$$

In most practical situations, the vibrating structure will be flexible, unlike the rigid piston assumed in the previous discussion. For such flexible structures, the vibration velocity at different locations of the structure will be different, depending on its geometry and structural dynamics. It is as shown in Figure 9.2.

Accounting for all factors related to the structural dynamics and the geometry of a vibrating structure by σ_{rad}, the radiation ratio, Equation 9.5 can be modified as

$$\pi_{rad} = <u^2>_t R_{rad}\sigma_{rad} \tag{9.6}$$

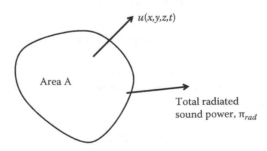

FIGURE 9.2
Sound radiated by a general flexible structure.

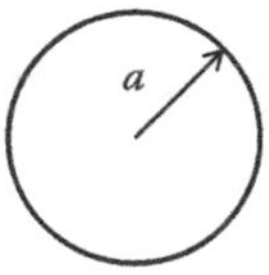

FIGURE 9.3
Spherical monopole source.

9.2 Geometric Radiation Ratio of a Pulsating Sphere

Consider a sphere of radius a whose external surface is sinusoidally pulsating at a velocity u_a, as shown in Figure 9.3. Let p_a be the corresponding dynamic pressure at the radius a.

From Equation 2.72 (see Chapter 2), the pressure velocity relationship on the sphere surface is given by

$$\frac{p_a}{u_u} = \rho_o c_o \left(\frac{jk_o a}{1 + jk_o a} \right) \tag{9.7}$$

where k is the wave number.

A new parameter, Z_{rad}, can be defined as

$$Z_{rad} = \frac{Ap_a}{u_a} = A\rho_o c_o \left(\frac{jk_o a}{1 + jk_o a} \right) \tag{9.8}$$

Multiplying and dividing Equation 9.8 by $1 - jk_o a$

$$Z_{rad} = \frac{Ap_a}{u_a} = A\rho_o c_o \left(\frac{jk_o a}{1 + jk_o a} \right)\left(\frac{1 - jk_o a}{1 - jk_o a} \right) = \left[\frac{A\rho_o c_o}{1 + (k_o a)^2} \right][(k_o a)^2 + jk_o a] \tag{9.9}$$

The real part of Equation 9.9 is given by

$$R_a = \left[\frac{A\rho_o c_o}{1 + (k_o)^2} \right](k_o a)^2 \tag{9.10}$$

The imaginary part of Equation 9.9 is given by

$$X_a = \left[\frac{A\rho_o c_o}{1 + (k_o a)^2} \right] k_o a \tag{9.11}$$

Comparing Equations 9.4 and 9.10, the radiation ratio of a pulsating sphere is given by

$$\sigma_{rad} = \frac{(k_o a)^2}{1 + (k_o a)^2} \tag{9.12}$$

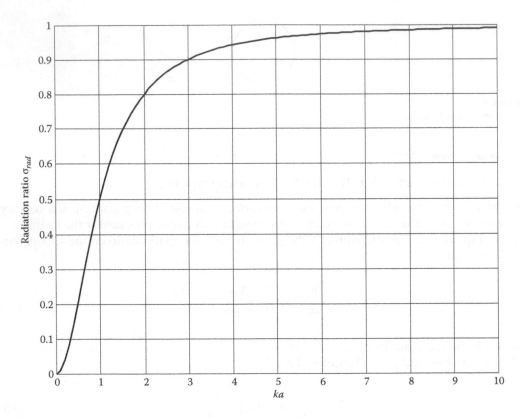

FIGURE 9.4
Radiation ratio of a pulsating sphere.

Hence, the radiation ratio of a pulsating sphere is dependent on the product of the wave number and its radius. Equation 9.12 is plotted in Figure 9.4.

Example 9.1

A sphere of 50 mm radius, pulsating with a certain frequency ω and vibrating with a peak amplitude of surface velocity of 40 mm/s, is used to produce sound in the far field with a radiation ratio of 70%.

a. Determine the frequency, dynamic pressure, particle velocity, and the phase difference between them at a distance of 10 m from the center of the sphere.
b. What will be the radiation ratio if the sphere pulsates at the same amplitude of velocity but at 2500 Hz frequency? Assume velocity of sound in air as 340 m/s and characteristic impedance of air as 400 rayl.

ANSWERS

a. From Equation 9.12, the product of wavenumber and monopole radius is given by

$$k_o a = \sqrt{\frac{\sigma_{rad}}{1 - \sigma_{rad}}} = \sqrt{\frac{0.7}{1 - 0.7}} = 1.52 \qquad (1)$$

From the given radius of the monopole, using Equation 1, the wave number of the airborne sound is given by

$$k_o = \frac{1.52}{50 \times 10^{-3}} = 30.55 \text{ rad/m} \tag{2}$$

From Equation 3.42 (see Chapter 3), the angular frequency of the airborne sound is given by

$$\omega = k_o c_o = 30.55 \times 340 = 10387 \text{ rad/s} (1653 \text{ Hz}) \tag{3}$$

From Equation 7.29 (see Chapter 7), the root mean square (rms) value of volume velocity is given by

$$Q_{rms} = \frac{4\pi a^2 U_a}{\sqrt{2}} = \frac{4\pi \times 0.05^2 \times 0.04}{\sqrt{2}} = 0.89 \times 10^{-3} \text{ m}^3/\text{s} \tag{4}$$

From Equation 7.31 (see Chapter 7), the rms value of dynamic pressure at a distance of 10 m from the monopole source is given by

$$p_d = \frac{Q_{rms}\omega\rho_o}{4\pi r\sqrt{1+(k_o a)^2}} = \frac{0.89 \times 10^{-3} \times 10387 \times 1.2}{4\pi \times 10\sqrt{1+(1.52)^2}} = 0.0485 \text{ Pa} \tag{5}$$

From Equation 3.103 (see Chapter 3), the rms value of particle velocity of at a distance of 10 m from the monopole source is given by

$$u = \frac{p_d\sqrt{1+(k_o r)^2}}{\rho_o c_o k_o r} = \frac{0.0485\sqrt{1+(30.55 \times 10)^2}}{400 \times 30.55 \times 10} = 0.12 \text{ mm/s} \tag{6}$$

b. The angular frequency corresponding to 2500 Hz, ω_1 is equal to $2\pi \times 2500 = 15708$ rad/s.

From Equation 3.42, the wave number of the airborne sound at 2500 Hz is given by

$$k_o = \frac{\omega}{c_o} = \frac{15708}{340} = 46.2 \text{ rad/s} \tag{7}$$

The radiating ratio of the monopole when it is producing sound at 2500 Hz is given by Equation 9.12 as

$$\sigma_{rad} = \frac{(k_o a)^2}{1+(k_o a)^2} = \frac{(46.2 \times 0.05)^2}{1+(46.2 \times 0.05)^2} = 84.2\% \tag{8}$$

Hence a pulsating sphere becomes an excellent radiator of sound at higher frequencies.

9.3 Radiation Ratio of an Infinite Vibrating Plate

9.3.1 Bending and Sound Wave Speeds and Wavelengths

Noise radiated due to vibration of a plate is of significant practical interest. Although most plates used in the construction of machines are finite plates with boundary conditions, the infinite plate helps in capturing the phenomenon of sound radiation using wave interaction of airborne and structure-borne sound.

As shown in Figure 9.5, the bending waves of the plate result in airborne sounds that are longitudinal waves. Let us assume that the plate is vibrating at a single frequency, f. Due to the fundamental property of force excitation of linear systems, the resulting airborne sound will also be of the same frequency.

The wavelength of bending waves is given by

$$\lambda_b = \frac{2\pi}{k_B} \tag{9.13}$$

where k_B is the bending wave number and is given by

$$k_B^4 = \frac{\rho_a (2\pi f)^2}{B} \tag{9.14}$$

where ρ_a is the density per unit area and B is the flexural rigidity of the plate.

By using the definition of longitudinal wave speed and the radius of gyration of the plate, using Equations 9.13 and 9.14 the bending wavelength can also be expressed as

$$\lambda_b = 1.35 \sqrt{\frac{hC_L}{f}} \tag{9.15}$$

where h is the plate thickness and C_L is the longitudinal wave speed for the plate material.

Critical frequency is the frequency at which the bending wave speed in the plate is equal to the speed of sound in the surrounding air. Based on the relationship between wavelength, frequency, and wave speeds in both the air and plate, it can be proved that if the wave speeds in the plate and air are equal, then the corresponding wavelengths are also equal.

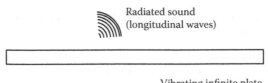

Radiated sound
(longitudinal waves)

Vibrating infinite plate
(bending waves)

FIGURE 9.5
Vibrating plate resulting in sound.

Let f_c be the critical frequency. The wavelength of sound in air at the critical frequency is given by

$$\lambda_o^c = \frac{c_o}{f_c} \tag{9.16}$$

From Equation 9.15, the bending wavelength at the critical frequency is given by

$$\lambda_b^c = 1.35 \sqrt{\frac{hc_L}{f_c}} \tag{9.17}$$

From Equations 9.16 and 9.17, the critical frequency is given by

$$f_c = \frac{c_o^2}{1.82 hc_L} \tag{9.18}$$

The ratio of the bending wavelength and acoustic wavelength at any frequency is given by

$$\frac{\lambda_b}{\lambda_o} = 1.35 \sqrt{\frac{hc_L f}{c_o^2}} \tag{9.19}$$

Using the value of c_o from Equation 9.18, Equation 9.19 becomes

$$\frac{\lambda_b}{\lambda_o} = 1.35 \sqrt{\frac{hc_L f}{1.82 f_c hc_L}} = \sqrt{\frac{f}{f_c}} \tag{9.20}$$

9.3.2 Sound–Structure Interaction of an Infinite Plate

Before we discuss sound–structure interaction of an infinite plate, it would be relevant to discuss the trace wave number. As shown in Figure 9.6, consider a wave propagating in a direction Θ, with a wavelength λ and wavenumber k. The projection of the wave number along the x-axis, $k_x = k \cos\Theta$, is called the trace wave number, and $\Lambda_x = \lambda/\cos\Theta$ is the trace wavelength.

Consider an infinite plate as shown in Figure 9.7, executing bending vibrations at a frequency f (Hz); let λ_b be the corresponding wavelength. The sound waves (airborne sound) produced by the plate vibration are assumed at an angle θ to the vertical and have a wavelength λ_o. Since the plate is infinite in the x-direction, the spatial variation of vibration is only along the x-direction; the spatial variation of dynamic pressure and particle velocity of the airborne waves are, however, along the x- and y-directions.

First, let us consider the dynamic pressure of sound waves. Let k_o be the wave number of the sound wave. Since the direction of propagation is at angle θ with respect to the vertical, the wave number can be resolved in the x and y-directions as shown in Figure 9.8.

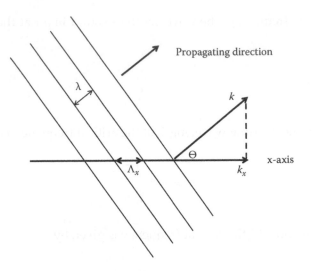

FIGURE 9.6
Plane wave trace wave number and trace wavelength.

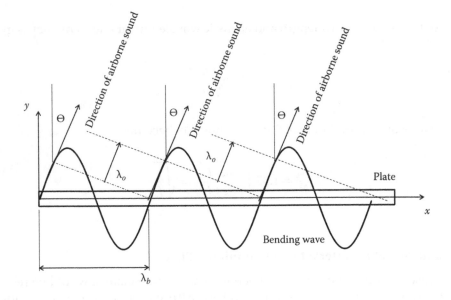

FIGURE 9.7
Direction and wavelength of sound waves and bending waves.

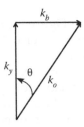

FIGURE 9.8
Resolution of sound wave number along coordinate directions.

The dynamic pressure of sound waves radiated by the plate in two dimensions is given by

$$\tilde{p}(x,y,t) = Pe^{j(\omega t - k_x x - k_y y)} \tag{9.21}$$

The wave number of sound waves in terms of its x and y components is given by

$$k_o = \sqrt{k_x^2 + k_y^2} \tag{9.22}$$

Since the plate and its surrounding air are in constant interaction, the spatial variation of dynamic pressure of sound waves along the x-direction must be the same as that of the spatial variation of the flexural plate vibration. Therefore, the wave number of the sound waves in the x-direction must be same as that of the bending wave number in the plate. It is given by

$$k_x = k_B \tag{9.23}$$

where k_B is the bending wave number of the plate vibration at the frequency f. As mentioned before, this frequency is the same for airborne sound and plate vibration; that is, a plate vibrating at a certain frequency will produce sound at the same frequency.

The dynamic pressure Equation 9.21 now becomes

$$\tilde{p}(x,y,t) = Pe^{j(\omega t - k_B x - k_y y)} \tag{9.24}$$

From Equation 7.20 (Chapter 7), the particle velocity of air due to sound waves is given by

$$\vec{u}(x,y,t) = -\frac{1}{j\omega\rho_o}\vec{\nabla}\tilde{p} \tag{9.25}$$

Using Equation 9.24, computing its gradient and substituting in Equation 9.25

$$\vec{u}(x,y,t) = \frac{P(k_B\hat{i} + k_y\hat{j})e^{j(\omega t - k_B x - k_y y)}}{\omega\rho_o} \tag{9.26}$$

The wave-boundary matching condition is that the component of acoustic particle velocity in the y-direction on the plate surface should be equal to the plate velocity at that point.

Let the surface velocity of the plate be given by

$$\vec{u}_p = \hat{j}U_p e^{j(\omega t - k_B x)} \tag{9.27}$$

where U_p is the peak amplitude of plate velocity. The unit vector $\hat{j}$ indicates that the plate velocity of flexural vibration is always in the vertical direction.

From Equation 9.26, the acoustic particle velocity at $y = 0$ that is on the plate surface is given by

$$\vec{u}(x, 0, t) = \frac{P(k_B \hat{i} + k_y \hat{j}) e^{j(\omega t - k_B x)}}{\omega \rho_o} \tag{9.28}$$

The vertical component of particle velocity from Equation 9.28 is given by

$$\tilde{u}_y = \frac{P k_y e^{j(\omega t - k_B x)}}{\omega \rho_o} \tag{9.29}$$

Equating Equations 9.27 and 9.29

$$U_p e^{j(\omega t - k_B x)} = \frac{P k_y e^{j(\omega t - k_B x)}}{\omega \rho_o} \tag{9.30}$$

Equation 9.30 simplifies to yield

$$U_p = \frac{P k_y}{\omega \rho_o} \tag{9.31}$$

The peak amplitude of dynamic pressure in terms of the amplitude of plate velocity is given by

$$P = \frac{U_p \omega \rho_o}{k_y} \tag{9.32}$$

From Equations 9.24 and 9.32

$$\tilde{p}(x, y, t) = \frac{U_p \omega \rho_o}{k_y} e^{j(\omega t - k_B x - k_y y)} \tag{9.33}$$

Putting $\omega = k_o c_o$ and $k_y = \sqrt{\left(k_o^2 - k_B^2\right)}$ in Equation 9.33

$$\tilde{p}(x, y, t) = \frac{U_p k_o c_o \rho_o}{\sqrt{\left(k_o^2 - k_B^2\right)}} e^{j\left(\omega t - k_B x - \sqrt{\left(k_o^2 - k_B^2\right)} y\right)} \tag{9.34}$$

Equation 9.34 can be simplified as

$$\tilde{p}(x, y, t) = \frac{U_p c_o \rho_o}{\sqrt{\left(1 - \left(\dfrac{k_B}{k_o}\right)^2\right)}} e^{j\left(\omega t - k_B x - \sqrt{\left(k_o^2 - k_B^2\right)} y\right)} \tag{9.35}$$

Based on the frequency of excitation, different types of sound waves are produced. Since the frequency of excitation also relates to the wave number and wavelength, the following conditions can be examined for various values of wave numbers and wavelengths of the plate vibration and sound waves.

The first condition is when the bending wave number of the plate is less than the wave number of the sound wave; this can also be expressed on whether the bending wavelength is greater than the sound wavelengths or whether the frequency of plate vibration is greater than the critical frequency. Mathematically, these conditions can be expressed as

$$k_B < k_o$$

or

$$\lambda_b > \lambda_o \qquad (9.36)$$

or

$$f > f_c$$

The conditions of Equation 9.36, when applied to the dynamic pressure of Equation 9.35 ensure that the dynamic pressure component in both the x- and y-directions will have an oscillatory component that will propagate in the form of plane waves. Since these plane waves can travel longer distances, they produce significant sound pressure levels and these levels are, of course, dependent on the nearness of the plate vibrating frequency close to the critical frequency. At the critical frequency, the largest sound pressure level is produced that is limited only by the damping of the plate. Since we have not accounted for damping in this study, it is sufficient to understand that the largest sound is produced by a plate vibrating at the critical frequency that is determined by the coincidence of bending wave speed with the speed of sound.

The second condition relates to when the bending wave number of the plate is more than the wave number of the sound wave; this can also be expressed on whether the bending wavelength is less than the sound wavelengths or whether the frequency of plate vibration is less than the critical frequency. Mathematically, these conditions can be expressed as

$$k_B > k_o$$

or

$$\lambda_b < \lambda_o \qquad (9.37)$$

or

$$f < f_c$$

For the conditions of Equation 9.37 when applied to the dynamic pressure equation of Equation 9.35 result in the third term of the exponential becoming imaginary and hence the dynamic pressure is produced only in the near field along the y-direction. Therefore, very insignificant levels of sound pressure are produced by infinite plates vibrating at frequencies less than the critical frequency.

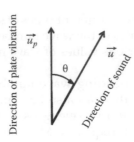

FIGURE 9.9
Direction of plate vibration velocity and sound particle velocity.

The direction of plate vibration and sound waves are shown in Figure 9.9. At the plate surface, the dynamic pressure in the direction of particle velocity, assuming plane wave propagation, is given by

$$p_{rms} = u_{rms} \rho_o c_o \tag{9.38}$$

The plate velocity and particle velocity of air are related by

$$u_{rms} \cos \theta = u_{p_{rms}} \tag{9.39}$$

From Equations 9.38 and 9.39

$$u_{p_{rms}} = \frac{p_{rms}}{\rho_o c_o} \cos \theta \tag{9.40}$$

Sound power radiated by the infinite plate is given by

$$\pi_{rad} = p_{rms} u_{p_{rms}} A \tag{9.41}$$

where A is the plate area.
From Equations 9.2 and 9.6, the sound power radiated by an infinite plate can also be expressed as

$$\pi_{rad} = \rho_o c_o A u_{p_{rms}}^2 \sigma \tag{9.42}$$

where σ is the radiation ratio. From Equations 9.41 and 9.42, the radiation ratio is given by

$$\sigma = \frac{p_{rms}}{\rho_o c_o \dfrac{p_{rms}}{\rho_o c_o} \cos \theta} = \frac{1}{\cos \theta} \tag{9.43}$$

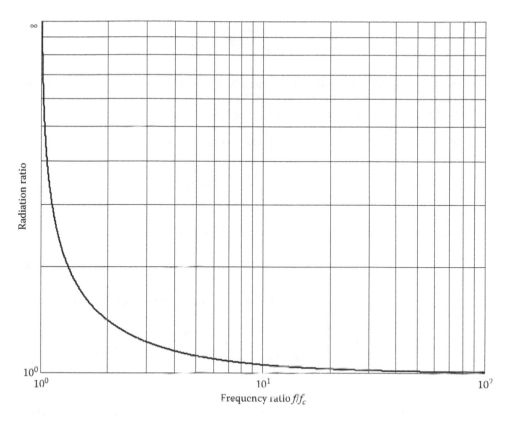

FIGURE 9.10
Radiation ratio of an infinite plate.

Using the angle relationship between the plate vibration wave number and the sound wave number, the radiation ratio, as shown in Figure 9.10, can be expressed as

$$\sigma = \frac{1}{\cos\theta} = \frac{k_o}{k_y} = \frac{1}{\sqrt{\left(1 - \left(\frac{k_B}{k_o}\right)^2\right)}} = \frac{1}{\sqrt{\left(1 - \frac{f_c}{f}\right)}} \tag{9.44}$$

See Cremer, Heckl, and Ungar (1987), Fahy and Gardonio (2007), Junger and Feit (1986), and Norton and Karczub (2003) for further details.

Example 9.2

A very large, flat, steel undamped plate, which can be considered as an infinite plate, is mechanically driven by an exciter perpendicular to its flat surface. Estimate the peak radiated sound pressure level due to the vibrating plate at 800 Hz and 10000 Hz. The plate is 1.5 mm thick and its peak surface velocity is 8 mm/s at both frequencies. Assume C_L = 5064 m/s for steel, and c_o = 340 m/s as speed of airborne sound, and density of steel = 7800 kg/m³.

From Equation 6.25 (Chapter 6), the critical frequency is given by

$$f_c = \frac{c_o^2}{1.82 h c_L} = \frac{340^2}{1.82 \times 1.5 \times 10^{-3} \times 5064} = 8362 \text{ Hz} \tag{1}$$

At the vibrating frequency of 800 Hz there is no sound radiation from the infinite plate since this frequency is much less than the critical frequency. At the vibrating frequency of 10000 Hz there is sound radiation from the infinite plate since this frequency is more than the critical frequency.

From Equation 3.42, the wave number of the airborne sound is given by

$$k_o = \frac{2\pi f}{c_o} = \frac{2\pi \times 10000}{340} = 184.80 \text{ rad/m} \tag{2}$$

The density per unit surface area of the plate is given by

$$\rho_a = \rho h = 7800 \times 0.0015 = 11.7 \text{ kg/m}^2 \tag{3}$$

From Equation 9.14, the plate bending wave number at 10000 Hz is given by

$$k_B = \left[\frac{(2\pi f)^2 \rho_a}{EI} \right]^{1/4} = \left[\frac{(2\pi \times 10000)^2 11.7}{\frac{200 \times 10^9 \times (1.5 \times 10^{-3})^3}{12}} \right]^{1/4} = 169.28 \text{ rad/m} \tag{4}$$

From Equation 9.35, the dynamic pressure resulting from vibration of the infinite plate is given by

$$P_{rms} = \frac{U_p \rho_o c_o}{\sqrt{2} \left[1 - \left(\frac{k_B}{k_o} \right)^2 \right]^{1/2}} = \frac{0.008 \times 400}{\left[1 - \left(\frac{169.28}{184.80} \right)^2 \right]^{1/2}} = 5.64 \text{ Pa} \tag{5}$$

The sound pressure level corresponding to the dynamic pressure is given by

$$L_p = 10 \log \left(\frac{P_{rms}}{p_{ref}} \right)^2 = 10 \log \left(\frac{5.64}{20 \times 10^{-6}} \right)^2 = 109 \text{ dB} \tag{6}$$

9.4 Sound–Structure Interaction of Vibrating Structures

We discussed sound radiation of infinite plates in the previous section using the wave approach. This has provided some understanding of how vibrating structures can generate sound. Unfortunately, we were able to figure out only sound radiated at frequencies greater than the critical frequency. Although many machinery noise problems arising out of impact excitation are related to sound radiation beyond the critical frequency of plates (as they contribute more toward A-weighted response), we must also be aware of the sound–structure interaction below the critical frequencies. Therefore, we will discuss the physics of sound–structure interaction of a general structure, without going into detailed mathematical treatment that can be obtained from references. The discussion in this section is based on Smith and Lyon (1965).

Sound generated by a vibrating structure and vibration generated by an incident sound wave form the basics of sound–structure interaction. Although sound generated by a vibrating structure is of interest to those trying to reduce machinery noise, both phenomena are equally important from a practical standpoint, as they are also complementary to each other. Hence, our discussion is aimed at a finite structure in a fluid medium that can sustain airborne-sound generation.

In our earlier discussion on the modal approach to vibration of continuous systems, we have seen that the total response of the system can be obtained in terms of the responses to uncoupled oscillators. We had, of course, not considered the presence of structural damping. But as long as structural damping is not significant, obtaining uncoupled single-degree-of-freedom (SDOF) systems is definitely possible. Even if there is a small coupling between them, it could be neglected without any difficulty at least in case of sound–structure interaction in air. With such a set of uncoupled oscillators, the net response can be computed by superposition of the response from each of them.

Obtaining the parameters of each of the uncoupled oscillators is fairly easy. The only difficulty would be to determine the driving force on each of them. When they were just vibratory systems with an external excitation, we could compute the generalized force in terms of the mode shape, excitation force, and generalized mass. But in the case of sound–structure interaction, the vibration of the structure encounters resistance from the surrounding medium, which is known as the radiation impedance. In addition, the structure can also be excited by airborne sound. Once this driving force is figured out, the solution to the rest of the problem is rather easy.

Consider a finite structure placed in a fluid medium that can sustain airborne sound generation. Many times we compute the natural frequencies and mode shapes of a structure assuming that the surrounding air does not have much influence on them. We can get away with such an assumption in most applications, including sound–structure interaction in air, where we consider that natural frequencies and the corresponding mode shapes are unlikely to be influenced by the surrounding air.

The general velocity response can be expressed as an infinite sum of the product of the modal velocity response of the uncoupled oscillators and the corresponding mode shape so the spatial variation of the response can be accounted for. Since kinetic energy is proportional to the square of velocity, it is proportional to the square of the mode shape function for a continuous system. Since a mode shape has an arbitrary scaling factor, it can be scaled in such a way that the modal mass for all the resonant frequencies will be equal to the total mass of the structure. Hence the total kinetic energy will be the product of the total mass of the structure times half the summation of the square of modal velocities, $T = \dfrac{M}{2} \sum_n v_n^2$, and the modal stiffness will be the product of the square of undamped natural frequency and the structural mass, $K_n = \omega_n^2 M$.

By assuming that damping forces are not coupled, the total power dissipated can be expressed as the sum of the product of the internal modal damping and square of the modal velocity, $\Pi = \Sigma R_{n,\text{int}}\, v_n^2(t)$. It has thus become necessary to differentiate internal damping, because in a sound–structure interaction, the radiation impedance will contribute toward the damping arising from the conversion of vibration into sound.

Once we have identified the elements of the resonators of a general structure, we must then identify the forces acting on them. We have the impedance felt by the resonators in converting vibration into sound and the forces acting on the resonator due to the sound field. Therefore, one is the dynamic pressure acting on the resonators in the absence of structural motion, known as blocked pressure, and the other is the dynamic pressure resulting due to the vibration of the structure, known as radiated pressure. The corresponding forces on the resonators can be respectively described as blocked force and radiation force; blocked force is the force acting on the resonator in the absence of structural motion and radiation force is the reaction experienced by the resonators from the fluid due to structural motion.

The blocked force depends on the frequency, magnitude of pressure, and direction of incidence. A simple example is the doubling of pressure due to reflection of a plane wave with a large surface, but it could take more complex forms for different applications. The relationship between blocked pressure and block force is related by the coupling parameter for each case and it will have the units of area.

The modal radiation force on the other hand is proportional to the modal velocity and the constant of proportionality between the modal radiation force and modal velocity is known as radiation impedance, and it is in general a complex number. However, in the specific case of a baffled piston, discussed in Chapter 7, we have seen that it has only the real part, which is equal to the product of the characteristic impedance of air and the area of the piston.

The imaginary part corresponds to a reactive term related to an inertial term, similar to a single-degree-of-freedom system. The equivalent mass is not constant but varies slowly with frequency, and the value of this mass is much smaller than the mass of the structure that is interacting with a surrounding fluid. The main effect of this mass is to reduce the modal resonance from its value in the vacuum, because of the inverse proportionality relationship of mass and undamped natural frequency. However, the reduction is considerably small for air, but could be higher for water surrounding the structure. A pulsating sphere also has an imaginary part of radiation impedance (see Chapter 7, Problem 7.19).

The real part of the radiation impedance is called radiation resistance. Although it could be smaller than the imaginary part, it cannot be neglected, as it is important in dissipating power through radiating sound. Therefore, one must compare it with internal mechanical resistance of the structure (see Chapter 10, Example 10.1). In many cases dissipation due to radiation can dominate and hence it cannot be neglected. For determining the net response of the system, the internal mechanical impedance must be added to the radiation impedance.

Once the radiated pressure and the corresponding particle velocities are computed, sound intensity can be obtained from them that can lead to the radiation factor. The aim of this book is not to present a detailed discussion on this procedure, which is available in reference texts. However, the preceding discussion will help in understanding the reciprocity relationship of sound–structure interaction that will be discussed in Chapter 10.

Since plates are commonly used in the construction of machinery, we shall discuss sound radiation of plates in the following section. This would help in computing the radiation factor of plates from references.

9.5 Bending Wave Sound Radiation from Finite Plates

Sound radiation from a finite structure due to vibration can be classified as (a) modal sound radiation at any frequency that could have both resonant and nonresonant frequencies and

(b) frequency-averaged sound radiation. Since a finite structure has natural frequencies and corresponding mode shapes, when it is subjected to broadband excitation, it results in the excitation of a large number of resonant modes. Therefore, sound radiation due to broadband excitation is dominated by resonant structural modes, whereas modal sound radiation that occurs due to specific frequencies can have both resonant and nonresonant frequencies. Finite structures have higher radiation ratios at higher frequencies. However, the sound radiated from them due to nonresonant frequencies is not large because the vibration amplitudes are not very high at higher frequencies due to large values of plate impedance at nonresonant frequencies. Therefore, sound radiated due to broadband excitation is more dominant than sound generated due to modal sound radiation.

The vibration generated due to acoustic excitation is an important phenomenon that is slightly different from that discussed in the preceding paragraph. The vibration response of structures due to acoustic excitation consists of a forced vibration response at the excitation frequency of the input and the free response due to the excitation of structural natural frequencies. Whereas the forced vibration response at the excitation frequency is due to the trace wavelength, the free response is associated with the interaction of the trace wave with the corresponding natural frequencies. In case of noise transmitted across partitions, the above phenomenon of acoustic excitation becomes important. Since the structural response is both resonant and forced, the transmission of sound could be due to either of the phenomena or both.

Now let us discuss sound radiation from free bending waves in finite plate structures in which frequency-averaged resonant sound radiation is of practical interest in designing quiet machines. Consider a rectangular plate of sides ℓ_1 and ℓ_2, thickness h, Young's modulus E, and density ρ that is simply supported on all sides. The natural frequencies of this plate are given by

$$f_{m,n} = \frac{\pi}{2}\left(\frac{m^2}{\ell_1^2} + \frac{n^2}{\ell_2^2}\right)\sqrt{\frac{Eh^2}{12\rho}} \tag{9.45}$$

where $m = 1, 2, 3, \ldots$ and $n = 1, 2, 3, \ldots$.

Using Equation 9.45, each vibrational mode of the plate can be represented in a two-dimensional lattice as shown in Figure 9.11. The nodal lines of this lattice divide the plate into smaller rectangular vibrating elements that act as individual volume sources and

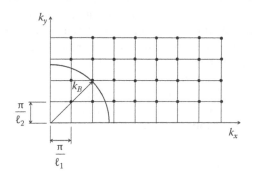

FIGURE 9.11
Wave number lattice of a rectangular plate.

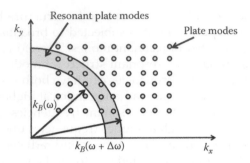

FIGURE 9.12
Wave number space of a flat plate.

hence each of these elements displaces surrounding fluid in their proximity. Therefore, the interaction of these vibrating elements with the surrounding air results in compression and rarefactions of the fluid medium, resulting in airborne sound; in addition, compressed or rarefied fluid media in the vicinity of the plate surface will also interact with each other. Hence, the sound power radiated is not simply related to the average velocity as in the case of an infinite plate. Due to boundary conditions, standing waves are generated and the sound power has to be thus related to the vibration modes within the excitation frequency bandwidth that are resonant. Hence, the radiation ratio of each of these modes will have to be accounted for in estimating the sound power.

The wave number diagrams, such as those shown in Figure 9.12, are useful in representing the vibration characteristics of a structure in relation to structure-borne and airborne sound. The quarter circle segment between bending wave numbers $k_B(\omega)$ and $k_B(\omega + \Delta\omega)$ contain resonance frequencies of the plate. Figure 9.12 shows how there can be a number of resonant frequencies of the plate within a frequency band. The frequency band can be computed based on the bandwidth of bending wave numbers of the excitation frequency bandwidth. Similar wave number diagrams can be used to explain the phenomenon of airborne sound radiation above and below the critical frequency.

9.5.1 Acoustically Slow Modes

Figure 9.13 shows the wave number diagram for resonant excitation below the critical wave number k_c. The wave number lattice of all the resonant modes of a rectangular plate are shown, but the quarter circle of radius less than the critical frequency is the focus of our discussion. It is assumed that the force excitation is within a bandwidth $\Delta\omega$ corresponding to wave numbers k_{B_1} and k_{B_2} that are less the critical bending wave number k_c. The airborne sound generated due to this vibration will have wave numbers between k_1 and k_2. The resonant modes of the plate within the shaded region can be further divided as edge modes and corner modes, which will be discussed later. Edge modes are more efficient than corner modes. However, all the plate modes that are below the critical frequency are known as acoustically slow or subsonic modes.

9.5.2 Acoustically Fast Modes

Figure 9.14 shows the resonant modes of a plate when subjected to a band-limited excitation at frequencies higher than the critical frequency. All the resonant structural modes are shown

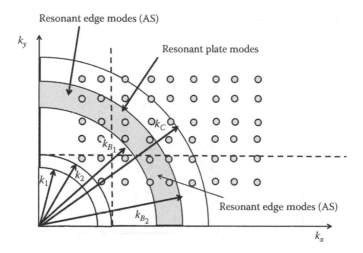

FIGURE 9.13
Wave number diagram for resonant excitation below the critical frequency (acoustically slow modes/subsonic modes).

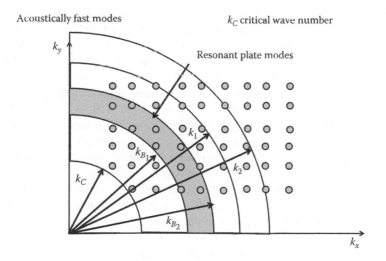

FIGURE 9.14
Wave number diagram for resonant excitation above the critical frequency (acoustically fast modes/supersonic modes).

within the region of bending wave numbers k_{B_1} and k_{B_2}. These wave numbers are less than the acoustic wave numbers k_1 and k_2. They are called acoustically fast or supersonic speeds, as the bending wave speeds are greater than the speed of sound in this band. Under these conditions, the plate efficiently radiates airborne sound. We have earlier seen in the case of infinite plates that airborne sound is produced only due to the acoustically fast modes.

9.5.3 Force Excitation at a Frequency

Figure 9.15 shows the resonant modes of a plate excited at a forcing frequency corresponding to a wavenumber k_B of the plate. The vibration response will therefore be the

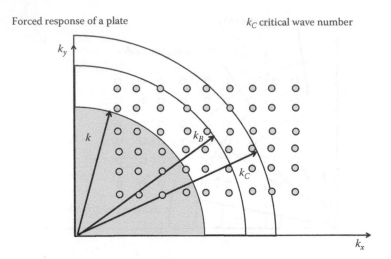

Forced response of a plate k_C critical wave number

FIGURE 9.15
Wave number diagram for force excitation of a plate at an excitation frequency.

superposition of all the modes at the forcing frequency. As discussed in Chapter 1, the response could be mass controlled or damping controlled or stiffness controlled, depending on the relative location of the excitation frequency with respect to the resonant frequency. Since a plate has many resonant frequencies, the mass-controlled region of one resonant frequency could be overlapping with the stiffness-controlled region of the next resonant frequency. It should be noted that only those resonant frequencies of the plate within the forcing frequency wave number k will couple with the forcing frequency and radiate sound. In addition, the excitation frequency could be from either an incident airborne sound field or vibration excitation below the critical frequency.

We know that vibration excitation generates very little sound from an excitation frequency below the critical frequency. However, when the excitation is from an incident sound field that has sufficient strength to excite light structures, then the nonresonant forced modes matching with the wavelengths of the incident sound can allow efficient transmission of sound. Under such a forced condition, the response is mass controlled and hence the plate mass and not stiffness or damping controls the transmission of sound at these frequencies. This forms the basis for limped mass law that is used to determine the transmission loss of partitions (see transmission loss section in Chapter 6).

Thus we can summarize the discussion as follows. Mechanical excitation of plates results in a significant portion of most of the radiated sound produced by resonant plate modes above the critical frequency. With acoustic excitation, however, excited by an incident airborne sound field, transmit sound very efficiently when the wavelengths of the forced modes match with the wavelength of the incident sound field. This phenomenon is important in case of aerospace structures that are subjected to high levels of acoustic excitation.

9.5.4 Corner Radiation of a Finite Plate

Figure 9.16 shows the vibration mode of a plate. For this mode it is assumed that the bending wavenumbers k_x and k_y are greater than the corresponding acoustic wavenumber k at the same frequency. Hence, the wavelengths λ_x and λ_y are smaller than the acoustic wavelength λ. This results in alternate square divisions volume velocity vibrating out of

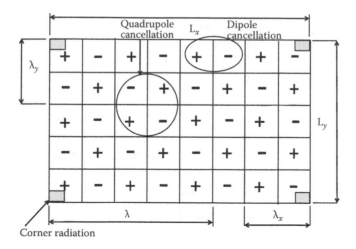

FIGURE 9.16
Corner radiations for a finite plate.

phase with each other. Depending on their relative location on the plate, they either form quadrupoles or dipoles or cancel each other; hence, very little sound is radiated from them. The only sources that are not canceled are the monopole sources at the corners, and since monopoles are the most efficient sound radiators, only corners of the plate radiate sound more efficiently in comparison to the edges for the aforementioned conditions.

Whether the aforementioned monopoles at the corners will radiate sufficient sound depends on the length of the plate with respect to the acoustic wavelength. If the length of the plate is less than the acoustic wavelength, the corner monopoles interact with each other depending on whether the mode numbers are even or odd. For odd values of m and n they vibrate in phase and behave like a combination of monopoles in phase that resembles the doubling effect discussed earlier. For m even and n odd, the adjacent pairs will be out of phase and behave like a dipole. For both m and n even, all four corners vibrate out of phase with each other and they behave like a quadrupole. When the plate dimensions are higher than the acoustic wavelength, the four monopoles do not interact with each other and they independently radiate sound. From this discussion it is clear that corner radiation becomes significant when the corners vibrate in phase.

9.5.5 Edge Radiation of a Finite Plate

Figure 9.17 illustrates the situation when one of the bending wave numbers is less than the corresponding acoustic wave number, that is, when one of the bending wavelengths (along the x-direction in this case) is smaller than the acoustic wavelength and the other bending wavelength (along the y-direction in this case) is larger than the acoustic wavelength. This results in rectangular strips of volume velocities of the plate alternately vibrating out of phase with each other; the strips have a longer dimension in the y-direction in comparison to the y-direction. These strips vibrating out of phase with each other form dipoles in the central regions of the plate and cancel each other, but the edges along which the bending wavelength is larger will not cancel. Compression and rarefaction of the fluid along these edges results in the radiation of airborne sound. Hence, the edge modes are more efficient radiators than corner modes that have many conditions to be satisfied in order to become efficient radiators.

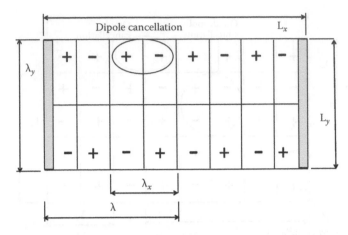

FIGURE 9.17
Edge radiation for a finite plate.

When the excitation frequency is increased toward the critical frequency, the cancellation in the central regions reduces since the separation between the sections reduces. The cancellation eventually stops above the critical frequency and the entire plate radiates airborne sound. These are now called surface modes and the bending wavelengths in both directions will be larger than the corresponding acoustic wavelengths. Hence, finite plates behave very similar to infinite plates in radiating airborne sound at frequencies greater than the critical frequency. See Fahy and Gardonio (2007), Junger and Feit (1986), and Norton and Karczub (2003) for further details.

9.6 Conclusions

The sound–structure interaction is a very important concept that will be required for estimating the sound pressure levels produced by vibrating objects. An important outcome of this concept is the radiation ratio, which is an objective measure of the sound radiating capability of a structure. Therefore, based on the discussions presented in this chapter, the radiation ratio of many practical structures can be computed. The coupling loss factor, which is one of the important parameters used in statistical energy analysis (SEA) is related to the radiation ratio, when vibro-acoustic systems are modeled using SEA. The main objective of this chapter was to provide the necessary background for discussing vibro-acoustics using SEA in the next chapter.

PROBLEMS

9.1. Consider an SDOF, as shown in the following figure, mounted flush with the surface of a reverberant room excited by an airborne sound source that generates broadband excitation. The mass of this SDOF, acting like a piston, whose front side only received the airborne sound is mounted as shown. The mass of the SDOF is 0.25 kg, the area of the piston is 50 cm², the undamped resonance frequency (f_n) is 400 Hz, and the loss factor of the SDOF (η) is 0.001. The airborne sound that is measured in the 400 Hz 1/3 octave band center frequency is 130 dB. Determine the radiation resistance, radiation loss factor, resistance ratio, spectral density of the airborne sound, and the response of the SDOF (displacement, velocity. and acceleration). Assume velocity of sound in air (c_o) is 340 m/s and the nominal density of air (ρ_o) is 1.2 kg/m³.

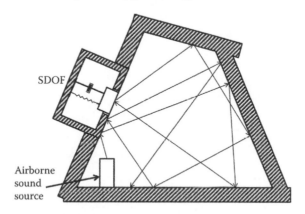

9.2. Consider Example 6.5 (Chapter 6). Compute the wavelengths of all resonant frequencies before 6680 Hz and classify the resonant frequencies as acoustically slow and acoustically fast modes.

9.3. Consider Problem 9.2. Determine all the resonant frequencies in the band of bending wave numbers 60 and 65 rad/m.

9.4. Consider Problem 9.2. Determine all the resonant frequencies in the band of bending wave numbers 85 and 90 rad/m.

9.5. If the plate of Problem 9.2 is excited into vibration corresponding to bending wave numbers 60 and 65 rad/m, discuss the qualitative nature of the airborne sound produced.

9.6. If the plate of Problem 9.4 is excited into vibration corresponding to bending wave numbers 85 and 90 rad/m, discuss the qualitative nature of the airborne sound produced.

9.7. Consider Example 6.5 (Chapter 6). Determine the radiation factor for this plate up to 7000 Hz. Assume that the plate is baffled.

10

Statistical Energy Analysis (SEA)

Statistical energy analysis (SEA) is a very important technique for solving vibro-acoustic problems at frequencies approximately above 500 Hz. It would be difficult to discuss SEA in isolation since it needs the basic knowledge of vibration, acoustics, wave theory, random vibration, vibration of continuous systems, and vibro-acoustics, which were discussed in the preceding nine chapters; hence, the discussion now as the last chapter of this book. Therefore, unless one has read through all the previous chapters, it will be difficult to make sense of this chapter.

This chapter will be discussed in the following sequence. First, the historic development of SEA will be given in a detailed discussion on why an entirely new approach like SEA had to be formulated. Then, the physics of SEA formulation is presented with the help of the thermal analogy of heat transfer between two bodies. This is then followed by deriving basic equations of SEA from a single-degree-of-freedom (SDOF) system that is extended to power transfer between two systems and extended to multisubsystem models to obtain a matrix formulation. Since SEA is a system modeling approach for a system of interconnected subsystems, the basic rules for formulating subsystems are presented with examples.

Continuous systems, such as beams, plates, and shells, are used to construct most machines and structures. Therefore, the basic properties of these continuous systems, including modal density, drive point mobility, conductance, and input power, are linked to SEA modeling and some of them were discussed earlier. In addition, SEA modeling also requires information about loss factors of each system and coupling loss factors between connected subsystems; loss factors need to be measured but coupling loss factors can be computed for many connected system, otherwise they will have to be measured. Coupling loss factors for some simple systems are presented. Finally, a number of applications of SEA are presented with examples. Transient SEA is also discussed towards the end.

10.1 History

The launching of the artificial satellite Sputnik in 1957 initiated many significant changes in the scientific world, including the creation of NASA in the United States. This resulted in a race to launch satellites with the help of rocket launch vehicles, and at the time of launch, the high decibel sound could extend up to 200 dB closer to the rocket; the source of sound came from combustion and boundary layer effects. Such a high level of airborne sound induced high vibration levels due to vibro-acoustic interaction that rattled the entire launch vehicle structure and its subsystems. Although stiffeners on the outer structure could reduce the vibration levels to prevent their failure, the more sensitive electronics of the payload mounted on printed circuit board (PCB) cards invariably failed. This significantly affected the launch mission, as control and guidance of the vehicle were seriously affected by the failure of electronics. Since the vibration problem could not be sorted using the then-existing methods based on modal analysis, an alternate approach had to be

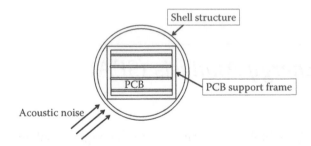

FIGURE 10.1
Vibration excitation of the PCBs due to acoustic excitation.

found. Hence, this event was responsible for the evolution of SEA, a very important tool in the study of vibro-acoustics. That is how vibro-acoustic analysis of the launch vehicle structure and its components using SEA became very important.

Figure 10.1 shows a simplified diagram of vibration excitation of PCBs due to acoustic noise at the time of launch. The failure of electronic systems mounted on the launch vehicles was the motivating reason for the development of SEA, and, hence, it is relevant for discussion. The cross section of a shell structure that encloses a frame on which PCBs are installed is shown in the figure. The shell structure is excited into vibration due to noise generated by the acoustics and boundary layer, which gets transferred to the frame and then to the PCB, and the cavity inside the shell structure has sound pressure levels limited by the transmission loss of the shell and this acoustic input can also directly excite the PCB into vibration. Both of these vibration and acoustic sources together can cause significant vibration levels in the PCB resulting in their possible failure.

The only known approach at that point of time for reducing the vibration of PCBs was the modal approach that was normally used to design machines from the viewpoint of reducing vibration at low frequencies. The same approach could not be scaled up in frequency for a launch vehicle since it had a light structure and a broadband excitation that made it necessary to account for a very large number of modes in the desired frequency range of 0 to 2000 Hz. The lightweight structures of these rockets were excited at higher frequencies due to broadband excitation of the rocket noise. Due to the larger influence of tolerances of the structure at higher frequencies, higher frequencies could not be accurately computed. In addition, the voluminous information regarding natural frequencies and mode shapes became unmanageable and the results obtained based on this approach were obviously not helpful. The computing power was also limited at that time. This kind of vibro-acoustic phenomenon that was never encountered before, and, therefore, the then-existing methods of using the modal approach at that time were not of much help. Therefore, a simplified approach of statistical averaging had to be evolved so they could obtain relevant information from the analysis that matches with the characteristic behavior of dynamic systems in the high frequency region.

10.2 Statistical Energy Analysis (SEA) Evolution

It is important to understand why high frequency excitations of structures within a rocket were excited, which was not commonly observed in any of the dynamic systems earlier.

Traditional machinery was rugged in construction and therefore, high frequency excitation was not an issue in the 1960s. Although high frequency excitation was an issue in jet aircraft design, aircraft structures were still tolerant to excitation since their structures were relatively strong and the acoustic and boundary layer excitation was also not very high. On the other hand, rocket structures had to be extremely light to accommodate higher payloads, and thin structures supported high frequency resonant vibration when subjected to broadband excitation. The high frequency characteristics of these structures are entirely different from the traditional structures that were robust, and, therefore, the traditional modal approach could not be scaled up in frequency. Hence, there was a need to develop a different analysis technique.

The SEA technique evolved as a technique for determining vibration response due to rocket noise of satellite launch vehicles during the early 1960s. In contrast to the modal approach that generated a lot of data that was not very accurate at high frequencies, a technique that could give a gross estimate of spatial and frequency average of vibration levels was more than sufficient than a detailed deterministic estimation of vibration at different locations. Thus, SEA, which was based on such a spatial and frequency averaging, became very important in solving problems that were beyond the reach of computers at that time. This is the historical context in which SEA evolved during the design of rocket launch vehicles in the 1960s.

Therefore, in order to overcome the difficulty of solving vibro-acoustic problems in the high frequency region, the SEA technique was formulated. R.H. Lyon is the originator of this technique. The basis for this technique is the power flow between two dynamic systems connected with each other and its dependence on the modal energy of each of them in a specified frequency band provided the basic platform to launch SEA. The advantage of this new technique was that parameters of this model depended only on the gross properties of the system and the response could be predicted only in terms of space-averaged and frequency-averaged values. This information was sufficient to design the dimensions of a particular component for satisfactory vibro-acoustics performance in the high frequency region. Thus SEA enabled a system-based modeling approach.

SEA was inevitable in the 1960s due to much less computing power at the time. Now, with the increased power of the computers, is SEA still relevant? At present, we have very powerful computers, both in terms of speed and memory, that can model complex structures having a large number of natural frequencies. However, the usefulness of SEA is still undiminished due to the following reasons: when a structure is subjected to broadband excitation, a large number of modes will have to be accounted for in the analysis. Information about natural frequencies and mode shapes of structures at higher frequencies will therefore be necessary. The natural frequencies and mode shapes become highly sensitive to structural detail at higher modes of vibration and the computer programs are rather inaccurate in computing natural frequencies and mode shapes corresponding to the higher modes of vibration. Therefore, one has to accurately describe geometry, construction and material property of the structures and also compute the natural frequencies and mode shapes accurately. Such an accurate description is practically impossible, since the above parameters can only be defined in terms of tolerance limits. Even assuming that the structure can be accurately described, by providing very close tolerance, and that the mode shapes and natural frequencies can be accurately computed, there is an additional problem of processing a large amount of data that are generated in the analysis. Therefore, SEA continues to be used in a large number of practical vibro-acoustic applications in the high frequency region.

Although SEA initially evolved as a simple technique to address the vibro-acoustic interaction of a small number of subsystems during its early stages of development, its scope has now been enhanced to cover most of the machinery that need reduction of vibration or noise. Having traced the origin of evolution of SEA due to rocket noise, it is important to note that it does not mean that SEA is only for rocket science. Today it is mainly used for dynamic systems on the ground for applications in consumer products such as automobiles, refrigerators, washing machines, and dishwashers, where customers demand both quietness and energy efficiency and require lightweight structures that need the application of SEA. The biggest users of SEA today can be found in the automotive industry wherein very complex SEA models having thousands of subsystems are being used to reduce cabin noise, which has become a very important parameter of customer satisfaction. Similarly, SEA is used for reducing cabin noise in airplanes. The exterior noise generated on war ships and submarines is an important aspect of survivability in water, and, hence, SEA is used for their reduction. Launching of satellites with much increased capacities now requires extensive application of SEA to ensure their satisfactory performance. SEA is also used in building acoustics and enclosure. Another important area that demands the use of vibro-acoustics through SEA is the concept of sound quality, which is required in many machines that are for consumer products. Based on jury tests, it will be required to maintain a certain sound pressure level in machines at selected frequency bands to maintain sound quality; such a tailored design of sound pressure levels of machines is only possible by using SEA.

Although SEA evolved during the early 1960s, room acoustics was the subject of interest much earlier, due to its application in auditorium acoustics. One of the important links between room acoustics and the evolution of SEA is that a large number of modes exist in room acoustics, but only a statistical view of their impact is taken. This type of statistical view is the cornerstone of SEA, and, therefore, the evolution of SEA is strongly linked to room acoustics. Hence, it is fair to say that the study of room acoustics that existed much before SEA was developed did provide the platform for developing SEA. Since there are millions of modes in room acoustics, it was not unusual to take a statistical viewpoint that was sufficient to explain the basic characteristics of sound in a room. This statistical viewpoint was conveniently extended to the evolution of SEA that included both acoustic and vibratory systems and has thus played an important role in developing SEA. In addition, the basic concepts of room acoustics are used in computing the energy of acoustic subsystems. However, it should be emphasized that vibratory systems that are used in SEA may not always guarantee a large number of modes that are common in room acoustics. This limits the lowest frequency below which SEA prediction becomes less accurate.

10.3 Basis for SEA

First, we must know why it is called SEA. Using this approach, we can obtain a statistical estimate of the response from a population of dynamic systems. For example, we can predict that the cabin noise of a particular brand of car as 65 ± 3 dB. In addition, we take a statistical view of the dynamic system and also the excitation. In addition, we use frequency-averaged and space-averaged information. Hence, the word *statistical* is used to cover all the above points. Those not familiar with SEA mistakenly relate it to statistics in the general sense, and, there is, of course, a subtle difference between statistical and statistics. The term *SEA* was coined for the following reasons. *Statistical* emphasizes that the systems being studied are drawn from populations of similar design construction, having known distribution of their dynamical parameters. This enables one to account for the manufacturing tolerance that exists in practical systems. *Energy* is the primary variable that is used for both vibration and acoustic systems from which other variables like acceleration, velocity, and sound pressure levels can be calculated. It is called *analysis* since it is only a framework of study.

SEA is particularly suited for vibration and noise studies of systems in the high frequency region. Typical examples are gearbox noise, diesel engine noise, sound transmission across partitions, jet noise, propeller noise, boundary layer turbulence, vibration resulting in satellites at microgravity due to the reaction of the wheels and coolers, and sound–structure interaction due to rocket noise. In the high frequency region, the vibrating structure and the acoustic environment have a very large number of modes that are contributing toward the response. The presence of a large number of modes is essential for the application of SEA, because the principle of SEA involves computation of time, space, and frequency averages. Such averages are meaningful only if there are a large number of modes. The presence of a large number of modes in the high frequency region poses the difficulty of computation of information regarding natural frequencies and mode shapes of these modes, which prompted the evolution of SEA in the first place. Therefore, the high frequency region is necessary for application of SEA, and SEA is inevitable in the high frequency region, as other methods do not work as well.

The number of resonance peaks of a structure, in a particular frequency band, generally increases with frequency (in beams they decrease with frequency). The resonance peaks will therefore be widely spaced at lower frequencies than at higher frequencies. Since the resonance peaks are rather crowded at higher frequencies, it results in more than one resonance within a half-power bandwidth. This is known as modal overlap. This translates in to difficulty of computing of natural frequencies and mode shapes in this region. But modal overlap is one of the favorable conditions for using SEA, which allows modal groups.

The basis for SEA is the interaction between various mode groups of vibro-acoustic systems. Although every mode group does not significantly interact with another mode group, all the possible mode groups can be initially accounted for in the modeling (especially when software is used), and later only mode groups that have significant interaction need to be considered. Many times, in machinery construction, similar systems are used, and hence there is always past information available from similar systems of existing machinery that can be used for modeling future machines. In addition, at the drawing board stage of designing a machine, the basic dimensions of various elements of which the machining is being constructed needs to be designed. The dimensions of the construction elements and the manner in which they are joined from the existing machines can be used to obtain an approximate estimate of the vibration and sound that will be produced by the machine using SEA.

10.4 Brief Literature Review

There were two landmark publications around 1960 that lead much of the research in SEA. Although both papers were independently published, they have many things in common that form the basis of SEA. The first paper by Lyon and Maidanik proposed that the flow of energy between two oscillators is proportional to the modal energy of each of them, which is the basis for deriving the power flow equation between two connected subsystems that could be extended to a multiple-subsystem model. The second paper by Smith proposes that there is equipartition of energy between a plate kept in a reverberant field as its damping is reduced (see Example 10.1). This is an important consideration in the acoustic excitation of launch vehicle structures that have low values of damping that result in equipartition of energy that simplifies many computations.

R.H. Lyon is credited as the originator of SEA. He is the coauthor of the NASA contractor report (Smith and Lyon 1965), which is perhaps the first reference material in this area. His textbook *Statistical Energy Analysis of Dynamical Systems* (1975) is the first known reference text in SEA, which was followed by *Theory and Application of Statistical Energy Analysis* (Lyon and DeJong 1995). These texts are considered to be the most authoritative reference material in this field, even today. Many of Lyon's papers form the basis of the SEA techniques, which have grown enormously over the past 50 years, both in terms of the number of publications and the number of useful applications. Another excellent reference textbook by R.H. Lyon, *Machinery Noise and Diagnostics* (1986), deals with many practical aspects of vibration and noise reduction using SEA.

It is interesting to note that the subject of random vibration evolved in the same period as SEA. Many developments in stochastic processes have benefited the study of SEA. Papers by Lyon and Eichler (1964) and Lyon and Maidanik (1962), which are considered the basis for SEA, apply the theory of stochastic processes extensively. Therefore, Chapter 4 on random vibration provides the necessary framework for studying SEA.

SEA has been mainly concerned with dynamic interaction of connected systems. Therefore, much of the earlier work on SEA was generally concerned with studying such an interaction between two oscillators. The outcome was later extended to subsystems, each of them having a large number of modes. The papers by Lyon and Eichler (1964) and Lyon and Maidanik (1962), which exclusively deal with the coupled oscillators, are considered to be the basis of SEA.

Spatial averaging of frequency response envelopes was been discussed by Langley (1994). The wave intensity technique (WVI) has been discussed by Langley and Bercin (1994), which deals with separation of reverberant and nonreverberant fields. Accounting for indirect coupling in physically disconnected subsystems was been discussed by Heron (1994). Without accounting for this indirect coupling, it could result in a significant error in many cases. One of the earliest successful applications of SEA was reported by Crocker and Price (1969) for determining transmission loss of partitions.

There have been many papers in the recent past presenting a critical review of SEA. Some of the important ones are by Fahy (1994), Woodhouse (1981a), and Hodges and Woodhouse (1986). A review of theoretical background of SEA has been presented in Woodhouse (1981b). These reviews help refine the existing SEA techniques and they address some of the important issues connected with their application.

10.5 Thermal Analogy of SEA

Since the concept of the flow of energy is fundamental to the study of SEA, the thermal analogy can be advantageously used to explain the basic concepts of SEA, as two connected bodies given some heat input in this model will be very similar to an SEA system. An SEA model of a vibro-acoustic system also must have at least two systems in order to define some of its parameters. Many of the practical systems will have two subsystems that are interconnected and at least one of them is excited by the input power. Hence, there is a very good analogy between statistical energy analysis and flow of heat between two bodies. The flow of heat is easy for conceptual visualization of power flow, and therefore the concepts involved in SEA can be better understood with such an analogy. Hence, before going to further details of mathematical formulation of SEA, the thermal analogy of SEA is discussed in this section (see Woodhouse 1981).

Figure 10.2 shows two bodies connected to each other. One of the bodies, body 1, is heated due to which its temperature increases. Some of the heat is lost to the surroundings, as heat and the rest is transferred to body 2; some heat might come back to system 1 from 2. When steady-state conditions are reached, there could be a net flow of heat from system 1 to 2. The temperatures of both the bodies are governed by the amount of heat radiated to the surroundings by each body and the amount of heat transferred between them.

The amount of heat transferred between them is quantified by the strength of coupling. That means if there is a greater flow of heat between them, they are strongly coupled; if the flow of heat between them is less, then they are weakly coupled. The amount of heat transferred depends on the material connecting the two bodies, which determines the amount of coupling. A good conductor of heat attached between the bodies provides strong coupling and an insulator provides weak coupling. Similarly, other parameters, such as mass, external surfaces, and surroundings, of each of the bodies influence the amount of heat radiated to the outside.

The aforementioned thermal model can be easily related to structural vibration. Let us consider a complex structure that consists of two coupled substructures, for example, two plates connected by welding. If one of the plates is applied with force excitation, some of the vibration energy is damped out in the first system and the rest flows in to the second plate depending upon the properties of interconnecting element. The exchange of energy between the two plates is due to the modes of vibration in the given frequency range in each system.

Detailed mathematics will follow in later sections, but at present we will relate the thermal model with a vibrating structure as follows. The temperature of the thermal system can be compared to the modal energy of a vibrating system, which is the total energy divided by the number of modes. The amount of energy that is present in the system after damping out and transferring to the neighboring substructure is used for increasing the

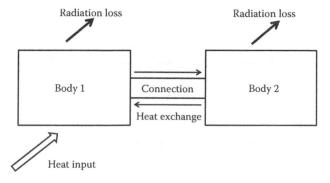

FIGURE 10.2
Heat flow between two bodies.

energy of the modes known as the modal energy, which will actually indicate the vibration level of the system.

The radiation loss of the thermal model can be compared to damping of the vibratory system. Energy is lost from a vibratory system due to damping and the total loss of energy can be computed if the frequency of vibration and damping factor of the material are known.

The strength of coupling of the thermal model can be compared to the mechanical coupling. Some materials allow vibration to be easily transmitted and some materials almost prevent the transmission of vibration. The amount by which they allow or disallow vibration is also dependent on the frequency of vibration. The strength of the mechanical coupling is quantified by the coupling loss factor, which is one of the most important parameters used in SEA.

The thermal capacity of a thermal model can be compared with the modal density of a vibratory model. Since each mode of vibration has definite energy stored in it, modal density, which is a measure of the number of modes in a frequency band, is related to the amount of energy stored in each system.

The aforementioned parameters of the vibratory model, which are analogous to the thermal model, are summarized in Table 10.1.

Depending on the type of material used to connect the two bodies and the extent of radiation loss from each of the bodies, temperatures of each of the bodies show a definite pattern as follows:

1. The temperature of the bodies when there is high radiation loss in addition to strong coupling is shown in Figure 10.3a. Heat from body 1 is lost as radiated heat and the rest flows to body 2 where it is further radiated into the surroundings. Due

TABLE 10.1

Analogy between Thermal Diffusion and SEA

Thermal Model	Vibratory Structure
Temperature	Modal energy
Radiative loss	Damping
Conductivity	Mechanical coupling
Thermal capacity	Modal density

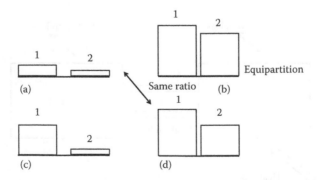

FIGURE 10.3
Temperature of two bodies shown in Figure 10.2. (a) High damping, strong coupling. (b) Low damping, strong coupling. (c) High damping, weak coupling. (d) Low damping, weak coupling.

to high radiation and strong coupling, the temperature of both bodies remains low, as shown in the figure.

2. The temperature of the bodies when there is low radiation from both of them but they are strongly coupled is as shown in Figure 10.3b. There is no significant loss of heat from body 1 due to radiation and most of the heat flows into body 2. Similarly, there is no significant loss of heat from body 2. Due to very low radiation and strong coupling, there is enough heat conserved within each of the bodies, which raises their temperature as shown in the figure. This is an example of equipartition of energy.

3. The temperature of the bodies when there is high radiation loss from both of them but are weakly coupled is shown in Figure 10.3c. Heat from body 1 is lost as radiated heat. Due to weak coupling very little heat flows into body 2 where it is further radiated in to the surroundings. Body 1 still has enough heat to increase its temperature. Due to these reasons, the temperature of body 1 is far greater than the temperature of body 2 as shown in the figure.

4. The temperature of the bodies when there is low radiation and low coupling is as shown in Figure 10.3d. There is no significant loss of heat due to radiation from both bodies. In addition, very little heat flows from body 1 to body 2. Due to very low radiation loss, temperatures of both the bodies increase as shown in the figure. However, the temperature of body 2 is definitely less than the case in which there was strong coupling between the bodies.

Now we can use Figure 10.3 to discuss how the mean square vibration level of two connected substructures depends on the strength of damping and coupling loss factors. The source of heat in Figure 10.2 becomes a source of vibration from a machine attached to one of the systems. The four aforementioned cases form a prototype for a typical vibration control problem. It is imperative that for any vibration control problem, the case presented in Figure 10.3a is the ideal situation. To reduce vibration, vibration has to be damped out in both substructures and both substructures must be completely isolated. Obviously, vibration of the source must be reduced in the first place, which is not always easy.

10.6 System Modeling

SEA basically is a system-based approach in which vibration and noise problems are formulated as an interaction between several subsystems connected to one another. It consists of averaging the response of each system, so the equations that are used for predicting the response depend mainly on the gross properties of the system: geometry, material properties, and the manner in which they are connected to other subsystems (e.g., welding and riveting). One of the SEA parameters that can be easily obtained is the modal density that has been discussed in earlier chapters. It represents a number of resonance frequencies, of a vibratory or acoustic system, in a given frequency band. The other important parameter is the coupling loss factor, which quantifies the extent of connection between subsystems that can be obtained from computation in some cases but has to be many times measured. Coupling loss factors of commonly used connections will be discussed in later sections. In this section, however, we will discuss how subsystems can be formed in SEA modeling and some examples of SEA subsystem modeling.

10.6.1 System Modeling Approach

In the system modeling approach of SEA, one can start with a very simple model and then progressively improve the model until desirable results are obtained. One can start with a very simple model with a minimum number of subsystems, and then the number of subsystems can be progressively increased based on the results obtained from the previous models. As explained in the preceding paragraph, SEA modeling requires certain input parameters, like damping loss factors and coupling loss factors. Hence, if one begins with a very complex model at the design stage, all of these parameters may not be readily available. If it is a new machine, the parameters relating to the existing machine may have to be used. Therefore, the approach of progressively improving the complexity of the model, depending on the extent of information required, will be a very practical approach.

10.6.2 Subsystems of SEA

The modal groups form the basis for forming subsystems of an SEA. In many cases, each physical group may have a single modal group. For example, in the case of an acoustic subsystem there is only one modal group of sound waves. Then it is very easy to visualize a subsystem. As a general case, many vibratory subsystems have more than one modal group, such as longitudinal, bending, and torsion. Each of these groups can exchange energy between them and with subgroups of other physical subsystems.

The example of a T-junction formed by three plates is shown in Figure 10.4 to illustrate the concept of multiple wave groups within a physical system. Each of these plates is assumed to support both longitudinal and bending modes. Thus, there are a total of six subsystems interacting with one another as shown in Figure 10.5. In practice, however, it may not be necessary to consider all of the subsystems of various mode groups. One can consider a few of them to begin with, like bending waves, and then progressively account for all of the other wave types. Because, bending waves are generally more dominant than others. Therefore, in many practical applications it may not be necessary to consider all the wave groups, thus making it easier for a beginner to identify subsystems that can be directly related to the physical model.

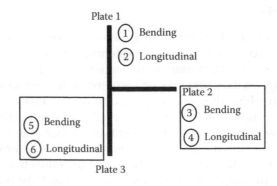

FIGURE 10.4
Three-plate junction.

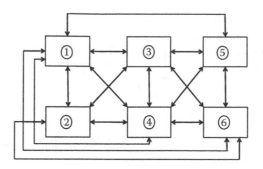

FIGURE 10.5
SEA model of the three-plate junction.

The following elements are generally used in SEA modeling. Derivation of modal density for some of the elements was discussed earlier.

1. Unstiffened panel
   a. Rectangular
   b. Circular
2. Torsionally vibrating beam
3. Transversely vibrating beam
4. Transversely vibrating hoop
5. Unstiffened cylinder
6. Stiffened cylinder
7. Flat stiffened panel
8. Flat honeycomb panel
9. Cylindrical honeycomb
10. Double curved honeycomb
11. 1-D acoustic cavity
12. 2-D acoustic cavity
13. 3-D acoustic cavity

If a combination of any of these elements is used, for example, cylinder stiffened with beam, the resulting modal density will be a sum of the modal densities of the cylinder and total length of the beam used for stiffening.

10.6.3 Vibration Excitation of Printed Circuit Boards due to Rocket Noise

We can now return to the original problem of failed PCBs shown in Figure 10.1 to demonstrate the model-based approach of SEA. Vibration excitation of the onboard electronics, which was the main reason for SEA formulation, can now be formally presented as an SEA model as shown in Figure 10.6. The acoustic excitation occurs close to the volume around the rocket and it can be defined in terms of an equivalent volume of reverberant space that causes the same excitation, and this definition helps in acoustic testing parts of launch

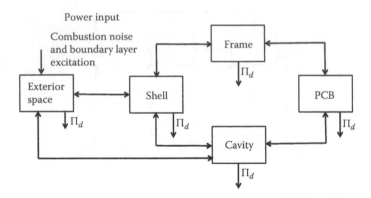

FIGURE 10.6
Vibro-acoustic model of PCB excitation due to rocket noise.

vehicle structures on the ground for the purposes of validation. As shown in the figure, acoustic excitation excites the shell into vibration that results in vibration of the frame and sound pressure levels in the cavity; the density of the shell structure also contributes to the sound pressure levels of the cavity, which is known as nonresonant transmission. The vibration of the frame and sound pressure levels in the cavity excite the PCBs into vibration. An appropriate design of this frame using SEA can minimize vibration in the PCBs. So this represents a typical vibro-acoustic problem. After going through the rest of the material in this chapter, one should be able to solve similar problems.

10.6.4 Vibro-Acoustic Model of an Exhaust Fan

An exhaust fan mounted on a large surface area wall is shown in Figure 10.7. The vibro-acoustic model of sound production by this exhaust fan in the room is shown in Figure 10.8. The fan is mounted on a frame that is mounted on the wall and all these elements radiate noise into the acoustic space of the room. Hence, all of these parameters need to be considered in the root cause analysis for noise reduction, apart from the fan noise characteristics. The fan is the source of noise and vibration, the wall and frame are vibratory systems, and the room air is an acoustic system. The fan directly radiates noise into the room. In addition, the fan also transmits vibration to the frame that in turn excites the wall into vibration. Since the wall is of a large surface area, it acts like a sounding board

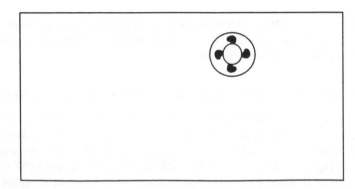

FIGURE 10.7
Exhaust fan mounted on wall.

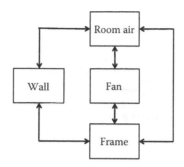

FIGURE 10.8
Vibro-acoustic model of the exhaust fan shown in Figure 10.7.

and radiates noise into the room. The frame also radiates some sound into the room due to vibration. Which of these phenomena contribute significantly to the total radiated noise depends on the construction details of these elements. For example, if the wall surface area is less or the wall does not get excited into vibration due to vibration isolation, its contribution will be significantly less. The wall can also have damping treatment or the room can be treated with acoustic absorption material to increase the room constant that can also reduce the sound pressure levels.

10.6.5 Vibro-Acoustic Model of Aircraft Cabin Noise

The SEA model of aircraft cabin noise is shown in Figure 10.9. Although the aircraft flies in free space, the reverberant effects of sound produced by it due to engine sound and boundary layer effects are significantly present in a typical volume around the aircraft. Therefore, it is assumed that an equivalent volume of external space represents an acoustic system that excites the fuselage into vibration, which in turns radiates sound into the cabin. In addition, sound is also transmitted into the cavity due to nonresonant transmission, wherein the sound reduction is mainly due to the structural mass and not due to its vibration. The transmission loss discussed in Chapter 6 is nonresonant transmission that is acceptable in many ground-based systems that have large structural mass partitions. However, for flexible structures that have significant vibration due to sound transmission, resonant transmission is more dominant that can be found only from measurement or by using SEA modeling. By appropriately designing the fuselage structure by increasing its transmission loss (resonant and nonresonant) and providing sufficient acoustic absorption within the cabin, it is possible to reduce sound pressure levels to an acceptable level.

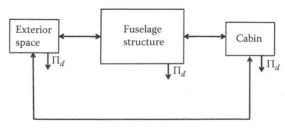

FIGURE 10.9
Vibro-acoustic model of aircraft.

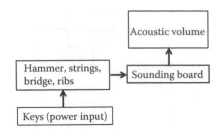

FIGURE 10.10

Vibro-acoustic model of a piano sound production.

10.6.6 Vibro-Acoustics of Piano Sound Production

Musical acoustics has also influenced the evolution of SEA, especially regarding the vibro-acoustic interaction. The piano is one such instrument that has all the elements of vibro-acoustics. Therefore, an example of a piano sound production can be used to illustrate this concept. A simplified vibro-acoustic model of sound production in a piano is shown in Figure 10.10. The inputs to the piano to generate various frequencies of sound are given through the hammer that strikes the strings that are attached to the sounding board. The sounding board that receives vibration from the string efficiently couples with the surrounding air to produce sound. It is the sounding board that plays a very important role in sound production and thus influences the quality of sound.

Although the principles of vibro-acoustics are similar in a music instrument and a machine, the objectives are different. We would like to produce sound efficiently in a music instrument as per the rules of music, whereas we would like to produce the least amount of sound in a machine. The sounding boards of a music instrument that produce sound due to vibration are clearly defined and designed for producing the desired sound, whereas the machines have multiple sounding boards that need to be clearly identified and designed properly to obtain a quiet machine. SEA does help in this process of identifying sounding boards of machines and the process of designing quiet machines is finally over when we discover the last sounding board, because, at each step, another dominant sounding board becomes a dominant sound source.

Now that we have discussed subsystem modeling in SEA, let us derive the basic SEA equations that will help in objective quantification of the system parameters that result in vibration and sound.

10.7 Power Flow Equations

10.7.1 Power Dissipated in a System

The power dissipated in a system can be derived based on the power dissipation in an SDOF; the power input at resonance of an SDOF system is given by

$$\Pi_d = F_{rms} v_{rms} \cos \phi \tag{10.1}$$

Since $F_{rms}/v_{rms} = R_m$ and $\cos\phi = 1$ at resonance, Equation 10.1 becomes

$$\Pi_d = R_m v_{rms}^2 \tag{10.2}$$

The equation of an SDOF is given by

$$m\ddot{y} + R_m\dot{y} + ky = f(t) \tag{10.3}$$

Equation 10.3 can be written in terms of damping factor and undamped natural frequency as

$$\ddot{y} + 2\zeta\omega_n\dot{y} + \omega_n^2 y = \frac{f(t)}{m} \tag{10.4}$$

where

$$R_m = 2\zeta\omega_n m \tag{10.5}$$

From Equations 10.2 and 10.5

$$\Pi_d = 2\zeta\omega_n m v_{rms}^2 \tag{10.6}$$

The time-averaged energy of an SDOF is given by

$$\langle E \rangle_t = m v_{rms}^2 \tag{10.7}$$

The loss factor can be defined as

$$\eta = 2\zeta \tag{10.8}$$

From Equations 10.6 through 10.8, the power dissipated at the resonant frequency of an SDOF is given by

$$\Pi_d = \omega_n \eta \langle E \rangle_t \tag{10.9}$$

Equation 10.9 can be generalized to a dynamic system whose power dissipation is of interest in a band of frequencies of center frequency ω_c. Because many practical systems have a number of resonant frequencies, the response is generally averaged over a frequency band of center frequency ω_c. Power dissipation by such a dynamic system is based on time-, space-, and frequency-averaged energy and is given by

$$\Pi_d = \omega_c \eta \langle E \rangle_{t,\omega,x} \tag{10.10}$$

The loss factor η of Equation 10.10 refers to the loss factor of the dynamic system within the band of frequencies of center frequency ω_c. $\langle E \rangle$ refers to the time-, frequency-, and space-averaged energy of the dynamic system in the band of center frequency ω_c.

10.7.2 Power Flow between Two Systems

Figure 10.11 shows the interaction between two dynamic systems. Each system is excited by a power input and dissipates a certain amount of power. Power flows from system 1 to 2 and vice versa due to dynamic interaction between these systems. The difference of these two power flows is the net power flow between the systems, which depends on the following SEA parameters: modal density of each system, damping loss factor of each system, and the coupling loss factor between the systems. The ratio of the energy and modal density is the modal energy. And the net power flow can be proved to be directly proportional to the difference of the modal energy of the two systems. The mathematical relationship between power flow and the aforementioned variables will be derived as follows.

Consider two dynamic systems, as shown in Figure 10.11 that are connected to each other as shown. Let us presume that we are interested in the response of each of these systems in a band of frequencies whose center frequency is ω_c. The following is given:

$\Pi_{1,in}$ is the power input to system 1.

$\Pi_{2,in}$ is the power input to system 2.

$\Pi_{d,1}$ is the power dissipated by system 1.

$\Pi_{d,2}$ is the power dissipated by system 2.

Π'_{12} is the power flowing from system 1 and 2.

Π'_{21} is the power flowing from system 2 to 1.

η_1 is the loss factor of system 1.

η_2 is the loss factor of system 2.

$n_1(\omega_c)$ is the model density of system1 at ω_c.

$n_2(\omega_c)$ is the model density of system 2 at ω_c.

$\langle E_1 \rangle_{t,\omega,x}$ is the space, time, and frequency average of system 1 in the frequency band ω_c.

$\langle E_2 \rangle_{t,\omega,x}$ is the space, time, and frequency average of system 2 in the frequency band ω_c.

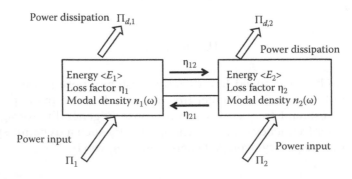

FIGURE 10.11
Two interconnected dynamic systems.

Using Equation 10.10, the power dissipated by both the systems are respectively given by

$$\Pi_{d,1} = \omega_c \eta_1 \langle E_1 \rangle_{t,\omega,x} \tag{10.11}$$

$$\Pi_{d,2} = \omega_c \eta_2 \langle E_2 \rangle_{t,\omega,x} \tag{10.12}$$

The power flowing between the systems 1 and 2 is similar to Equations 10.11 and 10.12. Therefore, we can assume that power flows from system 1 to 2 and also from system 2 to 1, and the net flow between the systems will be the difference between them.

The power that flows out of system 1 to 2 is assumed to be proportional to the product of center frequency and average energy of system 1 and is given by

$$\Pi'_{12} \alpha \omega_c \langle E_1 \rangle_{t,\omega,x} \tag{10.13}$$

Similarly, the power that flows out of system 2 to 1 is proportional to the product of center frequency and average energy of system 2 and is given by

$$\Pi'_{12} \alpha \omega_c \langle E_2 \rangle_{t,\omega,x} \tag{10.14}$$

The constants of proportionality in Equations 10.13 and 10.14 are given the name coupling loss factor. *Coupling* has been given since it concerns power flow between two systems and the *loss factor* is added to it because of its similarity to the loss factor used in the power dissipation Equations 10.11 and 10.12. In Equation 10.13, η_{12} is the coupling loss factor from system 1 to 2 that defines the proportionality constant, and in Equation 10.14, η_{21} is the coupling loss from system 2 to 1 that defines the proportionality constant.

Assuming that $\Pi'_{12} > \Pi'_{21}$, the net power flow between systems 1 and 2 is given by

$$\langle \Pi_{12} \rangle = \eta_{12} \omega_c \langle E_1 \rangle_{t,\omega,x} - \eta_{21} \omega_c \langle E_2 \rangle_{t,\omega,x} \tag{10.15}$$

Based on Lyon and Maidanik (1962), power flow between two dynamic systems is also proportional to the difference in modal energies of a dynamic system and is given by

$$\langle \Pi_{12} \rangle = \gamma \left[\frac{\langle E_1 \rangle_{t,\omega,x}}{n_1(\omega_c)} - \frac{\langle E_2 \rangle_{t,\omega,x}}{n_2(\omega_c)} \right] \tag{10.16}$$

where γ is the constant of proportionality, and if modal energy is defined as the ratio of total energy to the modal density, the net power transfer is related to the difference of modal energies of the two systems. Equation 10.16 gives a physical sense of power flow as being proportional to the difference between the modal energies. Therefore, it is the difference of the modal energies that can drive power flow between dynamic systems.

Comparing Equations 10.15 and 10.16, the following equations can be obtained

$$\frac{\gamma}{n_1(\omega_c)} = \eta_{12} \omega_c \qquad \frac{\gamma}{n_2(\omega_c)} = \eta_{21} \omega_c \tag{10.17}$$

From Equation 10.17, the following equation can be obtained:

$$n_1(\omega_c)\eta_{12} = n_2(\omega_c)\eta_{21} \tag{10.18}$$

Equation 10.18 that relates the product of modal density and coupling factors of the two systems is the well-known reciprocity relationship for the connected SEA subsystems. This equation is often useful in verifying modal densities and coupling loss factors obtained from experiments.

From Equations 10.16 and 10.17, the net power flow between two systems is given by

$$\langle \Pi_{12} \rangle = \omega_c n_1(\omega_c)\eta_{12} \left[\frac{\langle E_1 \rangle}{n_1(\omega_c)} - \frac{\langle E_2 \rangle}{n_2(\omega_c)} \right] \tag{10.19}$$

In addition to Equation 10.19, many times it is easier to obtain useful relationships by writing power balance equations between connected subsystems of an SEA model. The power balance for each subsystem works out as follows. For each subsystem, input power and power transferred from other system can be considered positive, and the power dissipated and power transferred to other subsystems can be considered negative. Their sum must be equal to zero or the absolute magnitudes of both of these groups must be equal for power balance. Using these power balance equations, the modal energy and subsequently energy from each system can be computed. Then the energy of each system can be used to obtain the actual vibration or acoustic response.

The damping loss factors will have to be measured. The coupling loss factors can be computed sometimes and some of them are presented in Section 10.10; otherwise, they will have to be measured by either considering two systems at a time or the entire system of interconnected subsystems. They can also be obtained from similar systems if the prototype is yet to be built. Power input will have to measured based on the force input and response. Modal density can be calculated for simple systems and it can be calculated for built-up systems also by considering them as a combination of several built-up systems. The finite element method (FEM) can also be used to compute the modal density of built-up systems.

The following sections discuss the ratio of energies of two systems with power input to only one of them. However, the power input information is avoided in this formulation.

10.7.2.1 Only the First System Is Excited

Some useful simplification results if only one of the systems of a two-subsystem model is excited, which is common in many practical systems. It is as shown in Figure 10.12, which is similar to Figure 10.11 in all aspects except that only the first system is excited.

The power balance for the first system shown in Figure 10.12 is given by

$$\Pi_{1,in} = \omega_c \langle E_1 \rangle \eta_1 + \omega_c \langle E_1 \rangle \eta_{12} - \omega_c \langle E_2 \rangle \eta_{21} \tag{10.20}$$

The power balance for the second system is given by

$$0 = \omega_c \langle E_2 \rangle \eta_2 + \omega_c \langle E_2 \rangle \eta_{21} - \omega_c \langle E_1 \rangle \eta_{12} \tag{10.21}$$

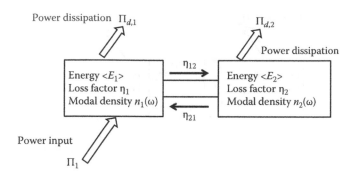

FIGURE 10.12
Only the first system is excited.

From Equation 10.21

$$\frac{\langle E_2 \rangle}{\langle E_1 \rangle} = \frac{\eta_{12}}{\eta_2 + \eta_{21}}$$

(10.22)

Since input power is always a difficult parameter to measure, Equation 10.21 was used to derive Equation 10.22.

Using the reciprocity of Equation 10.18, Equation 10.22 becomes

$$\frac{\langle E_2 \rangle}{\langle E_1 \rangle} = \frac{n_2(\omega_c)}{n_1(\omega_c)}\left[\frac{\eta_{21}}{\eta_2 + \eta_{21}}\right]$$

(10.23)

Equation 10.23 is more convenient since only one of the coupling loss factors was used, and modal densities of dynamic systems are anyway easily available.

10.7.2.2 Only the Second System Is Excited

The power balance for the first system shown in Figure 10.13 is given by

$$0 = \omega_c \langle E_1 \rangle \eta_1 + \omega_c \langle E_1 \rangle \eta_{12} - \omega_c \langle E_2 \rangle \eta_{21}$$

(10.24)

The power balance for the second system is given by

$$\Pi_{2,in} = \omega_c \langle E_2 \rangle \eta_2 + \omega_c \langle E_2 \rangle \eta_{21} - \omega_c \langle E_1 \rangle \eta_{12}$$

(10.25)

Since Equation 10.24 doesn't contain a power input, the ratio of energies is given by

$$\frac{\langle E_1 \rangle}{\langle E_2 \rangle} = \frac{\eta_{21}}{\eta_1 + \eta_{12}}$$

(10.26)

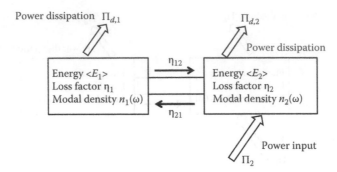

FIGURE 10.13
Only the second system is excited.

Using the reciprocity Equation 10.18, Equation 10.26 becomes

$$\frac{\langle E_1 \rangle}{\langle E_2 \rangle} = \frac{n_1(\omega_c)}{n_2(\omega_c)} \left[\frac{\eta_{12}}{\eta_1 + \eta_{12}} \right] \tag{10.27}$$

10.8 Matrix Approach to SEA Modeling

When there are a large number of interacting dynamic systems, it would be easier to express the SEA model in the form of a matrix. This matrix can relate power input to various subsystems and the resulting energy in each of them.

Figure 10.14 shows two interacting dynamic systems, both having power input to each of them. The power balance equation for system 1 is given by

$$\Pi_{1,in} = \Pi_{d_1} + \Pi'_{12} - \Pi'_{21} \tag{10.28}$$

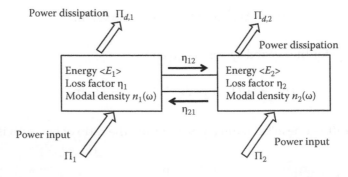

FIGURE 10.14
Two-system model excited by both systems.

Equation 10.28 can be expressed as

$$\Pi_{1,in} = \omega_c \langle E_1 \rangle \eta_1 + \omega_c \langle E_1 \rangle \eta_{12} - \omega_c \langle E_2 \rangle \eta_{21} \qquad (10.29)$$

Similarly, writing the power balance equation for system 2

$$\Pi_{2,in} = \Pi_{d_2} + \Pi'_{21} - \Pi'_{12} \qquad (10.30)$$

Equation 10.30 can be expressed as

$$\Pi_{2,in} = \omega_c \langle E_2 \rangle \eta_2 + \omega_c \langle E_2 \rangle \eta_{21} - \omega_c \langle E_1 \rangle \eta_{12} \qquad (10.31)$$

Equations 10.29 and 10.31 can be expressed in terms of a matrix as

$$\begin{bmatrix} \eta_1 + \eta_{12} & -\eta_{21} \\ -\eta_{12} & \eta_2 + \eta_{21} \end{bmatrix} \begin{Bmatrix} \langle E_1 \rangle \\ \langle E_2 \rangle \end{Bmatrix} = \begin{Bmatrix} \dfrac{\Pi_{1,in}}{\omega_c} \\ \dfrac{\Pi_{2,in}}{\omega_c} \end{Bmatrix} \qquad (10.32)$$

By multiplying and dividing $<E_1>$ and $<E_2>$ by the respective modal densities corresponding to the center frequency, Equation 10.32 becomes

$$\begin{bmatrix} n_1(\omega_c)(\eta_1 + \eta_{12}) & -n_2(\omega_c)\eta_{21} \\ -n_1(\omega_c)\eta_{12} & n_2(\omega_c)(\eta_2 + \eta_{21}) \end{bmatrix} \begin{Bmatrix} \dfrac{\langle E_1 \rangle}{n_1(\omega_c)} \\ \dfrac{\langle E_2 \rangle}{n_2(\omega_c)} \end{Bmatrix} = \begin{Bmatrix} \dfrac{\Pi_{1,in}}{\omega_c} \\ \dfrac{\Pi_{2,in}}{\omega_c} \end{Bmatrix} \qquad (10.33)$$

By using the reciprocity relationship of Equation 10.18, the off-diagonal elements of the matrix of Equation 10.33 can be written as

$$\begin{bmatrix} n_1(\omega_c)(\eta_1 + \eta_{12}) & -n_1(\omega_c)\eta_{12} \\ -n_2(\omega_c)\eta_{21} & n_2(\omega_c)(\eta_2 + \eta_{21}) \end{bmatrix} \begin{Bmatrix} \dfrac{\langle E_1 \rangle}{n_1(\omega_c)} \\ \dfrac{\langle E_2 \rangle}{n_2(\omega_c)} \end{Bmatrix} = \begin{Bmatrix} \dfrac{\Pi_{1,in}}{\omega_c} \\ \dfrac{\Pi_{2,in}}{\omega_c} \end{Bmatrix} \qquad (10.34)$$

Modal energies of subsystems become the unknown parameters in the linear system of Equation 10.34. If modal density, coupling loss factors, and power inputs are known, the modal energy of each subsystem can be obtained from which the corresponding energy of each subsystem can be obtained; the energy can then be used to determine the vibration or acoustic response in terms of acceleration/velocity or sound pressure level.

Equation 10.34 can be extended to multiple subsystems (Figure 10.15) as follows:

$$
\begin{bmatrix}
n_1(\omega_c)\left(\eta_1 + \sum\limits_{i\neq 1}^{N} \eta_{1i}\right) & -n_1(\omega_c)\eta_{12} & \cdot & -n_1(\omega_c)\eta_{1N} \\[2em]
-n_2(\omega_c)\eta_{21} & n_2(\omega_c)\left(\eta_2 + \sum\limits_{i\neq 1}^{N} \eta_{2i}\right) & \cdot & -n_2(\omega_c)\eta_{2N} \\[2em]
\cdot & \cdot & \cdot & \cdot \\[1em]
-n_N(\omega_c)\eta_{N1} & \cdot & \cdot & n_N(\omega_c)\left(\eta_N + \sum\limits_{i\neq 1}^{N} \eta_{Ni}\right)
\end{bmatrix}
$$

$$
\begin{Bmatrix}
\dfrac{\langle E_1 \rangle}{n_1(\omega_c)} \\[1.5em]
\dfrac{\langle E_2 \rangle}{n_2(\omega_c)} \\[1.5em]
\cdot \\[0.5em]
\dfrac{\langle E_N \rangle}{n_N(\omega_c)}
\end{Bmatrix}
=
\begin{Bmatrix}
\dfrac{\Pi_{1,in}}{\omega_c} \\[1.5em]
\dfrac{\Pi_{2,in}}{\omega_c} \\[1.5em]
\cdot \\[0.5em]
\dfrac{\Pi_{N,in}}{\omega_c}
\end{Bmatrix}
\tag{10.35}
$$

If the input power is known, the modal energies of the subsystems can be obtained from Equation 10.35.

A typical multiple subsystem model is shown in Figure 10.15. It is not necessary that all the subsystems are connected and if the loss factors of various subsystems and the coupling loss factors between them are unknown, multiple experiments can be conducted by exciting each subsystem at a time from which all the system parameters can be determined.

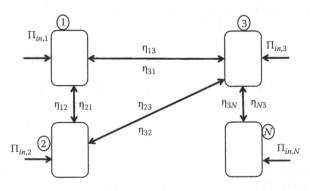

FIGURE 10.15
Multiple-subsystem model.

10.9 SEA Equations from Continuous Systems

Basic SEA equations can also be obtained from continuous systems. And the advantage of obtaining the equations from continuous systems is that they yield additional information that can be advantageously used for experimental measurement of modal density, coupling loss factors, and power input.

10.9.1 Averaging Normal Modes

The response of a continuous system, using the normal mode solution, is based on the nature of the force excitation, value of the natural frequencies, the mode shape functions evaluated at points of force excitation, and at points at where the response is required and damping. Due to difficulty in the exact determination of natural frequency and mode shapes, and since they are very sensitive to the tolerances of the system at high frequencies, the normal mode solution is unlikely to yield meaningful results in the high frequency region. However, suitable modifications to the normal mode solution can be used to derive the useful parameters of SEA.

Although extensive literature is available for applying the normal modes method for solving vibration problems at lower frequencies, certain practical difficulties arise in analyzing higher modes of vibration for broadband excitation. The higher modes of vibration are very sensitive to boundary conditions and the construction detail of the structure. Boundary conditions are fundamental in determining the normal modes and natural frequencies; exact boundary conditions are not always known in practice. It has been found from experiments that even with a simple beam of known boundary conditions, the discrepancy between the predicted and measured natural frequencies can be seen as early as the second mode of vibration. This discrepancy widens further at higher modes of vibration. In order to determine the response due to a high frequency broadband excitation, a large number of modes will have to be accounted for. With the discrepancy between the predicted and measured values, which widens at the higher modes, it may be difficult to predict the response of structures based on their theoretical models. Therefore, the theoretical models based on the normal modes method will be of little help in predicting vibration levels of structures subjected to broadband high frequency vibration. Designing of machinery and reduction of vibration and noise in existing machinery becomes a cut-and-try procedure, which is rather undesirable and also uneconomical.

In view of the preceding discussion, it is desirable to evolve a technique by which the normal modes and natural frequencies are averaged in some sense. This would result in equations that are used for predicting the response, depending mainly on the geometry, structural properties, and number of resonances in a given frequency band known as modal density. We are still left to seek information about modal density. Fortunately, modal density can be calculated for a number of simple structures like beams, plates, and shells, and it is independent of boundary conditions. Since there is no dependence on boundary conditions, it simplifies the modeling procedure and can predict vibration levels, which are close to measurement. We have already discussed that SEA is the inevitable option in the high frequency range.

The objective of this section, however, is to derive equations for the following parameters using a beam model: average velocity, drive point mobility, average conductance, and average input power; these parameters are similar for plates and shells. Averaging is carried out in the time, spatial, and frequency domains. Similar equations for the aforementioned parameters result for a plate, shell, and so on. As reported by Langley and Bercin (1994), the parameters can be derived in a general sense that is applicable to beams, plates, or shells. In order to make the derivation easily understandable, a beam has been chosen in this section; the modal solution for beam vibration was presented in Chapter 5. The parameters can be used for predicting the space, time, and frequency average response of structures modeled as a single beam, plate, or shell.

10.9.2 Average Velocity

The product of space- and time-averaged velocity and the total mass of the beam represents the total energy of the beam, which is equal to the sum of the energy of the individual modes.

The space- and time-averaged velocity of a beam is given by

$$\left\langle \overline{v^2} \right\rangle = \frac{1}{M} \sum_{n=1}^{\infty} m_n \left\langle \dot{q}_n^2 \right\rangle, \tag{10.36}$$

where M is the total mass of the beam, m_n the generalized mass, and $\dot{q}_n$ the generalized velocity for the nth mode. Substituting for the amplitude of the generalized velocity for harmonic excitation, Equation 10.36 becomes

$$\left\langle \overline{v^2} \right\rangle = \frac{1}{2M} \sum_{n=1}^{\infty} \frac{\omega^2 \phi_n^2(x_0) F^2}{m_n \left(\left(\omega_n^2 - \omega^2 \right)^2 + \eta^2 \omega_n^4 \right)}, \tag{10.37}$$

where $\phi_n(x_0)$ is the mode shape at the location of a harmonic force excitation of peak amplitude F, ω_n the natural frequency corresponding to the nth mode, ω the excitation frequency, and η the damping factor. Averaging over all possible locations of the excitation force can be termed equivalent to computing the space, time average velocity, averaged over of an ensemble of systems, in which the location of the excitation force is a random variable (Cremer, Heckl, and Ungar 1987). Since mode shapes are relative responses, they can be normalized with respect to the generalized mass to yield: $\phi_n^2(x_o)/m_n = 1/M$. Based on this assumption, we obtain the following equation:

$$\left\langle \overline{v^2} \right\rangle_{ens} = \frac{1}{2M^2} \sum_{n=1}^{\infty} \frac{\omega^2 F^2}{\left(\left(\omega_n^2 - \omega^2 \right)^2 + \eta^2 \omega_n^4 \right)}, \tag{10.38}$$

If the preceding space-, time-, and ensemble-averaged velocity is averaged over a frequency band $\Delta\omega$, between the resonant mode numbers N_1 and N_2, we get the following equation:

$$\left\langle \overline{v^2} \right\rangle_{ens,\Delta\omega} = \frac{|F|^2}{2M^2} \frac{\pi}{2\eta} \frac{1}{\Delta\omega} \sum_{n=N_1}^{N_2} \frac{1}{\omega_n}, \tag{10.39}$$

If ω is the center frequency of the frequency band $\Delta\omega$, and if ΔN represents the number of modes in this frequency band, Equation 10.39 becomes

$$\langle \overline{v^2} \rangle_{ens,\ \omega} = \frac{|F|^2}{2M^2} \frac{\pi}{2\eta\omega} \frac{\Delta N}{\Delta\omega} \approx \frac{|F|^2}{2M^2} \frac{\pi}{2\eta\omega} n(\omega)$$

(10.40)

where $n(\omega)$ modal density (modes/radian/s) in the frequency band is given by

$$n(\omega) = \frac{\Delta N}{\Delta\omega}$$

(10.41)

and $n(f)$ modal density (modes/Hz) in the frequency band is given by

$$n(f) = \frac{\Delta N}{\Delta f}$$

(10.42)

See section 5.6 on modal density.

10.9.3 Drive Point Mobility

Mobility (also known as admittance, discussed in Chapter 1) is defined as the ratio of velocity to force. The single mode drive point mobility for the nth mode, which is the ratio of the amplitude of generalized velocity to the amplitude of generalized force, due to a harmonic force of amplitude F acting at a location x_0, is given by

$$Y_n(\omega) = \frac{\dot{q}_n}{Q_n} = \frac{\omega\omega_n^2\eta}{\left[\left(\omega_n^2 - \omega^2\right)^2 + \left(\eta\omega_n^2\right)^2\right]} + j\frac{\omega\left(\omega_n^2 - \omega^2\right)}{\left[\left(\omega_n^2 - \omega^2\right)^2 + \left(\eta\omega_n^2\right)^2\right]}$$

(10.43)

By averaging Equation 10.43 over all possible locations of the excitation force, that is, by averaging an ensemble of beams in which the location of the force could be considered random, we obtain the following equation:

$$\overline{Y}(x_0,\omega) = \frac{1}{M}\sum_{n=1}^{\infty} Y_n(\omega) = G(\omega) + jS(\omega)$$

(10.44)

The real part of Equation 10.44, $G(\omega)$ represents conductance and the imaginary part $S(\omega)$ susceptance. Averaging conductance over a frequency interval $\Delta\omega$, the ensemble-averaged, frequency-averaged conductance is given by

$$\langle G \rangle_{ens,\Delta\omega} = \frac{\pi}{2M}\frac{\Delta N}{\Delta\omega} = \frac{\pi}{2M}n(\omega) = \frac{1}{4M\delta f}$$

(10.45)

where ΔN represents the number of resonance modes in the frequency band $\Delta\omega$, $n(\omega)$ the modal density in modes per radian, and δf the average separation frequency of the resonances in hertz. It is important to note that the average conductance depends mainly on the mass and number of resonance frequencies in a particular frequency band. The average spacing of the resonance frequencies, $\overline{\delta f}$, is given by

$$\overline{\delta f} = \frac{1}{n(f)} \tag{10.46}$$

where $n(f)$ is the modal density in hertz and it is related to modal density in radians per second by

$$n(f) = 2\pi n(\omega) \tag{10.47}$$

10.9.4 Input Power

Every dynamic system has the capability to accept a certain power from an external source that is mainly governed by the amplitude of force and the velocity response at the point of force application at each frequency. It is well known from Chapter 1 that an SDOF system accepts maximum power at an excitation frequency equal to the undamped resonance frequency and, hence, has the least impedance corresponding to this frequency. Therefore, input power to the SEA model of a dynamic system is an important parameter for determining the energy of various subsystems. As a general case, all the subsystems of the SEA model can be excited, but practically only a few of the subsystem are subject to the input power. Equation 1.96, derived in Chapter 1 for an SDOF system, is an important equation that can be used for computing input power to any vibratory system; a similar equation in terms of sound pressure and particle velocity was used to compute the sound power of an acoustic source. Although it was derived similar to the famous equation of alternating current (ac) circuits VI cos ϕ, it is much more powerful in computing the input power of any dynamic system and can also be expressed in terms of the admittance function and modal density for vibratory systems that are useful in the application of SEA.

From Equation 1.96, the input power for a vibratory system can be written as

$$\Pi_{in} = \frac{1}{2} \mathrm{Re}\left\{\tilde{F}\tilde{v}^*\right\} \tag{10.48}$$

From Equation 1.114, admittance can be expressed in terms of a real and complex number as follows:

$$\overline{Y}_m = G + jB \tag{10.49}$$

The real part of admittance G is known as conductance and the imaginary part B known as susceptance. Since admittance is the ratio of velocity and force, the power input of Equation 10.48 can be written as

$$\Pi_{in} = \frac{|F|^2}{2} \mathrm{Re}\left\{\overline{Y}_{in}\right\} \tag{10.50}$$

The input power to the beam system is given by

$$\Pi_{in} = \frac{1}{2}|F|^2 \langle G \rangle_{ens,\Delta\omega} = F_{rms}^2 \langle G \rangle_{ens,\Delta\omega} \tag{10.51}$$

where F is the amplitude of the force in the frequency band $\Delta\omega$. Substituting the average driving point conductance from Equation 10.45, the average power input is given by

$$\Pi_{in} = \frac{1}{2}|F|^2 \frac{\pi}{2M} \frac{\Delta N}{\Delta\omega} = \frac{|F|^2}{8M\delta f} \tag{10.52}$$

Substituting the equation for space-, frequency-, and ensemble-averaged velocity for a point excitation, the power input is given by the equation

$$\Pi_{in} = M\omega_c \eta \langle v^2 \rangle = \langle E \rangle_{t,\omega,x}\, \omega_c \eta \tag{10.53}$$

where $\langle E \rangle$ is the total energy of the system. The right-hand side of Equation 10.53 represents the power dissipation from the system in the absence of flow to another system from power balance considerations. Equation 10.53 is similar to Equation 10.9 obtained earlier.

Example 10.1

To demonstrate how energy can be shared between two vibratory systems for differing values of damping, consider a resonator attached to a large thin plate as shown in Figure 10.16. The plate is assumed to have a diffuse reverberant vibration field that is assumed to be almost uniform throughout the plate of root mean square (rms) velocity v_p (except at the point of attachment) and v_b is assumed to be the velocity at the point of attachment of the resonator. Determine the ratio of energies of the resonator and the plate (see Smith and Lyon 1965).

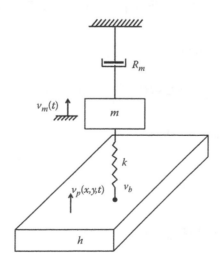

FIGURE 10.16
SDOF excited by plate vibration.

The force on the spring due to the difference in the velocity at its ends (assuming $v_m > v_b$) is given by

$$f = k \int (v_b - v_m) dt \tag{1}$$

Combining the damping force with the spring force defined in Equation 1, the differential equation for the resonator becomes

$$k \int (v_b - v_m) dt - R_m v_m = m \frac{dv_m}{dt} \tag{2}$$

From Equations 1 and 2

$$f = m \frac{dv_m}{dt} + R_m v_m \tag{3}$$

Differentiating Equation 2 with respect to time

$$k(v_b - v_m) - R_m \frac{dv_m}{dt} = m \frac{d^2 v_m}{dt^2} \tag{4}$$

The velocity v_b is the difference between the plate velocity and the velocity due to the reaction of the spring on the plate, and is given by

$$v_b = v_p - f \bar{G} \tag{5}$$

where $\bar{G}$ is the average conductance of the plate.
From Equations 3, 4, and 5

$$k \left[v_p - \left(m \frac{dv_m}{dt} + R_m v_m \right) \bar{G} - v_m \right] - R_m \frac{dv_m}{dt} = m \frac{d^2 v_m}{dt^2} \tag{6}$$

Rearranging terms of Equation 6

$$m \frac{d^2 v_m}{dt^2} + (R_m + k \bar{G} m) \frac{dv_m}{dt} + v_m k (1 + \bar{G} R_m) = k v_p \tag{7}$$

Dividing Equation 7 by m throughout

$$\frac{d^2 v_m}{dt^2} + \left(\frac{R_m}{m} + k \bar{G} \right) \frac{dv_m}{dt} + v_m \frac{k}{m} (1 + \bar{G} R_m) = \frac{k}{m} v_p \tag{8}$$

By substituting $\dfrac{R_m}{m} = \eta \omega_n$ and $\dfrac{k}{m} = \omega_n^2$ in Equation 8

$$\frac{d^2 v_m}{dt^2} + \frac{dv_m}{dt} \left(\eta \omega_n + \omega_n^2 \bar{G} m \right) + \omega_n^2 \left(1 + \bar{G} R_m \right) v_m = \omega_n^2 v_p \tag{9}$$

The coupling loss factor between the oscillator and the plate can be defined as

$$\eta_{coup} = \omega_n \bar{G} m \tag{10}$$

By substituting Equation 10 in Equation 9

$$\frac{d^2 v_m}{dt^2} + \omega_n \frac{dv_m}{dt}(\eta + \eta_{coup}) + \omega_n^2 (1 + \eta\eta_{coup}) v_m = \omega_n^2 v_p \tag{11}$$

From Equation 11, the steady-state transfer function of the oscillator is given by

$$\bar{H}(\omega) = \frac{\omega_n^2}{\omega_n^2(1 + \eta\eta_{coup}) + j\omega\omega_n(\eta + \eta_{coup}) - \omega^2} \tag{12}$$

From Equation 4.158 (Chapter 4), the velocity spectral density of the resonator mass vibration is given by

$$E\left[v_m^2(t) \right] = \int_{-\infty}^{\infty} \left| \bar{H}(\omega) \right|^2 S_{v_p}(\omega) d\omega \tag{13}$$

The spectral density of the plate vibration within a frequency range of $\Delta\omega$ is given by

$$S_{v_p} = \frac{\langle v_p^2 \rangle}{\Delta\omega} \tag{14}$$

From Equations 12, 13, and 14

$$E\left[v_m^2(t) \right] = \frac{\omega_n^4 \langle v_p^2 \rangle}{2\Delta\omega} \int_{-\infty}^{\infty} \left| \frac{1}{\omega_n^2 \left(1 + \eta\eta_{coup}\right) + j\omega\omega_n \left(\eta + \eta_{coup}\right) - \omega^2} \right|^2 d\omega \tag{15}$$

From the standard integral used in Example 4.7 (Chapter 4),

$$I_2 = \int_{-\infty}^{\infty} \left| \frac{B_0 + i\omega B_1}{A_0 + j\omega A_1 - \omega^2 A_2} \right|^2 d\omega = \frac{\pi\{A_0 B_1^2 + A_2 B_0^2\}}{A_0 A_2 A_1} \tag{16}$$

Comparing Equations 14 and 15

$$B_0 = 1, \quad B_1 = 0$$
$$A_0 = \omega_n^2(1 + \eta\eta_{coup}) \approx \omega_n^2, \quad A_1 = \omega_n(\eta + \eta_{coup}) \tag{17}$$
$$A_2 = 1$$

From Equations 16 and 17, the mean square response of the oscillator mass is given by

$$\left\langle v_m^2 \right\rangle_t = \frac{\pi}{2} \left(\frac{\omega_n}{\eta + \eta_{coup}} \right) \frac{\left\langle v^2 \right\rangle_t}{\Delta\omega} \tag{18}$$

From Equation 10.45, the average conductance of the plate is given by

$$\bar{G} = \frac{\pi}{2m_p} n(\omega) \tag{19}$$

where m_p is the plate mass and $n(\omega)$ its modal density. From Equations 10 and 19, the coupling loss factor is given by

$$\eta_{coup} = \frac{\pi \omega_n n(\omega)}{2} \left(\frac{m}{m_p} \right) \tag{20}$$

From Equations 18 and 20, the mean square velocity of the resonator mass is given by

$$\left\langle v_m^2 \right\rangle_t = \frac{\pi}{2} \left(\frac{\omega_n}{\eta_o + \dfrac{\pi \omega_n n(\omega)}{2} \left(\dfrac{m}{m_p} \right)} \right) \frac{\left\langle v^2 \right\rangle_t}{\Delta\omega} \tag{21}$$

When the resonator damping becomes negligible, Equation 21 becomes

$$m \left\langle v_m^2 \right\rangle_t = \frac{m_p \left\langle v^2 \right\rangle_t}{n(\omega)\Delta\omega} \tag{22}$$

Equation 22 clearly demonstrates the concept of equipartition of energy between two interacting systems. Whereas the left-hand side of the equation represents the total energy of the resonator, the right-hand side represents the total energy of the plate per mode of the plate.

When the resonator damping is comparable to the coupling loss factor, by extending the analogy of Equation 10.23, the total energy of the resonator can be expressed in terms of the total energy of the plate as

$$m \left\langle v_m^2 \right\rangle_t = \frac{m_p \left\langle v^2 \right\rangle_t}{n(\omega)\Delta\omega} \left(\frac{\eta_{coup}}{\eta_o + \eta_{coup}} \right) \tag{23}$$

Equation 23 shows that the internal damping of the resonator must be comparable to the damping caused by coupling to a structure to reduce its vibration. It also establishes that even if the resonator has no damping at all, its maximum energy is limited to the average modal energy of the structure to which it is attached. So the objective should be to increase internal damping in the resonator or minimize the modal energy of the structure to reduce vibration of the resonator.

10.10 Coupling Loss Factor

Coupling loss factors between the subsystems of an SEA model are the most important parameters that cannot be easily obtained. Sometimes, depending on the complexity of the connection, they can be calculated and there is vast literature available. Otherwise, the coupling loss factors can be obtained from experimental measurements. Since this is an introductory text, we shall only discuss the coupling loss factor between simple, connected systems in the following sections. The discussion in this section is based on Lyon and DeJong (1995).

10.10.1 Point-Connected Subsystems

Consider two one-dimensional systems of different areas of cross section that are semi-definite and joined as shown in Figure 10.17. It is assumed that there is reverberant energy, $\langle E_1 \rangle$, in subsystem 1, which also has the source of excitation, consisting of incident and reflected waves at the junction and a part of the energy is transmitted to the subsystem 2. It is assumed that both systems have very little damping, which ensures reverberant vibration field in both of them.

From Equation 10.13, power transmitted from system 1 to 2 in the frequency band of center frequency f_c is given by

$$\Pi'_{12} = 2\pi f_c \eta_{12} \langle E_1 \rangle \tag{10.54}$$

If some power π'_{21} is transmitted from subsystem 2 back to 1 is incoherent and if there is modal overlap, the net power transmitted from system 1 to 2 is given by

$$\Pi_{12} = \Pi'_{12} - \Pi'_{21} \tag{10.55}$$

In terms of the transmission coefficient, the transmitted power from system 1 to 2 is given by

$$\Pi_{tran} = \tau_{12} \Pi_{inc} \tag{10.56}$$

In terms of the input impedances of two subsystems, the transmission coefficient between systems 1 and 2 is given by

$$\tau_{12,\infty} = \frac{4R_1 R_2}{\left| Z_1 + Z_2 \right|^2} \tag{10.57}$$

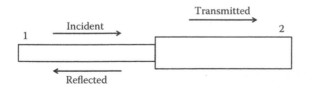

FIGURE 10.17
Two one-dimensional systems connected together.

where Z_1 and Z_2 are the respective input impedances of the two subsystems, and R_1 and R_2 are the respective real parts of the input impedances.

The reflected power, incident power, and the reflection coefficient are given by

$$\Pi_{ref} = \left| r \right|^2 \Pi_{inc} \tag{10.58}$$

From Equations 10.56 and 10.58, since the sum of transmitted and reflected power should be equal to the incident power, the reflection coefficient and the transmission coefficient are given by

$$\left| r \right|^2 = 1 - \tau_{12,\infty} \tag{10.59}$$

The number of reflections per second of the reverberant energy in system 1 is given by

$$n = \frac{c_{g1}}{L_1} \tag{10.60}$$

where c_{g1} is the group velocity of system 1.

The rate of change of reverberant energy in system 1 is therefore given by

$$\left\langle \dot{E}_1 \right\rangle = \frac{\left\langle E_1 \right\rangle c_{g1}}{L_1} \tag{10.61}$$

The rate of change of energy in system 1 (Equation 10.61) must be equal to the total power of system 1 consisting of the sum of incident and reflected power, and is given by

$$\left\langle \dot{E}_1 \right\rangle = \Pi_{inc} + \Pi_{ref} \tag{10.62}$$

From Equations 10.61 and 10.62, the reverberant energy of system 1 is given by

$$\left\langle E_1 \right\rangle = \frac{L_1}{c_{g1}} (\Pi_{inc} + \Pi_{ref}) \tag{10.63}$$

From Equation 10.58, Equation 10.63 becomes

$$\left\langle E_1 \right\rangle = \frac{L_1}{c_{g1}} (2 - \tau_{12}) \Pi_{inc} \tag{10.64}$$

From Equations 10.54 and 10.56

$$\Pi'_{12} = \Pi_{trans} = \tau_{12,\infty} \Pi_{inc} \tag{10.65}$$

From Equations 10.54 and 10.64

$$\eta_{12} = \frac{c_{g1}}{2\pi f_c L_1} \frac{\tau_{12,\infty}}{(2-\tau_{12})} \tag{10.66}$$

The preceding equations are meant for semi-indefinite systems.
 In terms of the average separation frequency, Equation 10.66 becomes

$$\eta_{12} = \frac{\overline{\delta f_1}}{\pi f_c} \frac{\tau_{12,\infty}}{(2-\tau_{12,\infty})} \tag{10.67}$$

From Equation 10.67, the modal coupling loss factor $\left(\beta_{12} = \frac{f_c \eta_{12}}{\overline{\delta f_1}} \right)$ is given by

$$\beta_{12} = \frac{1}{\pi} \frac{\tau_{12,\infty}}{(2-\tau_{12,\infty})} \tag{10.68}$$

10.10.2 Point-Connected Systems with Damping

Consider two point-connected systems as shown in Figure 10.18. Two sets of experiments are conducted on this system. Figure 10.18a shows the experiment in which a sinusoidal force, F_1, is applied at a location of system 1 and the velocity at the junction is measured as v_j. In addition the velocities v_1 and v_2 of the respective systems are measured at points a and b. Figure 10.18b shows the experiment in which a force, F_j, is applied at the junction

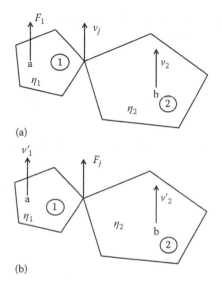

(a)

(b)

FIGURE 10.18
Point connected systems with damping at the junction. (a) First experiment, (b) reciprocal experiment.

and the velocities v_1' and v_2' at the same locations were also measured; both the force and velocity measurements are rms values.

If G_a is the conductance at point a and R_{2j} the input resistance of system 2 at the junction, the power balance for the first experiment for system 1 is given by

$$F_1^2 G_a = 2\pi f_c \eta_1 \langle E_1 \rangle + v_j^2 R_{2j} \tag{10.69}$$

The power balance for system 2 for the first experiment is given by

$$v_j^2 R_{2j} = 2\pi f_c \eta_2 \langle E_2 \rangle \tag{10.70}$$

By substituting Equation 10.70 in 10.69

$$F_1^2 G_a = 2\pi f_c \left(\eta_1 \langle E_1 \rangle + \eta_2 \langle E_2 \rangle \right) \tag{10.71}$$

Dividing Equation 10.70 by 10.71

$$\frac{v_j^2}{F_1^2} = \frac{G_a}{R_{2j}} \frac{\eta_2 \langle E_2 \rangle}{\left(\eta_1 \langle E_1 \rangle + \eta_2 \langle E_2 \rangle \right)} \tag{10.72}$$

The velocity at the junction for the reciprocal experiment is given by

$$v_j'^2 = \frac{F_j'^2}{\left| Z_{1j} + Z_{2j} \right|^2} \tag{10.73}$$

This equation is similar to Equation 1.66 (Chapter 1) and the sum of the junctions is considered as the total impedance at the junction.

The power balance for system 1 due to the reciprocal experiment can be written as follows:

$$v_j'^2 R_{1j} = 2\pi f_c \eta_1 E_1 = 2\pi f_c \eta_1 M_1 \langle v_1'^2 \rangle \tag{10.74}$$

The velocities at point a can be obtained from the reciprocal experiments as

$$\frac{\langle v_j^2 \rangle}{F_1^2} = \frac{\langle v_1'^2 \rangle}{F_j'^2} \tag{10.75}$$

From Equations 10.72 and 10.75

$$\langle v_1'^2 \rangle = \frac{F_j'^2 G_a}{R_{2j}} \frac{\eta_2 \langle E_2 \rangle}{\left(\eta_1 \langle E_1 \rangle + \eta_2 \langle E_2 \rangle \right)} \tag{10.76}$$

From Equations 10.74 and 10.76

$$v_j'^2 R_{1j} = \frac{F_j'^2 G_a}{R_{2j}} \frac{2\pi f_c M_1 \eta_1 \eta_2 \langle E_2 \rangle}{\left(\eta_1 \langle E_1 \rangle + \eta_2 \langle E_2 \rangle \right)} \tag{10.77}$$

From Equations 10.73 and 10.77

$$R_{1j} = \frac{\left| Z_{1j} + Z_{2j} \right|^2 G_a}{R_{2j}} \frac{2\pi f_c M_1 \eta_1 \eta_2 \langle E_2 \rangle}{\left(\eta_1 \langle E_1 \rangle + \eta_2 \langle E_2 \rangle \right)} \tag{10.78}$$

Rearranging Equation 10.78 and multiplying by 4 on both sides,

$$\frac{4 R_{1j} R_{2j}}{\left| Z_{1j} + Z_{2j} \right|^2} = \frac{8\pi f_c G_a \eta_1 M_1}{\left(1 + \dfrac{\eta_2 \langle E_2 \rangle}{\eta_1 \langle E_1 \rangle} \right)} \tag{10.79}$$

By averaging Equation 10.79 over frequencies, the left-hand side of the equation is equal to the frequency averaged transmission coefficient, given by

$$\overline{\tau}_{12} = 8\pi f_c \eta_1 M_1 \left\langle \frac{G_a}{\left(1 + \dfrac{\eta_2 \langle E_2 \rangle}{\eta_1 \langle E_1 \rangle} \right)} \right\rangle \tag{10.80}$$

From Equation 10.22, Equation 10.80 becomes

$$\overline{\tau}_{12} = \frac{8\pi f_c \eta_1 M_1 \overline{G}_a}{1 + \dfrac{\eta_1}{\eta_2} \dfrac{\eta_2 + \eta_{21}}{\eta_{12}}} \tag{10.81}$$

By using the Equation 10.45 for average conductance, Equation 10.81 becomes

$$\overline{\tau}_{12} = \frac{8\pi f_c \eta_1 M_1}{1 + \dfrac{\eta_1}{\eta_2} \dfrac{\eta_2 + \eta_{21}}{\eta_{12}}} \frac{1}{4 M_1 \overline{\delta f_1}} = \frac{2\pi f_c \eta_1}{1 + \dfrac{\eta_1}{\eta_2} \dfrac{\eta_2 + \eta_{21}}{\eta_{12}}} \frac{1}{\overline{\delta f_1}} \tag{10.82}$$

From Equation 10.18 the reciprocity relationship between coupling factors and average frequency separation is given by

$$\eta_{12} \overline{\delta f_2} = \eta_{21} \overline{\delta f_1} \tag{10.83}$$

From Equations 10.82 and 10.83

$$\overline{\tau}_{12} = \frac{2\pi f_c \eta_1}{1 + \dfrac{\eta_1}{\eta_2} \dfrac{\left(\eta_2 + \dfrac{\eta_{12}\overline{\delta f_2}}{\delta f_1}\right)}{\eta_{12}}} \frac{1}{\overline{\delta f_1}}$$

(10.84)

Equation 10.84 can be simplified to obtain the coupling loss factor as

$$\eta_{12} = \frac{\overline{\delta f_1}\,\overline{\tau}_{12}}{2\pi f_c - \overline{\tau}_{12}\left(\dfrac{\delta f_1}{\eta_1} + \dfrac{\delta f_2}{\eta_2}\right)}$$

(10.85)

By expressing Equation 10.85 in terms of modal overlap factors

$$\eta_{12} = \frac{\overline{\delta f_1}}{f_c} \frac{\overline{\tau}_{12}}{2 - \overline{\tau}_{12}\left(\dfrac{1}{\pi\beta_1} + \dfrac{1}{\pi\beta_2}\right)}$$

(10.86)

where the modal overlap factors are given by $\beta_1 = \dfrac{f_c \eta_1}{\delta f_1}$ and $\beta_2 = \dfrac{f_c \eta_2}{\delta f_2}$.

Equation 10.86 shows that the coupling loss factor depends not only on the impedance properties of the junctions but also on the damping of individual subsystems. When both the modal factors become equal to $2/\pi$, the coupling loss factor becomes the same as that derived for semidefinite systems using the wave approach.

10.10.3 Line-Connected Systems

Figure 10.19 shows two plates that are connected along a length L_j. Consider an incident wave inclined at angle θ and the incident power is $L_j \cos \theta$ times the power per unit width of the wave. The power transmitted is $\tau(\theta)$ times the incident power and the transmission coefficient is angle dependent.

Equation 10.66 can be slightly modified for the preceding two-dimensional system to obtain its coupling factor as

$$\eta_{12}(\theta) = \frac{c_{g1}}{2\pi f_c A_1} \frac{\tau_{12} L_j \cos \theta}{(2 - \tau_{12})}$$

(10.87)

Multiplying and dividing the right side of the equation by k_1, the wave number, Equation 10.87 becomes

$$\eta_{12}(\theta) = \frac{c_{g1}}{2\pi f_c k_1 A_1} \frac{\tau_{12} k_1 L_j \cos \theta}{(2 - \tau_{12})}$$

(10.88)

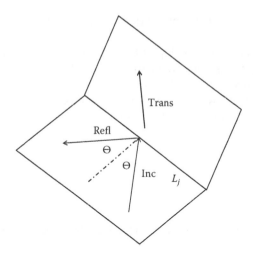

FIGURE 10.19
Two plates connected along a line.

Since the average separation frequency of two-dimensional systems is given by $\overline{\delta}\overline{f_1} = \dfrac{c_{g1}}{k_1 A_1}$, Equation 10.88 becomes

$$\eta_{12}(\theta) = \frac{\overline{\delta}\overline{f_1}}{2\pi f_c} \frac{\tau_{12} k_1 L_j \cos\theta}{(2-\tau_{12})} \tag{10.89}$$

Equation 10.89 has to be integrated over 0 to $\pi/2$ to obtain an average coupling loss factor for the line junction.

10.10.4 Types of Coupling Loss Factors

There are three types of coupling that are considered in SEA modeling:

1. Structure–structure coupling
2. Structure–acoustic coupling
   a. Unstiffened panel coupled to an acoustic cavity
   b. Flat stiffened panel coupled to an acoustic cavity
   c. Unstiffened cylinder coupled to an acoustic cavity
   d. Stiffened cylinder coupled with an acoustic cavity
   e. Flat honeycomb panel coupled to an acoustic cavity
3. Acoustic–cavity/acoustic–cavity coupling
   a. Acoustic cavity (3-D) coupled to another acoustic cavity (3-D)

Structure–structure coupling factors in general will have to be obtained from experiments, whereas structure–acoustic coupling and acoustic–cavity/acoustic cavity coupling can be obtained from computation. In case experimental results are not available, coupling loss factors for structure–structure interaction can be computed using transmission coefficients along the junction.

Since the coupling loss factors are dependent on the direction of power flow, if they are known in any of the directions, the other can be computed by using the reciprocal relationship relating modal density and coupling loss factors of connected systems. Measurement of SEA coupling loss factors was discussed by Cacciolati and Guyader (1994) and new methods of determining coupling loss factors were documented by Manik (1998).

10.11 Applications

Some applications of SEA using simple models are discussed in this section.

10.11.1 Plate–Shell Vibratory System

A flat plate attached to a cylindrical shell is a well-known structural element that is used in many applications. The coupling loss factors between them and the loss factors of each of them will be useful in determining the response of both the elements when each of them are subjected to excitation. One method of finding all the SEA parameters of the two-system model is to conduct two sets of experiments by driving the plate first and then by driving the shell. In both cases the space-averaged response in various frequency bands can be measured and be used to obtain the SEA parameters by writing the power balance equation for each experiment.

Example 10.2

Consider the structure shown in Figure 10.20. The aluminum plate is 4 mm thick and is 4×4 m. The cylinder is 3 m long, has a mean diameter of 2 m, and a 2 mm wall thickness. The following damping information is available in the 315 Hz octave band: the internal loss factor of the plate, η_1, is 0.0034, and the internal loss factor of the cylinder, η_2, is 0.002. The plate vibration velocity is 30 mm/s and the cylinder rms vibration velocity is 10 mm/s. Also $E = 80$ GPa, $\rho = 2800$ kg/m³, $v^{(1)}_{rms,1} = 30$ mm/s, and $v^{(1)}_{rms,2} = 10$ mm/s. Estimate the coupling loss factors, η_{12} and η_{21}, and the input power.

From Equation 2.38 (Chapter 2), the longitudinal wave speed is given by

$$c_L = \sqrt{\frac{E}{\rho}} = \sqrt{\frac{80 \times 10^9}{2800}} = 5345 \text{ m/s} \tag{1}$$

$$\text{Inner diameter } d_1 = 2\left(\frac{d_m}{2} - \frac{t}{2}\right) = 2\left(\frac{2}{2} - \frac{0.002}{2}\right) = 1.998 \text{ m}$$

$$\text{Outer diameter } d_2 = 2\left(\frac{d_m}{2} + \frac{t}{2}\right) = 2\left(\frac{2}{2} + \frac{0.002}{2}\right) = 2.002 \text{ m}$$

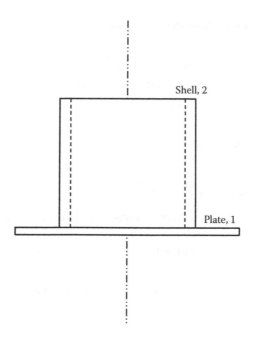

FIGURE 10.20
Shell and plate connected to each other.

The outer surface area of the shell is given by

$$A_s = \pi d_2 L = \pi \times 2.002 \times 3 = 18.8684 \text{ m}^2 \tag{2}$$

Excitation frequency $f = 315$ Hz.
From Equation 6.42 (Chapter 6), the ring frequency of the shell is given by

$$f_r = \frac{c_L}{\pi d_m} = \frac{5345}{\pi \times 2} = 850.72 \text{ Hz} \tag{3}$$

The frequency ratio, $f/f_r = 315/850.72 = 0.37$.
From Equation 6.38 (Chapter 6), the modal density of the shell is given by

$$\begin{aligned}
n_2(f) &= \frac{5A_s}{\pi c_L t} \left(\frac{f}{f_r} \right)^{0.5} \\
&= \frac{5 \times 18.8684}{\pi \times 5345 \times 0.002} (0.37)^{0.5} = 1.7093 \text{ modes/Hz}
\end{aligned} \tag{4}$$

From ESA (1996), the modal density of the shell is given by

$$n_1(f) = \frac{\sqrt{3} A_p}{c_L h} = \frac{\sqrt{3} \times 16}{5345 \times 0.004} = 1.2961 \text{ modes/Hz} \tag{5}$$

The total energy of the plate (system 1) is given by

$$
\begin{aligned}
E_1 = m_1 v_{rms,1}^2 &= (A_p h)\rho v_{rms,1}^2 \\
&= 16 \times 0.004 \times 2800 \times (30 \times 10^{-3})^2 \\
&= 0.1613 \text{ J}
\end{aligned}
\tag{6}
$$

The total energy of the shell (system 2) is given by

$$
\begin{aligned}
E_2 = m_2 v_{rms,2}^2 &= \frac{\pi\left(d_2^2 - d_1^2\right) L \rho v_{rms,2}^2}{4} \\
&= \frac{\pi(2.002^2 - 1.9980^2) 3 \times 2800 \times (10 \times 10^{-3})^2}{4} \\
&= 0.0106 \text{ J}
\end{aligned}
\tag{7}
$$

From Equation 10.23, for a two-system SEA model, the ratio of modal energy is given by

$$
\frac{\dfrac{E_2}{n_2(\omega)}}{\dfrac{E_1}{n_1(\omega)}} = \frac{\dfrac{0.0106 \times 2\pi}{1.7093}}{\dfrac{0.1613 \times 2\pi}{1.2961}} = 0.0496 = \frac{\eta_{21}}{\eta_2 + \eta_{21}}
\tag{8}
$$

Using the loss factor of system 2 (shell) given in the problem, the coupling loss factor from system 2 to 1 is given by

$$
\eta_{21} = \frac{0.0496}{1 - 0.0496}\eta_2 = \frac{0.0496}{1 - 0.0496} \times 0.002 = 1.044 \times 10^{-4}
\tag{9}
$$

Using the reciprocity Equation 10.18, the coupling loss factor from system 1 to system 2 is given by

$$
\eta_{12} = \eta_{21}\frac{n_2(f)}{n_1(f)} = \frac{1.044 \times 10^{-4} \times 1.7093}{1.2961} = 1.3774 \times 10^{-4}
\tag{10}
$$

By writing the power balance equation for system 1, the power input for the plate is given by

$$
\begin{aligned}
\Pi_{1,in} &= 2\pi f(\eta_{12}E_1 - \eta_{21}E_2 + \eta_1 E_1) \\
&= 2\pi \times 315(1.3774 \times 10^{-4} \times 0.1613 - 1.044 \times 10^{-4} \times 0.0106 + 0.0034 \times 0.1613) \\
&= 1.1271 \text{ W}
\end{aligned}
\tag{11}
$$

The following example illustrates how various SEA parameters of a dynamic system can be experimentally obtained by conducting reciprocal experiments consisting of exciting one subsystem at a time and using the power input and energy of subsystems in the matrix equation.

Example 10.3

Vibration experiments were conducted on the plate–shell system shown in Figure 10.20 in the 500 Hz octave band. When the plate was driven with an input power of 2.96 W in the first experiment, the average rms velocity of the plate was 30 mm/s and that of the shell was 10.8 mm/s. When the shell was driven with an input power of 0.42 W in the second experiment, the average rms velocity of the plate was 5.2 mm/s and that of the shell was 15 mm/s. Determine the damping loss factors and coupling loss factors of the system. (Superscripts are used for the experiment number and subscripts for the subsystem number. The plate is considered as subsystem 1 and the shell subsystem 2.)

The following is given: $E = 80$ GPa; $\rho = 2800$ kg/m³; area of the plate, $A_p = 4 \times 4 = 16$ m²; plate thickness, $h = 5$ mm; shell length, $L = 3$ m; inner diameter $d_1 = 1.49$ m; outer diameter $d_2 = 1.5$ m; $v_{rms,1}^{(1)} = 30$ mm/s; $v_{rms,2}^{(1)} = 10.8$ mm/s; $v_{rms,1}^{(2)} = 5.2$ mm/s; $v_{rms,2}^{(1)} = 15$ mm/s.

$$\Pi_1^{(1)} = 2.96 \text{ W}; \; \Pi_2^{(1)} = 0; \; \Pi_1^{(2)} = 0; \; \Pi_2^{(2)} = 0.42 \text{ W}$$

From Equation 2.38 (Chapter 2), the longitudinal wave speed is given by

$$c_L = \sqrt{\frac{E}{\rho}} = \sqrt{\frac{80 \times 10^9}{2800}} = 5345 \text{ m/s} \tag{1}$$

From Equation 6.38, the modal density of the plate is given by

$$n_1(f) = \frac{\sqrt{3}A_p}{c_L h} = \frac{\sqrt{3} \times 16}{5345 \times 0.005} = 1.0369 \text{ modes/Hz} \tag{2}$$

The mean diameter of the shell is given by

$$d_m = \frac{d_1 + d_2}{2} = \frac{1.49 + 1.5}{2} = 1.495 \text{ m} \tag{3}$$

The outer surface area of the shell is given by

$$A_s = \pi d_2 L = \pi \times 1.5 \times 3 = 14.137 \text{ m}^2 \tag{4}$$

Excitation frequency $f = 500$ Hz.
From Equation 6.42, the ring frequency of the shell is given by

$$f_r = \frac{c_L}{\pi d_m} = \frac{5345}{\pi \times 1.495} = 1138 \text{ Hz} \tag{5}$$

The frequency ratio, $f/f_r = 500/1134 = 0.44$.
The modal density of the shell is given by (ESA 1996)

$$n_2(f) = \frac{5A_s}{\pi c_L t}\left(\frac{f}{f_r}\right)^{0.5} \tag{6}$$

$$= \frac{5 \times 14.137}{\pi \times 5345 \times 0.005}(0.439)^{0.5} = 0.5578 \text{ modes/Hz}$$

The total energy of the plate (system 1) for the first experiment is given by

$$E_1^{(1)} = m_1 v_{rms,1(1)}^2 = (A_p h)\rho v_{rms,1}^2 (1)$$
$$= 16 \times 0.005 \times 2800 \times (0.030)^2 \qquad (7)$$
$$= 0.2 \text{ J}$$

The total energy of the shell (system 2) for the first experiment is given by

$$E_2^{(1)} = m_2 v_{rms,2(1)}^2 = \frac{\pi \left(d_2^2 - d_1^2\right) L \rho v_{rms,2(1)}^2}{4}$$
$$= \frac{\pi (1.5^2 - 1.49^2) 3 \times 2800 \times (0.0108)^2}{4} \qquad (8)$$
$$= 0.023 \text{ J}$$

The total energy of the plate (system 1) for the second experiment is given by

$$E_1^{(2)} = m_1 v_{rms,1(2)}^2 = (A_p h)\rho v_{rms,1(2)}^2$$
$$= 16 \times 0.005 \times 2800 \times (0.0052)^2 \qquad (9)$$
$$= 6.057 \text{ mJ}$$

The total energy of the shell (system 2) for the first experiment is given by

$$E_2^{(2)} = m_2 v_{rms,2(2)}^2 = \frac{\pi \left(d_2^2 - d_1^2\right) L \rho v_{rms,2(2)}^2}{4}$$
$$= \frac{\pi (1.5^2 - 1.49^2) 3 \times 2800 \times (0.015)^2}{4} \qquad (10)$$
$$= 0.0444 \text{ J}$$

By writing down the power balance equation for each experiment and subsystem

$$\Pi_1^{(1)} = \eta_1 \omega_c E_1^{(1)} + \eta_{12} \omega_c E_1^{(1)} - \eta_{21} \omega_c E_2^{(1)}$$
$$\Pi_2^{(1)} = \eta_2 \omega_c E_2^{(1)} + \eta_{21} \omega_c E_2^{(1)} - \eta_{12} \omega_c E_1^{(1)}$$
$$\Pi_1^{(2)} = \eta_1 \omega_c E_1^{(2)} + \eta_{12} \omega_c E_1^{(2)} - \eta_{21} \omega_c E_2^{(2)} \qquad (11)$$
$$\Pi_2^{(2)} = \eta_2 \omega_c E_2^{(2)} + \eta_{21} \omega_c E_2^{(2)} - \eta_{12} \omega_c E_1^{(2)}$$

By adding the equations of (11) and noting that $\Pi_2^{(1)} = 0$; $\Pi_1^{(2)} = 0$;

$$\Pi_1^{(1)} = \eta_1 \omega_c E_1^{(1)} + \eta_2 \omega_c E_2^{(1)}$$
$$\Pi_2^{(2)} = \eta_1 \omega_c E_1^{(2)} + \eta_2 \omega_c E_2^{(2)} \qquad (12)$$

These equations can be expressed as

$$
\begin{bmatrix} E_1^{(1)} & E_2^{(1)} \\ E_1^{(2)} & E_2^{(2)} \end{bmatrix} \begin{Bmatrix} \eta_1 \\ \eta_2 \end{Bmatrix} = \begin{Bmatrix} \Pi_1^{(1)}/\omega_c \\ \Pi_2^{(2)}/\omega_c \end{Bmatrix}
\tag{13}
$$

The loss factors of the two systems are given by

$$
\begin{Bmatrix} \eta_1 \\ \eta_2 \end{Bmatrix} = \begin{bmatrix} E_1^{(1)} & E_2^{(1)} \\ E_1^{(2)} & E_2^{(2)} \end{bmatrix}^{-1} \begin{Bmatrix} \Pi_1^{(1)}/\omega_c \\ \Pi_2^{(2)}/\omega_c \end{Bmatrix}
$$

$$
= \begin{bmatrix} 0.2 & 0.023 \\ 6.057 \times 10^{-3} & 0.0444 \end{bmatrix}^{-1} \begin{Bmatrix} 2.96/2\pi \times 500 \\ 0.42/2\pi \times 500 \end{Bmatrix}
\tag{14}
$$

$$
= \begin{Bmatrix} 0.0044 \\ 0.0024 \end{Bmatrix}
$$

Before we proceed to the computation of coupling loss factors, let us define the following modal energy ratios. The ratio of modal energies of the two systems when the plate was excited is given by

$$
x_1 = \frac{\dfrac{E_2^{(1)}}{n_2(\omega)}}{\dfrac{E_1^{(1)}}{n_1(\omega)}} = \frac{\dfrac{0.023 \times 2\pi}{0.5578}}{\dfrac{0.2 \times 2\pi}{1.0369}} = 0.212
\tag{15}
$$

The ratio of modal energies of the two systems when the shell structure was excited is given by

$$
x_2 = \frac{\dfrac{E_1^{(2)}}{n_1(\omega)}}{\dfrac{E_2^{(2)}}{n_2(\omega)}} = \frac{\dfrac{0.006057 \times 2\pi}{1.0369}}{\dfrac{0.0444 \times 2\pi}{0.5578}} = 0.0734
\tag{16}
$$

From Equation 10.22 and Equation 15, using the data for exciting the plate only, the coupling loss factor η_{21} becomes

$$
\eta_{21} = \frac{x_1 \eta_2}{1 - x_1} = \frac{0.212 \times 0.0024}{1 - 0.212} = 6.492 \times 10^{-4}
\tag{17}
$$

From Equation 10.27 and Equation 16, using the data for exciting the shell structure only, the coupling loss factor η_{12} becomes

$$
\eta_{12} = \frac{x_2 \eta_1}{1 - x_2} = \frac{0.0734 \times 0.0044}{1 - 0.0734} = 3.486 \times 10^{-4}
\tag{18}
$$

Using the values of loss factors and coupling loss factors, the reciprocity relationship of Equation 10.18 can be proved as follows:

$$n_1(\omega)\eta_{12} = \frac{1.0369 \times 3.486 \times 10^{-4}}{2\pi} = 5.75 \times 10^{-5}$$

$$n_2(\omega)\eta_{21} = \frac{0.5578 \times 6.492 \times 10^{-4}}{2\pi} = 5.76 \times 10^{-5}$$

(19)

10.11.2 Plate in a Reverberant Room

Consider a plate of surface area A_p and thickness h, having a surface density of ρ, placed in a reverberant room of volume V, as shown in Figure 10.21. A sound source is placed at a corner that generates a reverberant sound field inside the room and this sound excites the plate into vibration. This in turn would result in radiation of noise from the plate. If $\langle p^2 \rangle^{f_c}$, which is the mean square pressure corresponding to the octave band center frequency f_c, the energy of the reverberant room (Beranek 1971).
$E_1^{f_c}$ is given by the following relation:

$$\left\langle E_1^{f_c} \right\rangle = \frac{\langle p^2 \rangle^{f_c} V}{\rho_o c_o^2},$$

(10.90)

where ρ_o is the density of air and c_o the velocity of air at the corresponding temperature of the reverberant room. Similarly, if $\langle v^2 \rangle^{f_c}$ is the mean square velocity in the frequency band f_c, the energy of the plate, $E_2^{f_c}$, in the frequency band is given by (similar to Equation 10.7)

$$\left\langle E_2^{f_c} \right\rangle = \langle v^2 \rangle^{f_c} \rho_s A_p$$

(10.91)

The power balance for the plate is as shown in Figure 10.22. The power balance equation in the frequency band f_c can be written as

$$\Pi_{in}^{f_c} = \Pi_{rad}^{f_c} + \Pi_d^{f_c}$$

(10.92)

where $\Pi_{in}^{f_c}$ is the power input to the plate, $\Pi_{rad}^{f_c}$ is the power radiated by the plate, and $\Pi_d^{f_c}$ is the power dissipated in the plate.

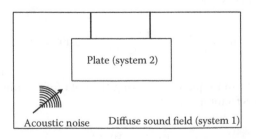

FIGURE 10.21
Plate in a reverberant room.

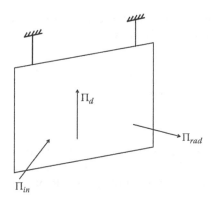

FIGURE 10.22
Power balance for the plate.

Since both sides of the plate are excited by acoustic excitation, from Equation 9.42 the radiated power in terms of radiation factor of the plate σ_{rad} is given by

$$\Pi_{rad}^{f_c} = 2\langle v^2 \rangle^{f_c} \rho_o c_o A_p \sigma_{rad} \tag{10.93}$$

The power radiated by the plate can also be expressed in terms of the energy of the plate, center frequency f_c, and the coupling loss factor η_{21} between systems 1 and 2. It is given by

$$\Pi_{rad}^{f_c} = \langle E_2^{f_c} \rangle (2\pi f_c) \eta_{21} \tag{10.94}$$

From Equations 10.51 through 10.55, the coupling loss factor can be obtained as

$$\eta_{21} = \frac{2\rho_o c_o \sigma_{rad}}{\rho_s (2\pi f_c)} \tag{10.95}$$

From Equation 10.23 the ratio of the energy of the two systems in the frequency band f_c is given by

$$\frac{\langle E_2^{f_c} \rangle}{\langle E_1^{f_c} \rangle} = \frac{n_2(f_c)}{n_1(f_c)} \frac{\eta_{21}}{\eta_{21} + \eta_2} \tag{10.96}$$

where $\eta_2^{f_c}$ is the loss factor of the plate in the frequency band f_c. From Equation 8.61 (Chapter 8), the modal density of the reverberant room $n_1(f_c)$ the frequency band f_c is given by

$$n_1(f_c) = \frac{(2\pi f_c)^2 V}{\pi c_o^3} \tag{10.97}$$

From Equation 6.38, the modal density of the plate $n_2(f_c)$ in the frequency band f_c is given by

$$n_2(f_c) = \frac{\sqrt{3} A_p}{c_L h} \tag{10.98}$$

From the preceding equations, the ratio of the mean square velocity and the mean square pressure in the frequency band f_c is given by

$$\frac{\langle v^2 \rangle^{f_c}}{\langle p^2 \rangle^{f_c}} = \left(\frac{\sqrt{3}\pi c_o}{\rho_s \rho_o (2\pi f_c)^2 h c_L} \right) \left(\frac{1}{1 + \dfrac{\eta_2 \rho_s 2\pi f_c}{2\rho_o \sigma_{rad}}} \right) \tag{10.99}$$

From Equation 10.99, if the velocity and pressure at various frequency bands are known, then the radiation factor of the plate can be determined. Once the radiation factor is determined at various frequency bands, the vibration levels of the plate can be determined for a given acoustic excitation.

Only in this example is the octave band center frequency explicitly written for all the parameters like energy, power input, power dissipated, coupling loss factors, loss factor, and modal density. Generally it should be understood that the parameters are frequency dependent and that all computations must be carried out in the octave or 1/3 octave band of center frequency f_c.

Example 10.4

Experiments were conducted in the 500 Hz octave band on a plate kept in a reverberant room. When only the sound source was input with a power of 63 W, the average sound pressure level in the room was 135 dB and the rms velocity of the plate was 30 mm/s. When only the plate was driven with an input power of 29 W, the average rms velocity of the plate was 40 mm/s and the average sound pressure level of the room was 110 dB. Determine the damping loss factors and coupling loss factors of the vibro-acoustic system.

The plate is of dimensions 10 × 10 × 0.01 m and E = 80 GPa and ρ = 2800 kg/m³. The room volume is 400 m³. Use air data from Example 10.1.

From Equation 2.38, the longitudinal wave speed is given by

$$c_L = \sqrt{\frac{E}{\rho}} = \sqrt{\frac{80 \times 10^9}{2800}} = 5345 \text{ m/s} \tag{1}$$

From Equation 10.98, the modal density of the plate is given by

$$n_1(f_c) = \frac{\sqrt{3} A_p}{c_L h} = \frac{\sqrt{3} \times 10 \times 10}{5345 \times 0.01} = 3.24 \text{ modes/Hz} \tag{2}$$

From Equation 8.61, the modal density of the acoustic modes within the volume is given by

$$n_2(f) = \frac{(2\pi f)^2 V}{\pi c_o^3} = \frac{(2\pi \times 500)^2 400}{\pi \times 340^3} = 31.97 \text{ modes/Hz} \tag{3}$$

The mass of the plate is given by

$$M_2 = \rho h A_p = 7800 \times 0.010 \times 10^2 = 2800 \text{ kg} \tag{4}$$

From Equation 3.57, the dynamic pressure corresponding to 135 dB sound pressure level is given by

$$p_{d_1} = \sqrt{10^{\frac{L_{p_1}}{10}} p_{ref}^2} = \sqrt{10^{\frac{135}{10}} (20 \times 10^{-6})^2} = 112.47 \text{ Pa} \tag{5}$$

From Equation 3.57, the dynamic pressure corresponding to 110 dB sound pressure level is given by

$$p_{d_1} = \sqrt{10^{\frac{L_{p_2}}{10}} p_{ref}^2} = \sqrt{10^{\frac{110}{10}} (20 \times 10^{-6})^2} = 6.32 \text{ Pa} \tag{6}$$

From Equation 8.26, the total energy of the acoustic volume for the first experiment is given by

$$E_1^{(1)} = \frac{\left(p_d^{(1)}\right)^2 V}{\rho_o c_o^2} = \frac{(112.47)^2 400}{1.2 \times 340^2} = 36.47 \text{ J} \tag{7}$$

The total energy of the plate for the first experiment is given by

$$E_2^{(1)} = M_2 \left(v_{rms}^{(1)}\right)^2 = 2800 \times (0.030)^2 = 2.52 \text{ J} \tag{8}$$

From Equation 8.26, the total energy of the acoustic volume for the second experiment is given by

$$E_1^{(2)} = \frac{\left(p_d^{(2)}\right)^2 V}{\rho_o c_o^2} = \frac{(6.32)^2 400}{1.2 \times 340^2} = 0.1153 \text{ J} \tag{9}$$

The total energy of the plate for the second experiment is given by

$$E_2^{(2)} = M_2 \left(v_{rms}^{(2)}\right)^2 = 2800 \times (0.04)^2 = 4.48 \text{ J} \tag{10}$$

The following equations can be written for a general two-system SEA model for two sets of experiments:

$$\begin{aligned} \Pi_1^{(1)} &= \eta_1 \omega_c E_1^{(1)} + \eta_{12} \omega_c E_1^{(1)} - \eta_{21} \omega_c E_2^{(1)} \\ \Pi_2^{(1)} &= \eta_2 \omega_c E_2^{(1)} + \eta_{21} \omega_c E_2^{(1)} - \eta_{12} \omega_c E_1^{(1)} \\ \Pi_1^{(2)} &= \eta_1 \omega_c E_1^{(2)} + \eta_{12} \omega_c E_1^{(2)} - \eta_{21} \omega_c E_2^{(2)} \\ \Pi_2^{(2)} &= \eta_2 \omega_c E_2^{(2)} + \eta_{21} \omega_c E_2^{(2)} - \eta_{12} \omega_c E_1^{(2)} \end{aligned} \tag{11}$$

$$\begin{bmatrix} E_1^{(1)} & 0 & E_1^{(1)} & -E_2^{(1)} \\ 0 & E_2^{(1)} & -E_1^{(1)} & E_2^{(1)} \\ E_1^{(2)} & 0 & E_1^{(2)} & -E_2^{(2)} \\ 0 & E_2^{(2)} & E_1^{(2)} & E_2^{(2)} \end{bmatrix} \begin{Bmatrix} \eta_1 \\ \eta_2 \\ \eta_{12} \\ \eta_{21} \end{Bmatrix} = \frac{1}{\omega_c} \begin{Bmatrix} \Pi_1^{(1)} \\ 0 \\ 0 \\ \Pi_2^{(2)} \end{Bmatrix} \tag{12}$$

Substituting values obtained earlier in Equation 12

$$\begin{bmatrix} 36.47 & 0 & 36.47 & -2.52 \\ 0 & 2.52 & -36.47 & 2.52 \\ 0.1153 & 0 & 0.1153 & -4.48 \\ 0 & 4.48 & -0.1153 & 4.48 \end{bmatrix} \begin{Bmatrix} \eta_1 \\ \eta_2 \\ \eta_{12} \\ \eta_{21} \end{Bmatrix} = \frac{1}{2\pi \times 500} \begin{Bmatrix} 63 \\ 0 \\ 0 \\ 29 \end{Bmatrix} \tag{13}$$

Solving the system of equations represented by Equation 13

$$\begin{aligned} \eta_1 &= 4.08 \times 10^{-4} \\ \eta_2 &= 0.002 \\ \eta_{12} &= 1.43 \times 10^{-4} \\ \eta_{21} &= 1.42 \times 10^{-5} \end{aligned} \tag{14}$$

10.11.3 Nonresonant Transmission between Two Rooms

Consider two rooms separated by a panel as shown in Figure 10.23. A noise source is kept in one of the rooms and it is required to determine the noise levels in the other room, assuming that the resonant modes of the panel do not participate in the noise transmission process.

From room acoustics, the ratio of the incident and transmitted sound powers is given by

$$\frac{W_1}{W_2} = \frac{1}{1 + \dfrac{A_2}{\tau S_{21}}} \tag{10.100}$$

where A_2 is the area of the second room, τ the sound transmission coefficient of the panel, and S_{21} the area of the panel.

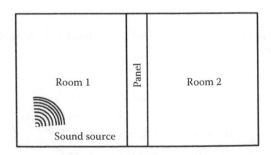

FIGURE 10.23
Nonresonant transmission between two rooms.

The area of the second room is given by

$$A_2 = \frac{8\pi f_c V_2 \eta_2}{c_o} \tag{10.101}$$

where V_2 is the volume of the second room and η_2 its loss factor.

From Equation 10.23, if the number of acoustic modes in both rooms is assumed to be constant, the energy ratio of the two rooms is given:

$$\frac{E_2}{E_1} = \frac{1}{1 + \dfrac{\eta_2}{\eta_{21}}} \tag{10.102}$$

From Equations 10.100 through 10.102 the coupling loss factor between the two rooms is given by

$$\eta_{21} = \frac{c_o \tau \eta S_{21}}{8\pi f_c V_2} \tag{10.103}$$

This example was specifically given to demonstrate that the coupling loss factors arising out of SEA analysis are in fact related to the physical system. In some simple systems they can be theoretically determined.

10.11.4 Resonant Sound Transmission

When a panel separates two rooms, at some frequencies the noise transmission might involve the modes of vibration of the panel and such a sound transmission is called resonant mode transmission. For this type of sound transmission, SEA is the only method of determining the transmission loss theoretically. Even when the complexity of the intermediate partition is increased, by using a double walled partition, SEA can still give very good results. The discussion in this section is based on Crocker and Price (1969).

The noise transmission between two rooms separated by a single-walled partition is as shown in Figure 10.24. The two rooms are named space 1 and space 3. The intermediate

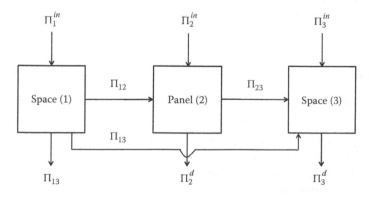

FIGURE 10.24
Resonant transmission between two rooms.

partition is denoted as panel 2. As a general case it is assumed that power can be input to all three systems and that there could be interaction between all three systems. Π_1^{in}, Π_2^{in}, and Π_3^{in} are the power input to the three systems; Π_1^d, Π_2^d, and Π_3^d are the power dissipated by each of the systems.

Π_{12} is the power flowing from space 1 to panel 2, Π_{23} is the power flowing from panel 2 to space 3, and Π_{13} is the power flowing from space 1 to space 3.

The power balance for each of the systems can be written as follows:

$$\Pi_1^{in} = \Pi_1^d + \Pi_{12} + \Pi_{13} \tag{10.104}$$

$$\Pi_2^{in} = \Pi_2^d - \Pi_{12} + \Pi_{23} \tag{10.105}$$

$$\Pi_3^{in} = \Pi_3^d - \Pi_{13} - \Pi_{23} \tag{10.106}$$

Based on Section 10.8, Equations 10.104 through 10.106 can be expressed in terms of the energy, modal density, loss factor, and coupling loss factor as follows (Crocker and Price 1969)

$$\Pi_1^{in} = \omega\eta_1 E_1 + \omega\eta_{12}n_1(\omega)\left(\frac{E_1}{n_1(\omega)} - \frac{E_2}{n_2(\omega)}\right) + \omega\eta_{13}n_1(\omega)\left(\frac{E_1}{n_1(\omega)} - \frac{E_3}{n_3(\omega)}\right) \tag{10.107}$$

$$\Pi_2^{in} = \omega\eta_2 E_2 - \omega\eta_{12}n_1(\omega)\left(\frac{E_1}{n_1(\omega)} - \frac{E_2}{n_2(\omega)}\right) + \omega\eta_{23}n_2(\omega)\left(\frac{E2}{n_2(\omega)} - \frac{E_3}{n_3(\omega)}\right) \tag{10.108}$$

$$\Pi_3^{in} = \omega\eta_3 E_3 - \omega\eta_{13}n_1(\omega)\left(\frac{E_1}{n_1(\omega)} - \frac{E_3}{n_3(\omega)}\right) - \omega\eta_{23}n_1(\omega)\left(\frac{E_2}{n_2(\omega)} - \frac{E_3}{n_3(\omega)}\right) \tag{10.109}$$

From the preceding equations it is possible to obtain the ratio of energy of space 1 and space 3 when only one of them is excited by a sound source. This in turn can be used for theoretically calculating the transmission loss of the partition at various frequencies.

10.12 Transient SEA

Shocks produced by pyrotechnic blasts in space vehicles and missiles, and during many manufacturing processes, result in high amplitude vibration at high frequencies. Since many manufacturing processes involve impulse excitation, the resulting vibration can produce high noise levels in the workplace in the sensitive frequency range of 1 to 4 kHz. The response to this impulsive vibration can be determined using transient SEA. Traditionally, transient SEA has been solved using a numerical technique in the time domain or by obtaining the shock spectrum in the frequency domain. The accuracy of numerical methods, however, depends on the discrete–time interval that has to be appropriately chosen. Many times, this time interval can become very small for predicting response at high frequencies, thus becoming computationally intensive. Since transient SEA equations are first-order matrix differential equations, the state-space approach can be easily used for solving them. By determining the fundamental matrix using the loss factors and coupling loss factors of various subsystems, the response can be obtained for a wide variety of excitation. The order of the matrix is the same as the number of subsystems of the SEA model. The state-space formulation is presented first and then an example problem of a two-system SEA model subjected to an impulse is solved using the state-space approach and compared with a numerical technique.

10.12.1 Basic Equations

Considering the dynamic energy in the resonant modes of a subsystem that changes with time, the power balance is given by

$$\Pi_{out} + \frac{dE}{dt} = \Pi_{in},$$ (10.110)

where Π_{out} is the power flowing out of the subsystem, either through internal damping or to other subsystems through coupling; $E(t)$ is the total dynamic energy of the resonant modes at any instant of time; and Π_{in} is the power input to the subsystem, either through an external excitation or from other subsystems through coupling.

Equation 10.110 can be cast in the standard form as follows:

$$\frac{dE}{dt} = \Pi_{in} - \Pi_{out}$$ (10.111)

Consider a single subsystem with an internal damping factor that has been initially excited with an energy E_o and then allowed to dissipate the energy. Hence $\Pi_{in} = 0$ and $\Pi_{out} = 2\pi f \eta E(t)$; we can assume that the dissipation is occurring at a single frequency f. Equation 10.111 then becomes

$$\frac{dE}{dt} = -2\pi f \eta E(t)$$ (10.112)

Based on the initial energy as the initial condition, the solution to Equation 10.112 is given by

$$E(t) = E_o e^{-2\pi f \eta t}$$ (10.113)

SEA, however, is always concerned with two or more subsystems. Therefore, let us first develop transient SEA analysis for a two-subsystem model that can be easily extended to a larger number of subsystems.

A two-subsystem SEA model is shown in Figure 10.25, which is initially subjected to energies E_{10} and E_{20}, respectively, in the frequency band of center frequency f_c. After receiving initial energy, each of the systems dissipates energy due to damping and also exchanges energy with the other system through coupling.

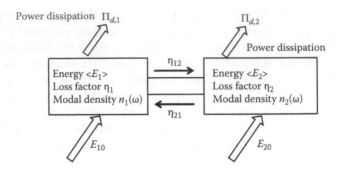

FIGURE 10.25
Energy exchange between two SEA subsystems due to initial energy.

Applying Equation 10.111 to each subsystem, the time variation of energy in each of them in the frequency band of center frequency f_c can be written as follows:

$$\frac{dE_1}{dt} = 2\pi f_c \left[-(\eta_{12} + \eta_1)E_1 + \eta_{21}E_2 \right]$$
$$\frac{dE_2}{dt} = 2\pi f_c \left[\eta_{12}E_1 - (\eta_{21} + \eta_2)E_2 \right] \tag{10.114}$$

Equation 10.114 can be cast in the matrix form as

$$\begin{Bmatrix} \dot{E}_1 \\ \dot{E}_2 \end{Bmatrix} = 2\pi f_c \begin{bmatrix} -(\eta_{12} + \eta_1) & \eta_{21} \\ \eta_{12} & -(\eta_{21} + \eta_2) \end{bmatrix} \begin{Bmatrix} E_1 \\ E_2 \end{Bmatrix} \tag{10.115}$$

10.12.2 Numerical Method

One of the ways of solving Equation 10.115 is to expand the energy vector in the Taylor series expansion as

$$\mathbf{E}(t + \Delta t) = \mathbf{E}(t) + \dot{\mathbf{E}}(t)\Delta t \tag{10.116}$$

From Equations 10.115 and 10.116

$$\mathbf{E}(t + \Delta t) = \mathbf{E}(t) + 2\pi f_c \begin{bmatrix} -(\eta_{12} + \eta_1) & \eta_{21} \\ \eta_{12} & -(\eta_{21} + \eta_2) \end{bmatrix} \begin{Bmatrix} E_1(t) \\ E_2(t) \end{Bmatrix} \Delta t \tag{10.117}$$

By assuming a reasonable time step in Equation 10.117, the decay of energy for an SEA model of known parameters can be obtained.

10.12.3 State-Space Formulation

Equation 10.115 can be written as

$$\left\{ \begin{array}{c} \dot{E}_1 \\ \dot{E}_2 \end{array} \right\} = \mathbf{A} \left\{ \begin{array}{c} E_1 \\ E_2 \end{array} \right\} \tag{10.118}$$

where

$$\mathbf{A} = 2\pi f_c \begin{bmatrix} -(\eta_{12} + \eta_1) & \eta_{21} \\ \eta_{12} & -(\eta_{21} + \eta_2) \end{bmatrix} \tag{10.119}$$

Equation 10.119 can be used to obtain the fundamental matrix, $\Phi(t) = e^{\mathbf{A}t}$ of the state-space formulation. Using this fundamental matrix, the energy decay after an initial impact can be accurately determined. When the number of subsystems becomes very large, standard subroutines that are available for state-space analysis can be directly used. The fundamental matrix can be obtained by using any of the following methods: (a) using eigenvalues and eigenvectors, (b) using the Cayley-Hamilton theorem, and (c) using the Laplace transform (see Manik 2012).

Example 10.5

Consider a two-subsystem SEA model with the following parameters: $\eta_1 = 0.01$, $\eta_2 = 0.02$, $\eta_{12} = 0.1$, and $\eta_{21} = 0.15$. Initial energy of E_0 at 100 Hz is given to the first system at $t = 0$ and then removed from the system. Determine the rate of energy decay in both systems using the state-space formulation, and use the numerical method to compare the decay of the first subsystem obtained using the state-space formulation for $\Delta t = 0.005$ and 0.009.

1. By using the parameters of the SEA model in Equation 10.119

$$\mathbf{A} = 2\pi \times 100 \begin{bmatrix} -0.11 & 0.15 \\ 0.1 & -0.17 \end{bmatrix} \tag{10.120}$$

The eigenvalues of this matrix, expressed in the form of an exponential of diagonal matrix, can be written as

$$e^{\mathbf{A}t} = \begin{bmatrix} e^{-8.7366t} & 0 \\ 0 & e^{-167.1925t} \end{bmatrix} \tag{10.121}$$

2. The modal vectors of the these eigenvalues can be expressed in the matrix form as

$$\Psi = \begin{bmatrix} 0.8420 & -0.6929 \\ 0.5394 & 0.7210 \end{bmatrix} \tag{10.122}$$

3. From Equations 10.121 and 10.122, the fundamental matrix is given by

$$\Phi(t) = \Psi e^{\Lambda t} \Psi^{-1} = \begin{bmatrix} 0.8420 & -0.6929 \\ 0.5394 & 0.7210 \end{bmatrix} \begin{bmatrix} e^{-8.7366t} & 0 \\ 0 & e^{-167.1925t} \end{bmatrix} \begin{bmatrix} 0.8420 & -0.6929 \\ 0.5394 & 0.7210 \end{bmatrix}^{-1} \tag{10.123}$$

4. Equation 10.123 simplifies to

$$\Phi(t) = \begin{bmatrix} (0.6190e^{-8.7366t} + 0.3810e^{-167.1925t}) & 0.5948(e^{-8.7366t} - e^{-167.1925t}) \\ 0.3965(e^{-8.7366t} - e^{-167.1925t}) & (0.3810e^{-8.7366t} + 0.6190e^{-167.1925t}) \end{bmatrix} \tag{10.124}$$

The fundamental matrix of Equation 10.124 can be used to determine the energy levels in both systems for any initial condition given to the two subsystems.

5. Let us assume that E_0 is supplied only to the first subsystem at $t = 0$. The energy levels in both systems can be obtained as

$$\left\{ \begin{array}{c} E_1 \\ E_2 \end{array} \right\} = \begin{bmatrix} 0.6190e^{-8.7366t} + 0.3810e^{-167.1925t} & 0.5948(e^{-8.7366t} - e^{-167.1925t}) \\ 0.3965(e^{-8.7366t} - e^{-167.1925t}) & (0.3810e^{-8.7366t} + 0.6190e^{-167.1925t}) \end{bmatrix} \left\{ \begin{array}{c} E_0 \\ 0 \end{array} \right\} \tag{10.125}$$

The energy dissipation in the two-subsystem SEA model after the first system is injected with energy E_0 is shown in Figure 10.26. There is an exponential decay of energy of the first system, whereas the second system's energy increases to a maximum of little over 30% of the initial energy. Then there is an exponential decay; the amount of energy acquired by the second subsystem will depend on the coupling loss and loss factors of both the systems. Therefore, by using the preceding analysis it would be possible to control the energy transferred to a target subsystems by controlling the various SEA parameters.

The energy dissipation of the first system is obtained using Equation 10.117 for $\Delta t = 0.009$ and 0.005 s with that obtained using the state-space approach. The comparison is shown in Figure 10.27. It is seen from Figure 10.27 that the value of the time step will significantly influence error with the numerical method. Hence the state-space formulation provides an accurate method of computing the response with fewer computations, which is beneficial for SEA models with a larger number of subsystems.

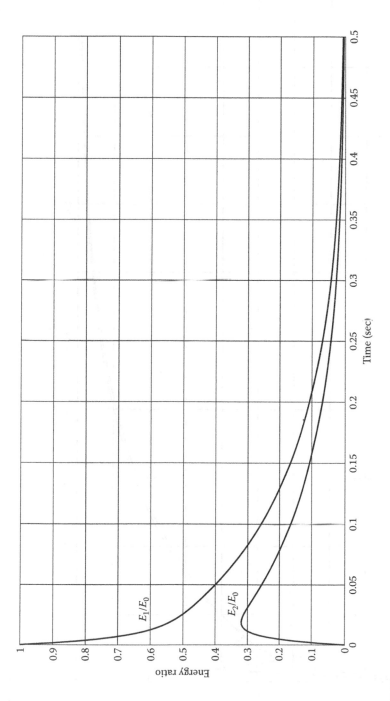

FIGURE 10.26

Energy dissipation in a two-subsystem SEA model using state-space approach (first system injected with initial energy).

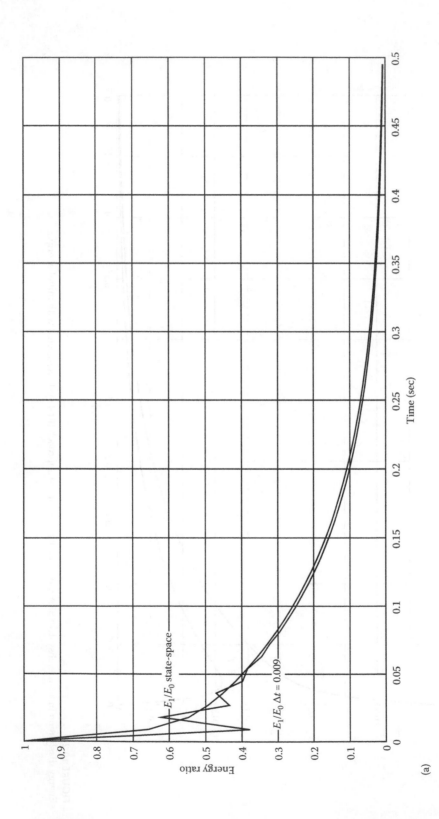

FIGURE 10.27
Comparison of energy ratios of the two-subsystem SEA model obtained from the numerical method and state-space approach: (a) $\Delta t = 0.009$.
(Continued)

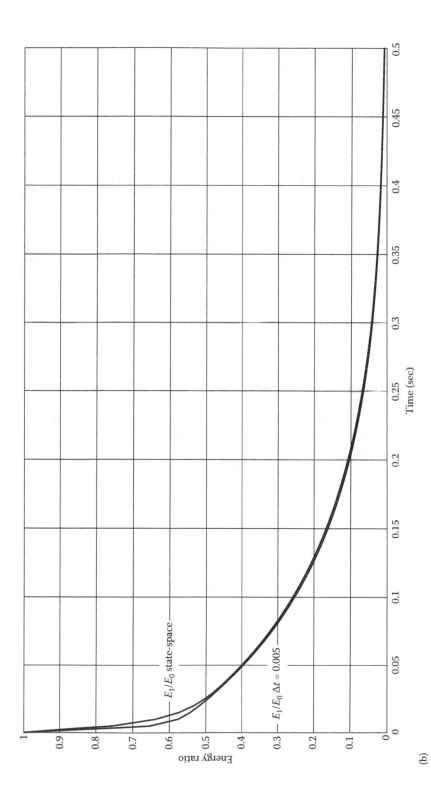

FIGURE 10.27 (CONTINUED)
Comparison of energy ratios of the two-subsystem SEA model obtained from the numerical method and state-space approach: (b) $\Delta t = 0.005$.

10.13 Conclusions

It is very clear from the review presented in this chapter that SEA can be a valuable tool for the design, in the medium to high frequency region. The modeling technique is very simple and the results are very easy to interpret. In addition, they give a physical feel for the designer to make suitable modifications in the design to reduce vibration or noise. This will lead to a better design from the vibration and noise point of view. Transient SEA can be used to reduce noise from impact-excited machines.

Optimization is a very important design tool in the hands of designers, application of which results in better quality products, with higher reliability at a lower price. Since SEA equations are simple algebraic equations for steady excitation, which closely predict the measured response of structures, optimization methods when applied to structures modeled through SEA will yield better results. There is good scope for work in this direction.

Finally, although the finite element method is not suitable for vibration analysis in the high frequency region, it can still provide inputs to SEA, such as modal density. Therefore, FEM, SEA, and optimization will have to be suitably integrated to derive maximum benefit for solving vibration and noise problems.

PROBLEMS

10.1. A portion of a rocket launch vehicle weighing 2000 kg was subjected to acoustic excitation in a test facility. The following results were obtained in the 2000 Hz octave band: sound pressure level 150 dB and velocity level 120 dB (ref: 10^{-9} m/s). If the volume of the room is 800 m³, assuming $c_o = 344$ m/s and $\rho_o = 1.2$ kg/m³, determine the modal density of the structure. Assume that the internal damping of the structure is much less than the coupling loss factor between the structure and air.

10.2. Consider Table 6.2 that gives the number of modes of an aluminum plate of 1 m² area and 3 mm thickness in the octave bands from 31.5 to 8000 Hz center frequencies. Determine the average conductance and power input from a 5 N (rms) force in the octave bands.

10.3. A steel plate and a steel beam are welded together as shown in the following figure. Assume $E = 200$ GPa and $\rho = 7800$ kg/m³. If the structural damping in both elements is much less than the respective coupling loss factors, plot a graph of the ratio of their velocities on a logarithmic scale.

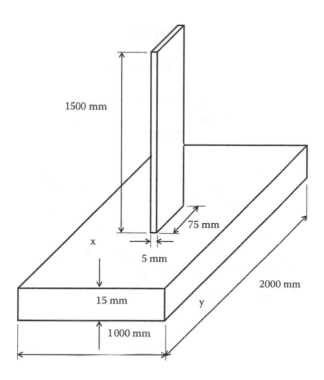

10.4. Consider Problem 10.3. Assume the beam has a damping factor of 0.02 and the plate has a damping factor of 0.01. If the energy ratios are 10% less than those of the equipartitioned condition due to the presence of damping, for power input to either the beam or plate, determine the coupling loss factors between the beam and plate. In addition, determine the power input to the plate to obtain a reverberant energy of 5 J in the plate and the corresponding rms value of force.

10.5. Consider Example 10.1. It has been assigned numerical values as shown in the following figure. If the plate is constructed of steel ($E = 200$ GPa, $\rho = 7800$ kg/m^3), plot the ratio of the energy of the discrete system mass to the total energy of the plate for the octave bands of center frequencies from 31.5 to 8000 Hz. In addition, determine the energy ratio at 1 kHz for the thickness of the plate from 2 to 20 mm.

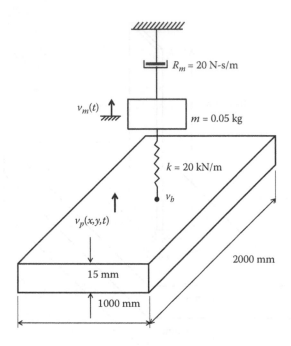

10.6.   Refer to Example 10.4. Determine the radiation efficiency of the plate at 500 Hz.

10.7.   Consider the following figure that represents the vibro-acoustic model of PCB excitation due to rocket noise. For the given input of 100 W at 500 Hz to the equivalent acoustic space around the rocket, determine the energy and modal energy of all the systems. In addition, determine the percentage decrease in the energy of the PCB if the coupling loss factor between the frame and PCB is reduced to 0.001 by providing vibration isolation (hint: extend Equation 10.32) to the five subsystem model and determine the energy and modal energy of all the systems. Use the reciprocity relationship to determine the remaining coupling loss factors.

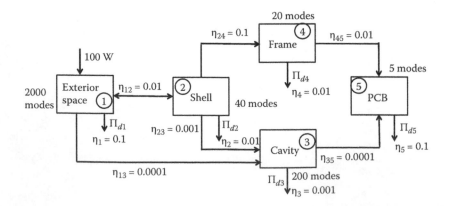

10.8. Consider the following figure that represents the vibro-acoustic model of a fan mounted on a frame that is in turn mounted on a wall (see Section 10.6.4). Determine the sound pressure level in the room in the 500 Hz octave band. If the room is acoustically treated to reduce the damping factor to 0.001, determine the new sound pressure level and percentage change in energy of other systems.

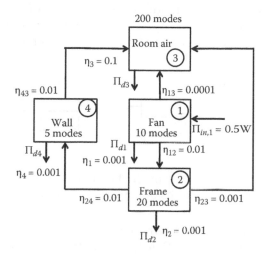

10.9. Consider a two-system SEA model shown in Figure 10.11. If the power input to the second system is one-third the power input to the first system, determine the modal energy ratio in terms of other system parameters.

10.10. Determine the mode number of a thin-walled steel cylinder of $L = 4$ m, outer diameter 0.5 m. and thickness 3 mm in the 1/3 octave band audio frequency range.

10.11. Consider a hollow beam of 30 mm square section, 1 mm thick, and 5 meters long. Determine the mode count for bending, torsion, and longitudinal waves in the 1/3 octave band audio frequency range.

10.12. An acoustic test facility of 800 m³ volume is maintained at 25°C. Determine the number of acoustic modes in the 1/3 octave band audio frequency range.

10.13. As shown in the following figure, two square plates perpendicular to each other are welded along their edges for a length of 500 mm, 100 mm, and 10 mm. Plot the coupling loss factor of the joint at 1/3 octave band audio frequencies.

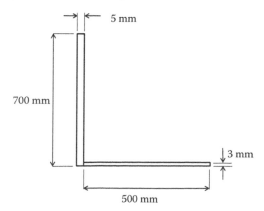

10.14. Consider the nonresonant transmission between two rooms (Figure 10.23). Assuming the volume of the second room as 100 m³, the area of the partition (steel plate) as 10 m², thickness of 10 mm, and a damping factor of 0.01 for the second room, determine the nonresonant coupling loss factor for 1/3 octave band frequencies in the audio frequency range.

10.15. As shown in the following figure, two steel beams of lengths L_1 and L_2 have areas of cross sections A_1 and A_2 connected by a spring of stiffness K. If only the first beam is excited for longitudinal vibration, determine the coupling loss factor.

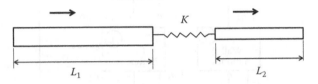

10.16. Consider Equation 10.32. If only the first system is excited, determine the modal energy of both systems in terms of power input, loss factors, and coupling loss factors.

10.17. A structure shown in the following figure is formed using a 3 mm steel plate (E = 200 GPa, ρ = 7800 kg/m³). Determine the modal density of this structure in the audio frequency range and verify it using an FEM program.

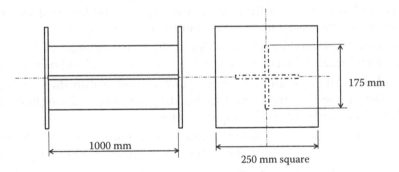

10.18. Consider an SDOF kept in a reverberant room (see the following figure). Determine the ratio of mean square velocity response of the SDOF mass to the mean square dynamic pressure in the reverberant room.

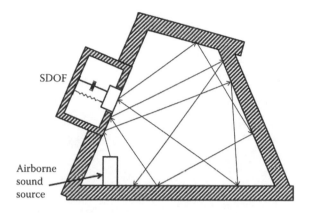

10.19. Refer to Problem 10.13 in which the plates are welded over a length of 500 mm. The 5 mm plate is excited in flexural vibration. This results in reverberant vibration (rms) of 25 mm/s in the 5 mm plate and 10 mm/s in the 3 mm plate. Determine the damping factors of both plates by using the theoretical value of their coupling loss factors and the power input to the system.

10.20. Refer to Problem 10.13 in which the plates are welded over a length of 500 mm. If this welded plate structure is kept in the acoustic volume of 25 m³ and tested for acoustic excitation, the following results were obtained at 500 Hz: sound pressure level 150 dB, and vibration velocity of 130 dB (ref: 10^{-9} m/s) for the 5 mm thick plate and 120 dB (ref: 10^{-9} m/s) for the 3 mm plate. Determine the radiation efficiency of the plates.

Bibliography

Works Cited

Barron, R.F., *Industrial Noise Control and Acoustics*, Marcel Dekker, New York, 2003.

Bell, L.H., *Industrial Noise Control*, Marcel Dekker, New York, 1982.

Beranek, L., *Noise Reduction*, McGraw-Hill, New York, 1960.

Beranek, L.L., *Noise and Vibration Control*, McGraw-Hill, New York, 1971.

Blackstock, D.T., *Fundamentals of Physical Acoustics*, Wiley-InterScience, New York, 2000.

Cacciolati, C., and Guyader, J.L., Measurement of SEA coupling loss factors using point mobilities, *Philosophical Transactions of the Royal Society, London, Series A*, 346, 465–475, 1994.

Cremer, L., Heckl, M., and Ungar, E.E., *Structure-Borne Sound*, Springer-Verlag, Berlin, 1987.

Crocker, M.J., and Price, A.J., Sound transmission using statistical energy analysis, *Journal of Sound and Vibration*, 9, 469–486, 1969.

ESA, European Space Agency, Structural Acoustic Design Manual, ESA PSS-03-204, 1996.

Fahy, F., and Gardonio, P., *Sound and Structural Vibration*, Academic Press, Oxford, 2007.

Fahy, F.J., Statistical energy analysis: A critical overview, *Philosophical Transactions of the Royal Society London, Series A*, 346, 431–447, 1994.

Harris, C.M. (ed), *Handbook of Acoustical Measurements and Noise Control*, Chapter 14, Measurement of Sound intensity, McGraw-Hill, New York, 1991.

Heron, K.H., Advanced statistical energy analysis, *Philosophical Transactions of the Royal Society, London, Series A*, 346, 501–510, 1994.

Hodges, C.H., and Woodhouse, J., Theories of noise and vibration in complex structures, *Reports on Progress in Physics*, 49, 107–170, 1986.

Junger, M.C., and Feit, D., *Sound, Structures, and Their Interaction*, MIT Press, Cambridge, MA, 1986.

Kinsler, L.E., and Frey, A.R., *Fundamentals of Acoustics*, 2nd ed., John Wiley & Sons, New York, 1962.

Langley, R.S., and Bercin, A.N., Wave intensity analysis of high frequency vibrations, *Philosophical Transactions of the Royal Society, London, Series A*, 346, 488–499, 1994.

Langley, R.S., Spatially averaged frequency response envelopes for one- and two-dimensional structural components, *Journal of Sound and Vibration*, 178(4), 483–500, 1994.

Lyon, R.H., and DeJong, R.G., *Theory and Application of Statistical Energy Analysis*, 2nd ed., Butterworth-Heinemann, Boston, 1995.

Lyon, R.H., and Eichler, E., Random vibration of connected structures, *Journal of the Acoustical Society of America*, 36, 1344–1354, 1964.

Lyon, R.H., and Maidanik, G., Power flow between linearly coupled oscillators, *Journal of the Acoustical Society of America*, 34, 623–639, 1962.

Lyon, R.H., *Machinery Noise and Diagnostics*. Butterworth, Boston, 1986.

Lyon, R.H., *Statistical Energy Analysis of Dynamical Systems: Theory and Applications*, MIT Press, Cambridge, MA, 1975.

Maidanik, G., Principle of supplementary of damping and isolation in noise control, *Journal of Sound and Vibration*, 77(2), 245–250, 1981.

Manik, D.N., A new method for determining coupling loss factors for SEA, *Journal of Sound and Vibration*, 211(3), 521–526, 1998.

Manik, D.N., *Control Systems*, Cengage Learning, New Delhi, 2012.

Norton, M., and Karczub, D., *Fundamentals of Noise and Vibration Analysis for Engineers*, Cambridge University Press, Cambridge, 2003.

Sabine, H.J., Notes on acoustic impedance measurement, *Journal of the Acoustical Society of America*, 14, 127, 1942.

Smith, P.W., Jr., and Lyon, R.H., *Sound and Structural Vibration*, NASA Contractor Report CR-160, March 1965.

Smith, P.W., Jr., Statistical models of coupled dynamic systems and the transition from weak to strong coupling, *Journal of the Acoustical Society of America*, 65(3), 695–698, 1979.

Szechenyi, E., Modal densities and radiation efficiencies of unstiffened cylinders using statistical methods, *Journal of Sound and Vibration*, 19(1), 65–81, 1971.

Watson, F.R., *The Absorption of Sound by Materials*, University of Illinois Engineering Experiment Station, Bulletin No. 172, November 1927.

Watters, B.G., The transmission loss of some masonry walls, *Journal of the Acoustical Society of America*, 31, 898–911, 1959.

Woodhouse, J., An approach to the background of statistical energy analysis applied to structural vibration, *Journal of the Acoustical Society of America*, 69(6), 1695–1709, 1981a.

Woodhouse, J., An introduction to statistical energy analysis of structural vibration, *Applied Acoustics*, 14, 455–469, 1981b.

Further Reading

Bies, D.A., and Hanson, C.H., *Engineering Noise Control*, E & FN SPON, London, 1996.

Dally, J.W., Riley, W.F., and McConnell, K.G., *Instrumentation for Engineering Measurements*, John Wiley & Sons, New York, 1984.

Digital Signal Analysis Using Digital Filters and FFT Techniques: Selected Reprints from Technical Review, Bruel & Kjaer, Nærum, Denmark, 1985.

Doebelin, E.O., *Measurement Systems*, McGraw-Hill International Edition, New York, 1990.

Ewins, D.J., *Modal Testing: Theory, Practice, and Application* (Mechanical Engineering Research Studies: Engineering Dynamics Series), 2nd ed., Research Studies, Philadelphia, 2000.

Fader, B., *Industrial Noise Control*, John Wiley & Sons, New York, 1981.

Fahy, F., *Foundations of Engineering Acoustics*, Academic Press, San Diego, CA, 2001.

Faulkner, L.L. (ed.), *Handbook of Industrial Noise Control*, Industrial Press, New York, 1976.

Hayt, W.H., and Kemmerly, J.E., *Engineering Circuit Analysis*, McGraw-Hill International Edition, New York, 1993.

Jacobsen, F., Sound intensity measurements, in *Handbook of Noise and Vibration Control*, edited by M. J. Crocker, chapter 45, John Wiley & Sons, Hoboken, NJ, 2007.

James, H.M., Nichols, N.B., and Phillips, R.S., *Theory of Servo Mechanisms* (MIT Radiation Lab Series), McGraw-Hill, New York, 1947.

Liquorish, A.D., Greenfield, J., Dudley, W., Lloyd, A.J.R., Taylor, R.M., Covell, G.A.B., and Warring, R.H., *Handbook of Noise and Vibration Control*, Trade and Technical Press, England, 1979.

Maekawa, Z., and Lord, P., *Environmental and Architectural Acoustics*, E & FN SPON, London, 1994.

Mark, W.D., and Crandall, S.H., *Random Vibration*, Academic Press, San Diego, CA, 1973.

Meirovitch, L., *Elements of Vibration Analysis*, McGraw Hill, New York, 1986.

Meirovitch, L., *Methods of Analytical Dynamics*, McGraw Hill, New York, 1970.

Moser, M., *Engineering Acoustics: An Introduction to Noise Control*, Springer, Berlin, 2003

Muller, P.C., and Schiehlen, W.O., *Linear Vibrations*, Martin Nijhoff, Dordrecht, 1988.

Nelson, P.A., and Elliott, S.J., *Active Control of Sound*, Academic Press, London, 1992.

Newland, D.E., *An Introduction to Random Vibration and Spectral Analysis*, Longman, London, 1984.

Nigam, N.C., *Introduction to Random Vibration*, MIT Press, Cambridge, MA, 1983.

Norton, M.P., *Fundamentals of Noise and Vibration Analysis for Engineers*, Cambridge University Press, Cambridge, 1989.

Rabiner, L.R., and Gold, B., *Theory and Application of Digital Signal Processing*, Prentice-Hall India, New Delhi, 1992.

Skudrzyk, E., *Simple and Complex Vibratory Systems*, Pennsylvania State University Press, University Park, PA, 1968.

Soedel, W., *Vibration of Shells and Plates*, Marcel Dekker, New York, 2004.

Sound Research Laboratories, *Noise Control in Industry*, E & FN Spon, London, 1991.

Ventsel, E., and Krauthammer, T., *Thin Plates and Shells: Theory, Analysis and Applications*, 3rd ed., Marcel Dekker, New York, 2001.

Yang, S.J., and Ellison, A.J., *Machinery Noise Measurement*, Clarendon Press, Oxford, 1985.

Index

Page numbers followed by f and t indicate figures and tables, respectively.

Printed and bound by CPI Group (UK) Ltd, Croydon, CR0 4YY

24/10/2024

01778293-0020